Lecture Notes in Mathematics

Lecture Notes in Mathematics

For information about Vols. 1–1090 please contact your bookseller or Springer-Verlag.

Vol. 1091: Multifunctions and Integrands. Proceedings, 1983. Edited by G. Salinetti. V, 234 pages. 1984.

Vol. 1092: Complete Intersections. Seminar, 1983. Edited by S. Greco and R. Strano. VII, 299 pages. 1984.

Vol. 1093: A. Prestel, Lectures on Formally Real Fields. XI, 125 pages. 1984.

Vol. 1094: Analyse Complexe. Proceedings, 1983. Edité par E. Amar, R. Gay et Nguyen Thanh Van. IX, 184 pages. 1984.

Vol. 1095: Stochastic Analysis and Applications. Proceedings, 1983. Edited by A. Truman and D. Williams. V, 199 pages. 1984.

Vol. 1096: Théorie du Potentiel. Proceedings, 1983. Edité par G. Mokobodzki et D. Pinchon. IX, 601 pages. 1984.

Vol. 1097: R.M. Dudley, H. Kunita, F. Ledrappier, École d'Éte de Probabilités de Saint-Flour XII – 1982. Edité par P.L. Hennequin. X, 396 pages. 1984.

Vol. 1098: Groups – Korea 1983. Proceedings. Edited by A.C. Kim and B.H. Neumann. VII, 183 pages. 1984.

Vol. 1099: C.M. Ringel, Tame Algebras and Integral Quadratic Forms. XIII, 376 pages. 1984.

Vol. 1100: V. Ivrii, Precise Spectral Asymptotics for Elliptic Operators Acting in Fiberings over Manifolds with Boundary. V, 237 pages. 1984.

Vol. 1101: V. Cossart, J. Giraud, U. Orbanz, Resolution of Surface Singularities. Seminar. VII, 132 pages. 1984.

Vol. 1102: A. Verona, Stratified Mappings – Structure and Triangulability. IX, 160 pages. 1984.

Vol. 1103: Models and Sets. Proceedings, Logic Colloquium, 1983, Part I. Edited by G.H. Müller and M.M. Richter. VIII, 484 pages. 1984.

Vol. 1104: Computation and Proof Theory. Proceedings, Logic Colloquium, 1983, Part II. Edited by M.M. Richter, E. Börger, W. Oberschelp, B. Schinzel and W. Thomas. VIII, 475 pages. 1984.

Vol. 1105: Rational Approximation and Interpolation. Proceedings, 1983. Edited by P.R. Graves-Morris, E.B. Saff and R.S. Varga. XII, 528 pages. 1984.

Vol. 1106: C.T. Chong, Techniques of Admissible Recursion Theory. IX, 214 pages. 1984.

Vol. 1107: Nonlinear Analysis and Optimization. Proceedings, 1982. Edited by C. Vinti. V, 224 pages. 1984.

Vol. 1108: Global Analysis – Studies and Applications I. Edited by Yu.G. Borisovich and Yu.E. Gliklikh. V, 301 pages. 1984.

Vol. 1109: Stochastic Aspects of Classical and Quantum Systems. Proceedings, 1983. Edited by S. Albeverio, P. Combe and M. Sirugue-Collin. IX, 227 pages. 1985.

Vol. 1110: R. Jajte, Strong Limit Theorems in Non-Commutative Probability. VI, 152 pages. 1985.

Vol. 1111: Arbeitstagung Bonn 1984. Proceedings. Edited by F. Hirzebruch, J. Schwermer and S. Suter. V, 481 pages. 1985.

Vol. 1112: Products of Conjugacy Classes in Groups. Edited by Z. Arad and M. Herzog. V, 244 pages. 1985.

Vol. 1113: P. Antosik, C. Swartz, Matrix Methods in Analysis. IV, 114 pages. 1985.

Vol. 1114: Zahlentheoretische Analysis. Seminar. Herausgegeben von E. Hlawka. V, 157 Seiten. 1985.

Vol. 1115: J. Moulin Ollagnier, Ergodic Theory and Statistical Mechanics. VI, 147 pages. 1985.

Vol. 1116: S. Stolz, Hochzusammenhängende Mannigfaltigkeiten und ihre Ränder. XXIII, 134 Seiten. 1985.

Vol. 1117: D.J. Aldous, J.A. Ibragimov, J. Jacod, Ecole d'Été de Probabilités de Saint-Flour XIII – 1983. Édité par P.L. Hennequin. IX, 409 pages. 1985.

Vol. 1118: Grossissements de filtrations: exemples et applications. Seminaire, 1982/83. Edité par Th. Jeulin et M. Yor. V, 315 pages. 1985.

Vol. 1119: Recent Mathematical Methods in Dynamic Programming. Proceedings, 1984. Edited by I. Capuzzo Dolcetta, W.H. Fleming and T. Zolezzi. VI, 202 pages. 1985.

Vol. 1120: K. Jarosz, Perturbations of Banach Algebras. V, 118 pages. 1985.

Vol. 1121: Singularities and Constructive Methods for Their Treatment. Proceedings, 1983. Edited by P. Grisvard, W. Wendland and J.R. Whiteman. IX, 346 pages. 1985.

Vol. 1122: Number Theory. Proceedings, 1984. Edited by K. Alladi. VII, 217 pages. 1985.

Vol. 1123: Séminaire de Probabilités XIX 1983/84. Proceedings. Edité par J. Azéma et M. Yor. IV, 504 pages. 1985.

Vol. 1124: Algebraic Geometry, Sitges (Barcelona) 1983. Proceedings. Edited by E. Casas-Alvero, G.E. Welters and S. Xambó-Descamps. XI, 416 pages. 1985.

Vol. 1125: Dynamical Systems and Bifurcations. Proceedings, 1984. Edited by B.L.J. Braaksma, H.W. Broer and F. Takens. V, 129 pages. 1985.

Vol. 1126: Algebraic and Geometric Topology. Proceedings, 1983. Edited by A. Ranicki, N. Levitt and F. Quinn. V, 423 pages. 1985.

Vol. 1127: Numerical Methods in Fluid Dynamics. Seminar. Edited by F. Brezzi, VII, 333 pages. 1985.

Vol. 1128: J. Elschner, Singular Ordinary Differential Operators and Pseudodifferential Equations. 200 pages. 1985.

Vol. 1129: Numerical Analysis, Lancaster 1984. Proceedings. Edited by P.R. Turner. XIV, 179 pages. 1985.

Vol. 1130: Methods in Mathematical Logic. Proceedings, 1983. Edited by C.A. Di Prisco. VII, 407 pages. 1985.

Vol. 1131: K. Sundaresan, S. Swaminathan, Geometry and Nonlinear Analysis in Banach Spaces. III, 116 pages. 1985.

Vol. 1132: Operator Algebras and their Connections with Topology and Ergodic Theory. Proceedings, 1983. Edited by H. Araki, C.C. Moore, Ş. Strătilă and C. Voiculescu. VI, 594 pages. 1985.

Vol. 1133: K.C. Kiwiel, Methods of Descent for Nondifferentiable Optimization. VI, 362 pages. 1985.

Vol. 1134: G.P. Galdi, S. Rionero, Weighted Energy Methods in Fluid Dynamics and Elasticity. VII, 126 pages. 1985.

Vol. 1135: Number Theory, New York 1983–84. Seminar. Edited by D.V. Chudnovsky, G.V. Chudnovsky, H. Cohn and M.B. Nathanson. V, 283 pages. 1985.

Vol. 1136: Quantum Probability and Applications II. Proceedings, 1984. Edited by L. Accardi and W. von Waldenfels. VI, 534 pages. 1985.

Vol. 1137: Xiao G., Surfaces fibrées en courbes de genre deux. IX, 103 pages. 1985.

Vol. 1138: A. Ocneanu, Actions of Discrete Amenable Groups on von Neumann Algebras. V, 115 pages. 1985.

Vol. 1139: Differential Geometric Methods in Mathematical Physics. Proceedings, 1983. Edited by H. D. Doebner and J. D. Hennig. VI, 337 pages. 1985.

Vol. 1140: S. Donkin, Rational Representations of Algebraic Groups. VII, 254 pages. 1985.

Vol. 1141: Recursion Theory Week. Proceedings, 1984. Edited by H.-D. Ebbinghaus, G.H. Müller and G.E. Sacks. IX, 418 pages. 1985.

Vol. 1142: Orders and their Applications. Proceedings, 1984. Edited by I. Reiner and K. W. Roggenkamp. X, 306 pages. 1985.

Vol. 1143: A. Krieg, Modular Forms on Half-Spaces of Quaternions. XIII, 203 pages. 1985.

Vol. 1144: Knot Theory and Manifolds. Proceedings, 1983. Edited by D. Rolfsen. V, 163 pages. 1985.

continued on page 193

Lecture Notes in Mathematics

Edited by A. Dold and B. Eckmann

1300

George B. Seligman

Constructions of Lie Algebras and their Modules

Springer-Verlag
Berlin Heidelberg New York London Paris Tokyo

Author

George B. Seligman
Department of Mathematics, Yale University
Box 2155 Yale Station, New Haven, CT 06520, USA

Mathematics Subject Classification (1980): Primary: 17B10, 17B20
Secondary: 15A66, 16A28, 17C40

ISBN 978-3-540-18973-2 Springer-Verlag Berlin Heidelberg New York
ISBN 978-0-387-18973-4 Springer-Verlag New York Berlin Heidelberg

2146/3140-543210

INTRODUCTION

Let F be a field of characteristic zero, and let $\mathcal{g}$ be a central simple Lie algebra over F. That is, $\mathcal{g}$ is a simple Lie algebra over F such that the centralizer in $\mathrm{End}_F(\mathcal{g})$ of ad $\mathcal{g}$ is the set of scalar multiplications by elements of F or, what is equivalent, such that $K \otimes_F \mathcal{g}$ is a simple Lie K-algebra for each field extension K of F. We are interested in constructing all irreducible $\mathcal{g}$-modules of finite dimension in terms related to structural descriptions of $\mathcal{g}$. These modules are finite-dimensional F-vector spaces M with a (right) action of $\mathcal{g}$ such that no proper F-subspaces are $\mathcal{g}$-stable.

Under additional conditions, namely those of isotropy and of possessing a reduced set of roots relative to a maximal split torus, this program has been carried out in [Se3]. Indeed, it is [fallaciously] claimed in Proposition I.3 of [Se3] that the constructions there yield a solution to the problem even when one deletes the requirement that $\mathcal{g}$ be central simple over F. The author thanks Drs. Grundhöfer and Löwen, of the Mathematisches Institut in Tübingen, for pointing out the fallacy. The interested reader will be able to see it readily from the "natural" representation of the real Lie algebra $\mathfrak{so}(1,3)$ on $\mathbb{R}^4$, as my critics did. Accordingly, no claims are made here concerning non-central cases.

The objective of these notes is to demonstrate the constructions of the irreducible modules in the isotropic cases with non-reduced root systems, and in all anisotropic cases where $[\mathcal{g}:F]$ is not among 14, 28, 52, 78, 133 or 248 - that is, wherever $\mathcal{g}$ is not "exceptional" - exceptional forms of D_4 being included under this rubric. (Indeed, the cases of dimensions 14 and 52 may be regarded as "solved" from our point of view, even when anisotropic - cf. §9.2.)

The program proceeds along the lines of [Se3], whose basic principles are recapitulated in Chapter 2. For purposes of motivation, and as an illustration of how the program is applied to construct all modules in anisotropic cases, I have treated algebras "of inner type A" in an introductory chapter.

Chapters 3 and 4 deal with the remaining classical algebras. In Chapter 3 the generalized even Clifford algebras of [Se5] are introduced, and are shown to play a central part in the construction of modules closely analogous to that of the symmetric powers of the "natural" central simple associative algebra of Chapter 1. Chapter 4 returns more strictly to the program of [Se3]: In the isotropic case, with t_0 a

*With support from NSF Grants DMS-8201333 and DMS-8512904.

maximal split torus and $\mathfrak{g}_0$ its centralizer, we give a finite family of fundamental $\mathfrak{t}_0$-weights and associated "admissible $\mathfrak{g}_0$-modules". These $\mathfrak{g}_0$-modules serve as generators in presentations of finite-dimensional irreducible $\mathfrak{g}$-modules with the fundamental $\mathfrak{t}_0$-weights as highest $\mathfrak{t}_0$-weights, and all irreducible $\mathfrak{g}$-modules of specified highest $\mathfrak{t}_0$-weight are obtained from these by a process of "Cartan multiplication" natural to our context.

The bulk of the remainder of the notes is devoted to exceptional algebras. Chapter 5 develops constructions of exceptional algebras from quadratic forms in low dimensions (at most 13), treating the question of existence somewhat more thoroughly than did [Sel], where conditions on the quadratic form were shown to be necessary. An alternate approach to the parametrization of these algebras would follow lines of Kantor [Ka] and of Allison [A4], at least in the case of relative rank one, using constructions involving a pair of composition algebras in place of our quadratic space and its spin module. A guiding principle has been to ground the ideas insofar as possible in a classical, "associative" context. In seeking to follow this principle, I have chosen to stress exclusively the quadratic form and its Clifford algebra. Chapter 6 gives case-by-case constructions of complete families of fundamental admissible $\mathfrak{g}_0$-modules, as in Chapter 4.

A similar pattern prevails in Chapters 7 and 8. Chapter 7 gives constructions for what we have dubbed "super-exceptional" Lie algebras, mainly of relative rank one, along the Kantor-Allison lines, but with a mild generalization that promises to yield some anisotropic algebras as well. In the terminology of Allison, the super-exceptional Lie algebras are parametrized by "structurable division algebras" with a one-dimensional space of skew elements, and we have relied on his classification of these. They involve as key ingredients a cubic Jordan algebra J over a quadratic extension field and a special semi-linear bijection of J. In all cases except that where J is the 27-dimensional exceptional Jordan algebra, we have given a somewhat more precise description of the semi-linear map, in order to study the modules in Chapter 8. This has also made it possible to exclude, on principles of the constructions, some cases for which the only previous basis for exclusion was the tables of Tits [T2].

A brief Chapter 9 indicates work where one or more of our restrictions will be found to be relaxed, or where our problems have been approached along other lines.

It is a pleasure to acknowledge the fact that a number of colleagues have supplied ideas and pre-publication access to their works. Among these, Bruce Allison, John Faulkner and David Saltman deserve special thanks.

Mary Ellen DelVecchio made of my irregular and often trying script a wonderfully clean first typed version, from which Caroline Curtis has prepared this text with speed, accuracy and good cheer. Its faults are mine.

Constructions of Lie Algebras
and their Modules

Table of Contents

CHAPTER 1: AN INTRODUCTORY EXAMPLE: $\mathfrak{sl}(n,D)$

§1. The Algebra and a Canonical Decomposition.

An example will serve to illustrate the context of the general considerations on modules presented here, and perhaps to motivate some of the hypotheses of the formal setting later on. The example we consider is the class of Lie algebras $\mathfrak{sl}(n,D)$, where D is a central associative division algebra over the field F of characteristic zero, with $[D:F] = d^2$, and where n is a positive integer (including $n = 1$). By $\mathfrak{sl}(n,D)$ will be understood the derived Lie algebra $[M_n(D), M_n(D)]$ of the algebra $M_n(D)$ of all n by n D-matrices, or, what is the same, the set of n by n D-matrices (a_{ij}) such that the reduced trace of Σa_{ii} is zero, or such that $\Sigma a_{ii} \in [D,D]$, the derived Lie algebra of D.

We fix an integer $r > 0$, and let $A = M_r(D)$. Let $\ell > 2$. Consider the Lie algebra $\mathfrak{sl}(\ell,A) = [M_\ell(A), M_\ell(A)]$, again an absolutely simple Lie algebra over F. Let E_{ij}, $1 \le i, j \le \ell$, denote the usual ℓ by ℓ matrix units in $M_\ell(A)$, so that every element x of $M_\ell(A)$ is uniquely expressible as

$$x = \sum_{i,j} a_{ij} E_{ij}, \qquad a_{ij} \in A.$$

The condition that $x \in \mathfrak{sl}(\ell,A)$ may then be expressed by

$$\sum_{i=1}^{\ell} \tau(a_{ii}) = 0,$$

where for $a \in A$ we have $a = \tau(a)1 + a_0$, $\tau(a) \in F$, $a_0 \in [AA]$; thus $\sqrt{[A:F]}\,\tau(a) = \operatorname{trd}(a)$, the <u>reduced trace</u> of a.

Map A into $\mathfrak{sl}(\ell,A)$ by sending $a \in A$ to

$$\varphi(a) = aE_{\ell\ell} - \frac{1}{\ell - 1} \sum_{i=1}^{\ell-1} \tau(a)\, E_{ii}.$$

Then if $a_0 \in [AA]$, $\varphi(a_0) = a_0 E_{\ell\ell}$, while $\varphi(1) = \frac{\ell}{\ell-1} E_{\ell\ell} - \frac{1}{\ell-1} I$. One checks at once that φ is a Lie morphism, and is injective.

Let $\mathfrak{t}_0$ be the commutative Lie subalgebra of $\mathfrak{sl}(\ell,A)$ consisting of F-combinations of the E_{ii}, so that $\mathfrak{t}_0$ has as basis the $T_i = E_{i+1,i+1} - E_{ii}$, $1 \le i < \ell$. Then $\varphi(A)$ is contained in the centralizer $\mathfrak{g}_0$ of $\mathfrak{t}_0$ in $\mathfrak{g} = \mathfrak{sl}(\ell,A)$, $\mathfrak{g}_0$ consisting of all $\Sigma_{i=1}^{\ell} a_i E_{ii}$ with $\Sigma\tau(a_i) = 0$. The subalgebra $\mathfrak{t}_0$ is split toral, the root-spaces (i.e., weight-spaces for non-zero weights in the adjoint representation) for $\mathfrak{t}_0$ in $\mathfrak{g}$ being the AE_{ij}, $i \neq j$. If α_i is the root to which $AE_{i,i+1}$ belongs, then the roots relative to $\mathfrak{t}_0$ are a system of roots of type $A_{\ell-1}$, with base $\alpha_1,\dots,\alpha_{\ell-1}$ and diagram

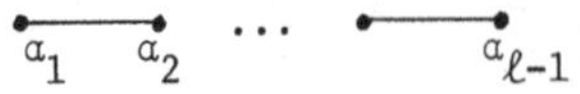

The fact that all roots are of the form $\pm(\alpha_i + \alpha_{i+1} + \dots + \alpha_j)$, $i \le j$, together with the linear independence of the α_i, shows that if α is a root, there is no integer $k > 1$ such that $k\,\alpha$ is a root. If $a \in A$ is invertible, then with $b = a^{-1}$ we have

$$[a\,E_{i+1,i},\ b\,E_{i,i+1}] = T_i.$$

That is, *there is a Zariski-open dense subset* $\mathscr{s}_i$ *of* $\mathcal{g}_{-\alpha_i}$, $1 \le i \le \ell$, *such that if* $Y \in \mathscr{s}_i$, *there is* $X \in \mathcal{g}_{a_i}$ *with* $[YX] = T_i$.

§2. Representation-theoretic Properties.

The conditions cited above have the following consequences for relations between the representation theory, over F, of $\mathcal{g}$ and that of $\mathcal{g}_0$ (see [Se5], §4; compare also [Se3], Chapter I):

Proposition 1.1. If M *is an irreducible (finite-dimensional)* $\mathcal{g}$*-module, then there is a unique* $\lambda \in t_0^*$ *such that the* t_0*-weight space* M_λ *of* M *contains a non-zero element* u *annihilated by* $n = \sum_{\alpha > 0} \mathcal{g}_\alpha$ $(= \sum_{i<j} AE_{ij})$. *The set of all such* u, *together with zero is an irreducible* $\mathcal{g}_0$*-submodule* W. *Each* $\lambda(T_i)$ *is a non-negative integer* b_i; *for all* $w \in W$, *for each* i $(1 \le i \le \ell)$, *and for all* $f_1,\dots,f_{b_i+1} \in \mathcal{g}_{-\alpha_i}$,

$$w\,f_1 \dots f_{b_i+1} = 0.$$

(Here we regard W as right $U(\mathcal{g})$-module.)

To recall briefly the grounds for Proposition 1, note that from the theory of representations of split 3-dimensional algebras $FX + FT_i + FY$ as in the last section, it follows that all T_i act diagonally in M with integral eigenvalues, hence that M is a sum of t_0-weight spaces M_λ. Combined with the fact that $M_\lambda \mathcal{g}_\alpha \subset M_{\lambda+\alpha}$, the form of the t_0-roots α and the linear independence of the α_i, this yields that there is a t_0-weight λ $(M_\lambda \ne 0)$ such that $M_\lambda\,\mathcal{g}_\alpha = 0$ for all positive roots α. Each M_μ, and in particular this M_λ, is a right $\mathcal{g}_0$-module.

If $M^{(0)}$ is any $\mathcal{g}_0$-submodule of M_λ, then the linear combinations of all $mY_{i_1} \dots Y_{i_s}$, $s \ge 0$, $m \in M^{(0)}$, $Y_{i_\nu} \in \mathcal{g}_{-\alpha_{i_\nu}}$, $1 \le i_\nu \le \ell$, form a $\mathcal{g}$-submodule of M, all t_0-weights of which are of the form $\lambda - \sum_{i=1}^{\ell} n_i\,\alpha_i$, n_i non-negative integers, the only elements of weight λ being those of $M^{(0)}$. It follows that $M^{(0)} = M_\lambda$, and that the t_0-weight λ is uniquely determined, either as the unique t_0-weight such that no $\lambda + \alpha_i$ is a t_0-weight or as the unique t_0-weight such that $M_\lambda\, g_\alpha = 0$ for all positive roots relative to t_0. Thus "W" is this M_λ.

From the theory of representations of 3-dimensional algebras, $b_i = \lambda(T_i)$ is a non-negative integer. If $Y \in s_i$, the Zariski-open dense subset of $g_{-\alpha_i}$ cited above, then

$$w\,Y^{b_i+1} = 0 \quad \text{for all} \quad w \in W. \tag{1}$$

By density, (1) holds for all $Y \in g_{-\alpha_i}$, all $w \in W$. Now $g_{-\alpha_i}$ is commutative, so the last relation of Proposition 1 follows by polarization of (1).

Proposition 1.2. *Conversely, let W be an irreducible finite-dimensional* (right) *g_0-module, on which each T_i acts as a non-negative integral scalar b_i; assume further that for each i and each pair of sequences $f_1, \ldots, f_{b_i+1}$ from $g_{-\alpha_i}$ and $e_1, \ldots, e_{b_i+1}$ from $g_{-\alpha_i}$, W is annihilated by the projection of*

$$f_1 \ldots f_{b_i+1}\, e_1 \ldots e_{b_i+1} \in U(g) \quad \text{on}$$

$U(g_0)$ determined by the direct decomposition

$$U(g) = U(g_0) \oplus (nU(g) + U(g)n^-).$$

Then there is a unique irreducible finite-dimensional g-module M whose highest t_0-weight space (in the sense of the preceding paragraph) *is g_0-isomorphic to W.*

The module M is constructed as the quotient of the right $U(g)$-module $W \otimes_{U(p)} U(g)$ by the submodule generated by all

$$w \otimes x_1^{(i)} \ldots x_{b_i+1}^{(i)}, \quad 1 \le i < \ell,$$

for all $w \in W$, $x_j^{(i)} \in g_{-\alpha_i}$.

(Here $n^- = \sum_{\alpha<0} g_\alpha = \sum_{i<j} AE_{ji}$, $p = g_0 + n$, a subalgebra, and W is made into a p-module by letting n annihilate W.)

A proof of Proposition 2 is to be found in [Se3], where its assertions are distributed among Propositions I.6 and I.7, together with Theorem I.1. More detail is given in Chapter 2 to follow. The idea is that the condition of annihilation by the projections in question guarantees that the submodule of $W \otimes_{U(p)} U(g)$ of the last paragraph, that one shows to be of finite codimension in this induced module, has only trivial intersection with W, hence is proper. That it is the unique maximal submodule follows by arguments that are quite commonplace in this subject.

For fixed $\lambda \in t_0^*$, with all $\lambda(T_i)$ non-negative integers, those irreducible g_0-modules $W = M_\lambda$ for irreducible g-modules of highest t_0-weight λ; or, by Proposition 2, those irreducible g_0-modules (of finite dimension) annihilated by the elements of $U(g_0)$ prescribed here, will be called λ-admissible g_0-modules.

From Proposition 2 it is clear that a g_0-isomorphism of λ-admissible g_0-modules (for fixed λ) extends uniquely to an isomorphism of the corresponding g-modules M.

The case $r=1$, where t_0 is a maximal split toral subalgebra of g, is considered in detail in Chapter V of [Se3]. Investigations along lines similar to that, dealing with other (non-reduced) isotropic algebras will be found in later chapters of this exposition. The theme of that approach, developed for the case where t_0 is maximal F-split, is given in Chapter I and II of [Se3], as follows: By appeal to "Cartan multiplication" and splitting, the set of λ's to be considered, in order to construct all irreducible finite-dimensional g-modules of any possible highest t_0-weight, is reduced to a finite set, where most of the b_i are zero. Then the corresponding relations, resulting from the fact that we have a g_0-module satisfying the conditions of Propositions 1 and 2, are interpreted in the particular setting to identify the λ-admissible g_0-modules for each λ in our finite set. A general form of this principle is given as Theorem 2.1, to follow.

A variant of this approach applies to cases that include some anisotropic ones, and has been used this way in [Se5]. In the case at hand (not treated in [Se5]), the variant takes advantage of the fact that $s\ell(r,D)$ embeds "in one corner" in $s\ell(\ell,A)$ as part of the algebra g_0, and uses the action of $s\ell(r,D)$ on the highest t_0-weight spaces M_λ of irreducible g-modules for special values of λ to construct all irreducible $s\ell(r,D)$-modules, even when $r = 1$.

§3. The Role of Symmetric Powers.

Let $\lambda_1,\ldots,\lambda_{\ell-1}$ be the basis for t_0^* dual to $T_1,\ldots,T_{\ell-1}$; we apply the above considerations to $\lambda = b\lambda_{\ell-1}$, where b is a fixed non-negative integer. Let W be a λ-admissible g_0-module, in the sense above. Let π be the projection of $U(g)$ on $U(g_0)$ of Proposition 2. Then W is annihilated by all

$$\pi((aE_{i+1,i})(b\, E_{i,i+1})) \quad \text{for} \quad i \le \ell-2.$$

Here $\pi((aE_{i+1,i})(bE_{i,i+1})) = [a\, E_{i+1,i},\ b\, E_{i,i+1}] = ab\, E_{i+1,i+1} - ba\, E_{ii} \in g_0$, and this element must annihilate W for all choices of a and b. As in [Se3], Chapter V, it follows that the effect on W of the general element

$$x = \Sigma_{i=1}^{\ell}\ a_iE_{ii} \quad \text{of} \quad g_0, \quad \tau(\Sigma\ a_i) = 0,$$

agrees with that of $(\Sigma_{i=1}^{\ell-1}\tau(a_i))\,E_{\ell-1,\ell-1} + a_\ell E_{\ell\ell}$, and with that of $(\frac{1}{\ell-1}\,\Sigma_{i=1}^{\ell-1}\tau(a_i))\,\Sigma_{j=1}^{\ell-1}E_{jj} + a_\ell E_{\ell\ell} = \varphi(a_\ell)$, where φ is as in §1. We are thus led to consider irreducible g_0-modules W where the action of x as above coincides with that of $\varphi(a_\ell)$, and where $\varphi(1) = \sum_{j=1}^{\ell-1}\frac{j}{\ell-1}\,T_j$ acts as does $T_{\ell-1}$, namely as scalar multiplication by b.

The final condition, that of annihilation of $\mathcal{W}$ by all

$$\pi(f_1 \ldots f_{b+1} e_1 \ldots e_{b+1})$$

for $f_i \in g_{-\alpha_{\ell-1}}$, $e_i \in g_{\alpha_{\ell-1}}$, has the following interpretation (as in [Se3], Chapter V): Consider $\mathcal{W}$ as right $U(g_0)$-module, and let ρ be the map $A \to \mathrm{End}_F(\mathcal{W})$ (acting on $\mathcal{W}$ on the right) defined by $w\rho(a) = w\varphi(a)$ for all $w \in \mathcal{W}$. Let $\rho_1(a_1;b_1) = \rho(a_1b_1)$, and for $k > 1$ define $\rho_k(a_1,\ldots,a_k;\ b_1,\ldots,b_k) \in \mathrm{End}(\mathcal{W})$ inductively, for $a_1,\ldots,a_k;\ b_1,\ldots,b_k \in A$, by

$$\rho_k(a_1,\ldots,a_k;\ b_1,\ldots,b_k) =$$
$$\sum_{i=1}^{k} \rho(a_ib_1)\,\rho_{k-1}(a_1,\ldots,\hat{a}_i,\ldots,a_k;\ b_2,\ldots,b_k)$$
$$-\sum_{i<j} \rho_{k-1}(a_1,\ldots,a_{i-1},a_ib_1a_j + a_jb_1a_i,\ldots,\hat{a}_j,\ldots,a_k;\ b_2,\ldots,b_k),$$

where the circumflex indicates omission. Then ρ_{b+1} is identically zero.

In particular, with all $a_j = a$ and all $b_j = 1$, it follows as in [Se3], Chapter III (see also [Se1]) that

$$\sum_{P \leftrightarrow (s_1,\ldots,s_{b+1})} \mathrm{sgn}\ C(P)\ |C(P)|\ \rho(a)^{s_1}\rho(a^2)^{s_2}\ldots\rho(a^{b+1})^{s_{b+1}} = 0 \tag{2}$$

for all $a \in A$. Here P runs over all partitions of $m+1$, with "$P \leftrightarrow (s_1,\ldots,s_{b+1})$" meaning that P has s_1 terms equal to one, s_2 terms equal to $2,\ldots,s_{b+1}$ terms equal to $b+1$. $C(P)$ is the conjugacy class in the symmetric group S_{b+1} corresponding to P, $|C(P)|$ its cardinality, and $\mathrm{sgn}\ C(P)$ is the sign of any member of $C(P)$. The identity for $\rho: A \to \mathrm{End}_F(\mathcal{W})$ is referred to as the $(b+1)$-th symmetric identity.

It has been noted that $\rho(1) = b\cdot\mathrm{id}$, and one sees at once that $\rho([ac]) = [\rho(a),\rho(c)]$ for $a,c \in A$. It follows by Theorem III.3 of [Se3] that $\mathcal{W}$ carries a unique structure of right module for the b-th symmetric power $S^b(A)$ of A, such that $\rho(a)$ is the action on $\mathcal{W}$ of

$$(a \otimes 1 \otimes \ldots \otimes 1) + (1 \otimes a \otimes 1 \otimes \ldots \otimes 1) + \ldots \tag{3}$$
$$+ (1 \otimes \ldots \otimes 1 \otimes a) \in S^b(A).$$

Conversely, if $\mathcal{W}$ is an irreducible right $S^b(A)$-module, one defines $\rho(a)$, for $a \in A$, to be the transformation of $\mathcal{W}$ given by the action of (3). For $x \in g_0$, $x = \Sigma_i a_i E_{ii}$, $\Sigma\tau(a_i) = 0$, define $\sigma(x)$ to be the endomorphism $\rho(a_\ell)$ of $\mathcal{W}$. Then $\mathcal{W}$ is an irreducible right g_0-module, with x acting by $\sigma(x)$, and $\mathcal{W}$ is annihilated by all $[aE_{i+1,i}, cE_{i,i+1}]$, $i < \ell-1$, as well as by all $\pi((aE_{\ell,\ell-1})^{b+1}(cE_{\ell-1,\ell})^{b+1})$, ([Se3], §V.3). From this last it follows by polarization that $\mathcal{W}$ is annihilated by all

$$\pi((a_1E_{\ell,\ell-1}) \cdots (a_{b+1}E_{\ell,\ell-1})(c_1E_{\ell-1,\ell}) \cdots (c_{b+1}E_{\ell-1,\ell})),$$

a_i, $c_j \in A$. From the definition of σ and the fact that $\rho(1) = b\cdot\mathrm{id}$, together with these observations, $\mathcal{W}$ is the highest $\mathfrak{t}_0$-weight space of a finite-dimensional irreducible $\mathfrak{g}$-module of highest $\mathfrak{t}_0$-weight λ. That is, we have established

Proposition 1.3. There is a one-one correspondence between (isomorphism classes of) irreducible right $S^b(A)$-modules and finite-dimensional irreducible right $\mathfrak{g}$-modules of highest $\mathfrak{t}_0$-weight $\lambda = b\lambda_{\ell-1}$.

§4. Splitting Behavior.

Recall that $A \cong M_r(D)$, D a central division algebra over F. We take E to be a maximal subfield of D, $[E:F] = d$, $[D:F] = d^2$, and we let K be a finite extension field of F (which we may assume galois over F) such that the K-algebra $E_K = K \otimes_F E$ is isomorphic to the product of d copies of K. Let f_i, $1 \le i \le d$, be orthogonal primitive idempotents in E_K, with sum 1. Then there are d by d matrix units f_{ij} in $D_K = K \otimes_F D$, forming a K-basis for D_K, with $f_{ii} = f_i$ and with $Kf_{ij} = f_iD_Kf_j$, $1 \le i,j \le d$. Let u_{ij}, $1 \le i,j \le r$, be a basic set of r by r matrix units in $A = M_r(D)$, forming a D-basis for $M_r(D)$, and let e_{ij}, $1 \le i,j \le rd$ be a K-basic set of rd by rd matrix units in $A_K = K \otimes_F A = M_r(D_K)$, with $e_{ij} = f_{i',j'}u_{i'',j''}$, when $i = (i'' - 1)d + i'$, $j = (j'' - 1)d + j'$, $1 \le i'',j'' \le r$, $1 < i',j' \le d$.

The diagonal r by r matrices with entries in E constitute a Cartan subalgebra $\mathfrak{h}$ for the Lie algebra A, and such matrices for which $\mathrm{tr}_{E/F}$ vanishes at the sum of the diagonal entries form a Cartan subalgebra $\mathfrak{h}'$ for $[AA]$. In A_K, $\mathfrak{h}_K$ has as K-basis the e_{ii}, $1 \le i \le rd$, while $\mathfrak{h}'_K$ has K-basis the $e_{i+1,i+1} - e_{ii}$, $1 \le i < rd$. Over K, these are splitting Cartan subalgebras for the respective Lie algebras. A Cartan subalgebra $\mathfrak{h}''$ for $\mathfrak{g} = \mathfrak{sl}(\ell,A)$ consists of those ℓ by ℓ diagonal A-matrices whose entries are r by r diagonal E-matrices, the E/F-trace of the sum of all $r\ell$ diagonal elements being zero. Thus $\mathfrak{t}_0 \subset \mathfrak{h}'' \subset \mathfrak{g}_0$, and $\mathfrak{h}''_K$ is a splitting Cartan subalgebra of $\mathfrak{g}_K$, with basis the elements $h_{(j-1)rd+i} = (e_{i+1,i+1} - e_{ii})E_{jj}$, $1 \le i < rd$, $1 \le j \le \ell$, and the elements $h_{jrd} = e_{11}E_{j+1,j+1} - e_{rd,rd}E_{jj}$, $1 \le j < \ell$. A basic set of positive roots $\beta_1,\ldots,\beta_{\ell rd-1}$ for $\mathfrak{g}_K$ relative to $\mathfrak{h}''_K$ has as root space corresponding to $\beta_{(j-1)rd+i}$, $1 \le i < rd$, $1 \le j \le \ell$, the space $Ke_{i,i+1}E_{jj}$, while the root-space for β_{jrd}, $1 \le j < \ell$, is $Ke_{rd,1}E_{j,j+1}$. The system is of type A:

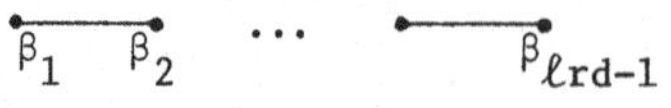

,

with $\beta_{(j-1)rd+i}\big|_{\mathfrak{t}_0} = 0$ if $1 \le i < rd$, $\beta_{jrd}\big|_{\mathfrak{t}_0} = \alpha_j$.

The fundamental weights μ_k, $1 \le k < \ell rd$, on h''_K, are given by

$$\mu_k = \frac{1}{\ell rd} \left(\Sigma_{s=1}^{k} s(\ell rd - k)\beta_s + \Sigma_{s=k+1}^{\ell rd-1} (\ell rd - s)k\, \beta_s\right).$$

From the restrictions above and the corresponding formulas for the λ_j, it follows that $\mu_k|_{t_0}$ is equal to $rd\,\lambda_j$, if $k = jrd$, and to $(rd-i)\lambda_{j-1} + i\,\lambda_j$ if $k = (j-1)rd + i$, $1 \le i < rd$, where $\lambda_0 = 0 = \lambda_\ell$. Thus if a dominant integral function $\Sigma_{s=1}^{\ell rd-1} q_s \mu_s$ (q_s non-negative integers) is to have restriction to t_0 equal to $\lambda = b\lambda_{\ell-1}$, it is necessary and sufficient that $q_s = 0$ for all $s < (\ell-1)rd$, while

$$\Sigma_{i=0}^{rd-1} (rd-i)q_{(\ell-1)rd+i} = b.$$

It follows (from the fact that n is contained in the sum of all positive root-spaces of g_K relative to h''_K) that <u>the g_K-irreducible constituents of M_K, where M is any irreducible g-module of highest t_0-weight λ, have highest h''_K-weights of the form</u>

$$\mu = \Sigma_{i=0}^{rd-1} p_i\, \mu_{(\ell-1)rd+i}, \tag{4}$$

<u>with</u>

$$\Sigma_{i=0}^{rd-1} (rd-i)p_i = b.$$

Conversely, if R is an irreducible g_K-module of highest weight μ as above, we may regard R as a (finite-dimensional) g-module, containing non-zero elements of t_0-weight λ annihilated by n. Now g_0 is a reductive subalgebra in the semi-simple Lie algebra g, the centralizer of t_0 in g, and the elements of $t_0 \subset h''_K$ act in R by semisimple transformations. Thus g_0 acts completely reducibly in R ([J3], Chapter III). An irreducible g_0-submodule of the set of elements of R of t_0-weight λ annihilated by n generates an irreducible g-submodule Q with this g_0-module W as highest t_0-weight space. Then $KQ = R$, as g_K-submodule of R, and KQ is a g_K-homomorphic image of $K \otimes_F Q = Q_K$, so this last module has an irreducible g_K-constituent isomorphic to R. That is, <u>every irreducible g_K-module of highest weight μ such that $\mu|_{t_0} = \lambda$ is an irreducible constituent of some Q_K, where Q is an irreducible g-module of highest t_0-weight λ.</u>

Upon K-linear extension, the embedding φ of §1 maps A_K into $(g_0)_K$, h_K into $(h'')_K$. With μ as above, restriction of μ to $\varphi(h_K)$ induces an element of h^*_K, say μ_0, which we now identify:

For $1 \le i \le rd$, $\varphi(e_{ii}) = e_{ii}E_{\ell\ell} - \frac{1}{\ell-1} \sum_{j=1}^{\ell-1} \tau(e_{ii})E_{jj}$, with $\tau(e_{ii}) = \frac{1}{\ell rd}$ since $\tau(1) = 1$, τ is invariant under K-automorphisms of A_K, and all e_{ii} are conjugate by automorphisms. That is,

$$\varphi(e_{ii}) = e_{ii}E_{\ell\ell} - \frac{1}{(\ell-1)rd}\sum_{j=1}^{\ell-1} E_{jj} ,$$

so that

$$\varphi(e_{i+1,i+1} - e_{ii}) = (e_{i+1,i+1} - e_{ii})E_{\ell\ell} .$$

The respective root-vectors for a fundamental set $\gamma_1,\ldots,\gamma_{rd-1}$ of roots for $[AA]_K$ relative to h'_K are the $e_{j,j+1}$, $1 \le j < rd$, with the unique element of $[Ke_{j,j+1}, Ke_{j+1,j}] \subset h'_K$ where γ_j takes on the value 2 being $e_{j+1,j+1} - e_{jj}$. If $\nu_1,\ldots,\nu_{rd-1}$ are the corresponding dominant integral functions on h'_K, we have, for μ as in (4),

$$\mu_0|_{h'_K} = \sum_{j=1}^{rd-1} \mu((e_{j+1,j+1} - e_{jj})E_{\ell\ell})\nu_j$$

$$= \sum_{j=1}^{rd-1} p_j \nu_j .$$

The determination of μ_0 is completed by noting that $\mu_0(1_A) = \mu(\varphi(1_A)) = \lambda(\varphi(1_A)) = b$.

It follows (cf. [Se5], §8) that if W is the highest t_0-weight space of an irreducible g-module of highest weight $\lambda = b\lambda_{\ell-1}$, then every irreducible $(g_0)_K$-constituent of W_K is an irreducible $[AA]_K$-module via φ. For each such constituent, the highest h'_K-weight ν is of the form $\nu = \Sigma_{i=1}^{rd-1} p_i \nu_i$ where, for some p_0, $b = \Sigma_{i=0}^{rd-1} p_i(rd-i)$. In particular, $\Sigma_{i=1}^{rd-1} p_i(rd-i) \le b$ and $\Sigma_{i=1}^{rd-1} p_i(rd-i) \equiv b$ (modulo rd). Combining this with the results of §3, we have:

<u>Proposition 1.4. Let W be an irreducible right $S^b(A)$-module; then W is an irreducible module for the Lie algebra $[AA]$, and all the irreducible $[AA]_K$-constituents of W_K have highest weights ν (relative to h'_K) satisfying $s(\nu) \le b$, $s(\nu) \equiv b \pmod{rd}$, where $s(\nu) = \Sigma_{i=1}^{rd-1}(rd-i)p_i$, if $\nu = \Sigma_{i=1}^{rd-1} p_i \nu_i$.</u>

§5. Classification of Irreducible [AA]-Modules.

It remains to prove a converse to Proposition 4, and to show the completeness of the $S^b(A)$-modules (as b varies) for the totality of all $[AA]$-modules. Thus suppose W is an irreducible $[AA]$-module (of finite dimension). Then W_K is a direct sum of $[AA]$-submodules, each isomorphic to W, so the same holds for all irreducible $[AA]_K$-submodules of W_K. Let ν be the highest h'_K-weight of one of these irreducible submodules (about which we shall say more in the proof of the theorem below). Let b be an integer such that $b \ge s(\nu)$, $b \equiv s(\nu) \pmod{rd}$.

Let $\mu \in h''^{*}_K$ be defined by $\mu = \Sigma_{i=0}^{rd-1} p_i \mu_{(\ell-1)rd+i}$, where $\nu = \Sigma_{i=1}^{rd-1} p_i \nu_i$, $p_0 rd = b - s(\nu)$, and let M be the irreducible g_K-module of the highest weight μ. By the above, the highest t_0-weight space of M is an irreducible $S^b(A_K)$-module, therefore an irreducible $[AA]_K$-module, and its highest h'_K-weight is $\mu|_{h'_K} = \nu$. That is, <u>every irreducible $[AA]_K$-module whose highest weight ν satisfies $s(\nu) \le b$,</u>

$s(\nu) \equiv b \pmod{rd}$ is an irreducible right $S^b(A_K)$ - $(= S^b(A)_K$-) module.

The main theorem is the following:

Theorem 1.1. Let $q = rd$, so that A is a central simple algebra over F, of characteristic zero, with $[A:F] = q^2$. For $b = 0,1,2,\ldots$, let $\mathcal{S}_b$ be the (finite) set of equivalence classes of irreducible [AA]-modules obtained as irreducible right $S^b(A)$-modules. Then for each j, $0 \le j < q$,

$$\mathcal{S}_j \subset \mathcal{S}_{j+q} \subset \mathcal{S}_{j+2q} \subset \ldots \subset \mathcal{S}_{\bar{j}} = \bigcup_{i=0}^{\infty} \mathcal{S}_{j+iq},$$

the $\mathcal{S}_{\bar{j}}$ are disjoint, and their union is the totality of finite-dimensional irreducible [AA]-modules.

Proof. To complete the proof, it suffices to show that if ν, ν' are highest weights of $[AA]_K$-irreducible constituents of $\mathcal{W}_K$, where $\mathcal{W}$ is an irreducible [AA]-module, then $s(\nu) \equiv s(\nu') \pmod q$. For then it suffices to take b equal to the largest of the $s(\nu)$'s associated with $\mathcal{W}_K$ in order to conclude, as in [Se3], Proposition II.2, iii), that $\mathcal{W}$ is isomorphic to an irreducible $S^b(A)$-module. The disjointness of the different $\mathcal{S}_{\bar{j}}$ follows from the incongruence (mod q) of the values $s(\nu)$, $s(\nu'')$ for highest h'_K-weights ν, ν'' associated with irreducible $[AA]_K$-constituents of modules in $\mathcal{S}_{\bar{j}}$, $\mathcal{S}_{\bar{j}''}$, for $j \neq j''$.

In fact, it follows from Galois-theoretical considerations as in [BT] or [T5] (see also [Se5], §10) that $\nu = \nu'$ in the setting above, i.e., that all $[AA]_K$-irreducible constituents of $\mathcal{W}_K$ are isomorphic. Here we assume K to be a finite Galois extension of F, with group G, splitting g as above. Then each $\sigma \in G$ acts on g_K so as to fix g, thus fixes each E_{ii} (in an evident extension to $M_\ell(A)_K$) and stabilizes each subspace $(K \otimes_F E)E_{ii}$, $(K \otimes_F A)E_{ii}$. inducing in $K \otimes_F E$ a permutation of the idempotents f_j and in $K \otimes_F A$ a permutation of the primitive idempotents $e_i = e_{ii}$. The fundamental roots $\gamma_1,\ldots,\gamma_{q-1}$ for $[AA]_K$ relative to h'_K are the restrictions to h'_K of the elements $\xi_2 - \xi_1,\ldots,\xi_q - \xi_{q-1} \in h^*_K$, where h_K is the K-subspace of A_K spanned by the e_i and $\{\xi_i\}$ is the basis for h^*_K dual to the basis $\{e_i\}$ for h_K.

If π_σ, for $\sigma \in G$, is the permutation of $\{1,\ldots,q\}$ such that $e_i^\sigma = e_{\pi_\sigma(i)}$, $1 \le i \le q$, it follows from the action of the Weyl group on the roots as in [Se3] §VII.6, that the unique $w_\sigma \in W$, the Weyl group of $[AA]_K$ relative to h'_K, such that

$$\gamma_1^{\sigma w_\sigma^{-1}}, \ldots, \gamma_{q-1}^{\sigma w_\sigma^{-1}}$$

is a permutation of $\gamma_1,\ldots,\gamma_{q-1}$, must result from the same permutation of the ξ_j, and therefore that $\gamma_i^{\sigma w_\sigma^{-1}} = \gamma_i$ for all i and all σ.

Now, as in [BT], if $\mathcal{R}$ is an irreducible $[AA]_K$-submodule of $\mathcal{W}_K$, one has

$\mathcal{W}_K = \sum_{\sigma \in G} R^{\sigma}$, where the action of G on $\mathcal{W}_K$ is that fixing $\mathcal{W}$, and each irreducible $[AA]_K$-submodule of $\mathcal{W}_K$ is isomorphic to some R^{σ}. Moreover, if ν is the highest weight η of R^{σ} has $\eta(h) = \nu(h^{\sigma\varphi_{\sigma}^{-1}})$ for all $h \in h'_K$, where φ_{σ} is an automorphism of $[AA]_K$ stabilizing h'_K and inducing the action of w_{σ}, and this means that $\eta = \nu^{\sigma^{-1} w_{\sigma^{-1}}^{-1}} = \nu$ by the above. This completes the proof of the assertion $\nu = \nu'$, and of the theorem.

If $\mathcal{W}$ is an irreducible $[AA]$-module in the class $S_{\bar{j}}$, let $c(\mathcal{W})$ be the class in the Brauer group $Br(^K/_F)$ of central simple F-algebras split by K of any simple ideal in $S^b(A)$. (These are all central simple algebras in the same Brauer class, namely $c(D)^j = c(D)^b$, as Saltman has noted - cf. the appendix below.) Then $c(\mathcal{W})$ has period e, where $e|d$. Thus the enveloping associative algebra, the image of $\mathcal{U}([AA])$ in $\mathrm{End}_F(\mathcal{W})$, is a central simple algebra, an ideal in some $S^b(A)$, $b \equiv j \pmod{rd}$, and has class $c(\mathcal{W}) = c(D)^j$ in the Brauer group. We thus have the

<u>Corollary. For each irreducible $[AA]$-module $\mathcal{W}$ in $S_{\bar{j}}$, the enveloping algebra is central simple, and its class $c(\mathcal{W})$ in the Brauer group is $c(D)^j$. Thus the enveloping algebras of all modules in $S_{\bar{j}}$ are similar central simple algebras.</u>

It is only when $r = 1$, $e = d$ - for example, where $A = D$, with d square free - that the $S_{\bar{j}}$ may alternatively be described as the class of those $\mathcal{W}$ with $c(\mathcal{W}) = c(D)^j$.

Appendix

Let A be a central simple associative algebra of finite dimension over the field F of characteristic zero. Let $\{u_i\}$ be a basis for A over F, $\{v_i\}$ the corresponding dual basis with respect to the trace form of the right regular representation: $\mathrm{Tr}(R_{u_i} R_{v_j}) = \delta_{ij}$, where $R_x : A \to A$ sends a to ax. Let $c = \sum_i u_i \otimes v_i \in A \otimes A$. Then it is shown in [Se3], IV.5, that the elements

$$s_j = \overbrace{1 \otimes \ldots \otimes 1}^{j-1} \otimes c \otimes \overbrace{1 \otimes \ldots \otimes 1}^{k-j-1} \in \otimes^k A, \quad 1 \leq j \leq k-1$$

generate a subalgebra S of $\otimes^k A$ isomorphic to the group algebra $F[S_k]$ of the symmetric group S_k, the element s_j corresponding to the transposition $(j,j+1)$. Furthermore, the centralizer of S in $\otimes^k A$ is the k-th symmetric power $S^k(A)$. A complete set of primitive central idempotents in S is again such a set in $S^k(A)$, in number equal to $p(k)$, the number of partitions of k. From the classical structure of $S^k(A)$ when A is split, it follows that <u>each of the minimal</u> (two-sided) <u>ideals in $S^k(A)$ is a central simple algebra</u>.

Some of these considerations, in particular the realization of the group algebra within $\otimes^k A$, have been treated in a more general context by Saltman [Sa]. He has also supplied a proof (unpublished) of the result which follows, in a more general form than that given here. This proof is entirely his, specialized to our situation:

Theorem 1.A. (Saltman): Each minimal ideal of $S^k(A)$ is a central simple algebra, whose Brauer class is the same as that of $\otimes^k A$.

Proof. Let $B = \otimes^k A$, $S \approx F[S_k]$, $S \subset B$ as above, and let $e_1,\ldots,e_n$ be the primitive central idempotents of S. Then $S^k(A) = C$ centralizes all the e_i, and $B = \sum_{i,j} \oplus e_i B e_j$. One sees at once that $C \subseteq \sum_i e_i B e_i$, and $S \subset \sum_i e_i B e_i$, because the decomposition yields that $\Sigma e_i B e_i$ is the centralizer of the $\{e_i\}$ in B. Clearly, then, the centralizer C of S in B is the centralizer of S in $\Sigma e_i B e_i$, and is the direct sum of the centralizers C_i of the individual Se_i in the individual $B_i = e_i B e_i$.

Now the left module Be_i is a sum of copies of the unique simple left B-module M_i, with $B \approx \mathrm{End}_D(M_i)$, the division algebra D being $\mathrm{End}_B(M_i)$ (and central). Thus $\mathrm{End}_B(Be_i)$ may be identified with a matrix algebra over D, and so is central simple of the same Brauer class as B^{op}. But a B-endomorphism of $Be_i = \sum_j e_j B e_i$ must send e_i to an element annihilated by all e_j for $j \neq i$, so to an element of $e_i B e_i$, and one sees that the result is an isomorphism of algebras (acting on the left) between $\mathrm{End}_B(Be_i)$ and $(e_i B e_i)^{op}$. That is $(e_i B e_i)^{op}$ and B^{op} have the same Brauer class, as do $e_i B e_i$ and B.

Now $e_i B e_i$ is central simple, contains $Se_i \approx M_n(F)$ for some n, and the centralizer in $e_i B e_i$ of Se_i is C_i. Thus $e_i B e_i \approx Se_i \otimes C_i$, C_i is central simple, and the class of C_i in the Brauer group is that of $e_i B e_i$, hence that of B (see, for example, [J5], Theorem 4.7). This completes the proof.

CHAPTER 2: GENERAL CONSIDERATIONS

§1. Structure of Irreducible Modules.

The general context in which the themes of Chapter 1 are developed is as follows: $\mathfrak{g}$ is a central simple Lie algebra over the field F of characteristic zero, and $\mathfrak{t}_0$ is a split toral subalgebra. Thus ad t is diagonalizable in F for each t in the commutative subalgebra $\mathfrak{t}_0$. It follows that one has a decomposition

$$\mathfrak{g} = \mathfrak{g}_0 + \sum_\alpha \mathfrak{g}_\alpha,$$

where $\mathfrak{g}_0$ is the centralizer of $\mathfrak{t}_0$ in $\mathfrak{g}$, and where α runs over the set of weights, other than zero, of the (right) adjoint representation of $\mathfrak{t}_0$ on $\mathfrak{g}$,

$$\mathfrak{g}_\alpha = \{x \in \mathfrak{g} \mid [xt] = \alpha(t)x, \text{ for all } t \in \mathfrak{t}_0\}.$$

We insist on the following conditions:

1) The $\mathfrak{t}_0$-weights α of $\mathfrak{g}$ shall be (combinatorially isomorphic to) a system of roots - not necessarily reduced - in the sense of Bourbaki, with a distinguished base $\alpha_1,\ldots,\alpha_\ell$ forming a basis for the dual space $\mathfrak{t}_0^*$.

2) There shall be a basis $T_1,\ldots,T_\ell$ for $\mathfrak{t}_0$, with $\alpha_i(T_i) = 2$, and subsets of root-spaces as follows: For each i, $1 \le i \le \ell$, there shall be an open Zariski-dense subset Y_i of $\mathfrak{g}_{-\alpha_i}$ such that for each $y \in Y_i$ there is $x \in \mathfrak{g}_{\alpha_i}$ with $[xy] = T_i$. If $\mathfrak{g}_{2\alpha_i} \neq 0$, there shall be an open dense subset Z_i of $\mathfrak{g}_{-2\alpha_i}$ such that if $y \in Z_i$ there is $x \in \mathfrak{g}_{2\alpha_i}$ with $[xy] = T_i$.

3) If $\mathfrak{g}_{2\alpha_i} \neq 0$, there shall be an open Zariski-dense subset X_i of $\mathfrak{g}_{-2\alpha_i}$ such that if $y \in X_i$, then $[y,\mathfrak{g}_{\alpha_i}] = \mathfrak{g}_{-\alpha_i}$.

4) For all i, $[\mathfrak{g}_{-\alpha_i},\mathfrak{g}_{-\alpha_i}] = \mathfrak{g}_{-2\alpha_i}$.

5) The Lie algebra $\mathfrak{g}$ is generated by $\mathfrak{g}_0$ and the $\mathfrak{g}_{\pm\alpha_i}$, $1 \le i \le \ell$.

By a result of Jacobson ([J2]; see also [J3], p. 102) $\mathfrak{g}_0$ is a reductive subalgebra of $\mathfrak{g}$. The fact that the $\mathfrak{t}_0$-weights are a system of roots means that we have a sense of positive root, and that $\mathfrak{n} = \mathfrak{n}^+ = \sum_{\alpha>0} \mathfrak{g}_\alpha$, $\mathfrak{n}^- = \sum_{\alpha<0} \mathfrak{g}_\alpha$ are nilpotent subalgebras of $\mathfrak{g}$, contained in the parabolic subalgebras

$$\mathfrak{p} = \mathfrak{p}^+ = \mathfrak{g}_0 + \mathfrak{n}^+, \ \mathfrak{p}^- = \mathfrak{g}_0 + \mathfrak{n}^-,$$

in which $\mathfrak{n}^+$ resp. $\mathfrak{n}^-$ are ideals. A maximal toral subalgebra $\mathfrak{h}$ of $\mathfrak{g}$ (not necessarily split) containing t_0 is then a Cartan subalgebra of g, as well as of g_0, and we may assume $t_0 \subseteq t \subseteq h$, where t is a maximal F-split toral subalgebra of g: (See [Sel], §I.1.)

For the greater part of our considerations, t_0 will be a maximal F-split toral subalgebra of g, and (as is implicit above) different from $\{0\}$. Then the conditions 1) - 5) are automatically satisfied - cf. [Sel], Chap. I. In fact, in 2) and 3) the Zariski-dense set in question may be taken to be the set of non-zero elements of the corresponding root-space. Moreover, each root-space g_α is an irreducible (right) g_0-module.

Returning to the general setting, let M be a (finite-dimensional, right) g-module. The condition 2) means that M is a module for a 3-dimensional subalgebra with basis x, T_i, y as there, so that t_0 acts F-diagonally in M, all eigenvalues of the T_i being integers. In fact, if $2\alpha_i$ is a root, then with $y \in Z_i$, $x \in g_{2\alpha_i}$, we have $2\alpha_i(\frac{1}{2}T_i) = 2$ and the representation theory of $< x,y,\frac{T_i}{2} >$ shows that all eigenvalues of such T_i are <u>even</u> integers. From the linear independence of the α_i and the ordering of the roots it is clear that each of n^+, n^- acts in M by nilpotent transformations. In particular, Engel's theorem yields that

$$M^+ = \{u \in M \mid \text{For all } x \in n,\ ux = 0\}$$

is a non-zero subspace of M, and a p-submodule.

In analogy with very classical settings, one has

<u>Proposition 2.1. Suppose M is irreducible. Then M^+ is a t_0-weight space, say of weight $\lambda \in t_0^*$, and is an irreducible g_0-module. The space M is linearly generated by all</u>

$$u\, e_{-\alpha_{i_1}} \cdots e_{-\alpha_{i_s}}, \tag{1}$$

$u \in M^+$, $s \geq 0$, $1 \leq i_j \leq \ell$.

Conversely if M^+ is an irreducible g_0-module, then M is an irreducible g-module.

<u>Proof</u>. Assume M irreducible, and partially order the elements of t_0^* as usual: $\lambda > \mu$ if $\lambda - \mu$ is a sum of positive roots. Let λ be minimal among the t_0-weights of M^+, and consider the linear span $M^{(0)}$ of the elements (1), where now u is constrained to be of weight λ. This space is clearly stable under the action of g_0 and of all $g_{-\alpha_i}$, and its stability under the action of all g_{α_i} follows by induction on s, using $M^+ g_{\alpha_i} = 0$, $[g_{-\alpha_j}\ g_{\alpha_i}] = 0$ if $i \neq j$, and $[g_{-\alpha_i}\ g_{\alpha_i}] \subseteq g_0$. By 5), $M^{(0)}$ is a g-submodule, $M^{(0)} = M$, and for every t_0-weight μ of M, $\mu \leq \lambda$, with the space M_λ of t_0-weight λ being contained in M^+. By choice of λ, $M^+ = M_\lambda$. The same procedure yielding $M^{(0)}$ could be applied starting with any

$\mathfrak{g}_0$-submodule of M^+ as the domain of our u's, yielding an $M^{(0)} = M$ whose only elements of M^+ are in the submodule. Thus M^+ is an irreducible $\mathfrak{g}_0$-module.

For the converse, the complete reducibility of M and the fact that if $M = M_1 \oplus M_2$, M_i $\mathfrak{g}$-modules, $M^+ = M_1^+ \oplus M_2^+$, M_i^+ $\mathfrak{g}_0$-modules make the assertion transparent.

§2. The Highest Weight $\mathfrak{g}_0$-Module.

Proposition 2.2. _Let M be irreducible as above, $M^+ = M_\lambda$. Then $\lambda(T_i)$ is a non-negative integer m_i for each i. For each i, each $u \in M^+$ and each sequence $f_{j_1},\ldots,f_{j_{m_i+1}}$ from $\mathfrak{g}_{-\alpha_i}$,_

$$u\, f_{j_1} \ldots f_{j_{m_i+1}} = 0. \tag{2}$$

Proof. The first assertion follows from the representation theory of the 3-dimensional algebra $< y(\in Y_i), x, T_i >$ as above, as does, if $m = m_i$, $uf^{m+1} = 0$ for each $u \in M^+$, $f \in Y_i$. By Zariski-density, this last holds for all $f \in \mathfrak{g}_{-\alpha_i}$. If $[\mathfrak{g}_{-\alpha_i}, \mathfrak{g}_{-\alpha_i}] = 0$ (2) follows by polarization.

Otherwise, $\mathfrak{g}_{-2\alpha_i} \neq 0$, m is even, say $m = 2r$, and $[\mathfrak{g}_{-2\alpha_i}, \mathfrak{g}_{-\alpha_i}] = 0 = [\mathfrak{g}_{-2\alpha_i}, \mathfrak{g}_{-2\alpha_i}]$, so that, as above,

$$u\, y_{j_1} \ldots y_{j_{r+1}} = 0$$

for all $u \in M^+$, all $y_j \in \mathfrak{g}_{-2\alpha_i}$. If $y \in X_i$, $f \in \mathfrak{g}_{-\alpha_i}$, and $[yx] = f$, where $x \in \mathfrak{g}_{\alpha_i}$, then $0 = uy^{r+1}x = (r+1)uy^r f$. It follows by the density of X_i that $uy_1 \ldots y_r f = 0$ for all $u \in M^+$, $y_j \in \mathfrak{g}_{-2\alpha_i}$, $f \in \mathfrak{g}_{-\alpha_i}$. Now let $s \leq r$, and assume proved that, for all $y_1,\ldots,y_s \in \mathfrak{g}_{-2\alpha_i}$, $f_1,\ldots,f_{m+1-2s} \in \mathfrak{g}_{-\alpha_i}$, $u \in M^+$,

$$u\, y_1 \ldots y_s\, f_1 \ldots f_{m+1-2s} = 0. \tag{3}$$

In particular we have for $y \in X_i$, $f \in \mathfrak{g}_{-\alpha_i}$, $[yx] = f$ as above, $0 = uy^s f^{m+2-2s} x$. The fact that $[fx] \in \mathfrak{g}_0$, and therefore normalizes each of $\mathfrak{g}_{-\alpha_i}$, $\mathfrak{g}_{-2\alpha_i}$, as well as stabilizing M^+, yields

$$0 = s\, u\, y^{s-1} f^{m+3-2s}$$

for all $u \in M^+$, $y \in X_i$, $f \in \mathfrak{g}_{-\alpha_i}$, hence by Zariski-density of X_i and polarization on y,

$$u\, y_1 \ldots y_{s-1} f^{m+3-2s} = 0 \quad \text{for all } y_1,\ldots,y_{s-1} \in \mathfrak{g}_{-2\alpha_i},\ u \in M^+,$$

$f \in g_{-\alpha_i}$. Now we may polarize on f, at the expense of possibly introducing commutators from $g_{-2\alpha_i}$, and therefore terms already known to be zero, to obtain a universal relation of the form (3) with s replaced by $s-1$. For $s = 0$ we obtain the desired conclusion, and the proposition is proved.

By taking a basis for the universal enveloping algebra $U(g)$ to consist of monomials in t_0-root vectors in n^+, a basis of g_0, and t_0-root vectors in n^- in that order, we obtain a direct decomposition of $U(g)$ into subspaces

$$U(g) = U(g_0) + n^+U(p^+) + U(p^-)n^- + n^+U(g)n^- \tag{4}$$

each subspace being a g_0-submodule under the adjoint action, and such that the centralizer of t_0 in $U(g)$ is contained in the sum of the first and last summands. Now if M, λ, i, and $m = m_i$ are as above, let $f_1,\ldots,f_{m+1} \in g_{-\alpha_i}$, and $e_1,\ldots,e_{m+1} \in g_{\alpha_i}$, and consider the action on M^+ of $z = f_1 \ldots f_{m+1}\, e_1 \ldots e_{m+1} \in U(g)$. This action is necessarily zero by Proposition 2, and z centralizes t_0. The remarks above then show that if π is the projection onto $U(g_0)$ consistent with the decomposition (4) of $U(g)$, $\pi(z)$ must annihilate M^+. For reference, we restate this observation as

__Lemma 2.1. Let__ M __be an irreducible__ g-__module__, $M^+ = M_\lambda$, __and fix__ i, $1 \le i \le \ell$, __so that__ $\lambda(T_i) = m$, __a non-negative integer. Let__ π __be the projection of__ $U(g)$ __on__ $U(g_0)$ __of the decomposition. Then for every sequence__ $f_1,\ldots,f_{m+1}$ __from__ $g_{-\alpha_i}$ __and every sequence__ $e_1,\ldots,e_{m+1}$ __from__ g_{α_i},

$\pi(f_1 \ldots f_{m+1}\, e_1 \ldots e_{m+1})$ __annihilates__ M^+.

The effect of these considerations is to yield information about the irreducible g_0-module M^+ associated with M. Namely, each T_i acts there by a non-negative integral scalar m_i, and the ideal in $U(g_0)$ generated by all the $\pi(z)$, formed as above for each i, annihilates M^+. Next we show that these conditions characterize the M^+ among irreducible g_0-modules.

Thus let R be an irreducible g_0-module such that for each i, T_i acts on R by a non-negative integral scalar m_i, even if $2\alpha_i$ is a root, and such that R is annihilated by all $\pi(z) \in U(g_0)$, where $z = f_1 \ldots f_{m_i+1}\, e_1 \ldots e_{m_i+1} \in U(g)$ are elements formed as above for each i. One makes R into a p-module by letting n annihilate R, and forms the induced right $U(g)$-module

$$I(R) = \mathrm{Ind}_p^g(R) = R \otimes_{U(p)} U(g).$$

This module is characterized among (not necessarily finite-dimensional) g-modules S as universal for p-morphisms of p-modules $R \to S$, and the canonical morphism $u \to u \otimes 1$ of R into $I(R)$ is injective.

Now let $I(R)^-$ be the g-submodule of $I(R)$ generated by all $u \otimes f_1 \dots f_{m_i+1}$ for $u \in R$, $1 \le i \le \ell$, $f_j \in g_{-\alpha_i}$.

Lemma 2.2. $I(R)^-$ is of finite codimension in $I(R)$.

Proof. When i is fixed, the linear combinations of all images $u \otimes f_1 f_2 \dots f_j$, $u \in R$, $f_\nu \in g_{-\alpha_i}$ constitute a finite-dimensional submodule of $S = I(R)/_{I(R)^-}$ for the Lie subalgebra $g_i = g_{-2\alpha_i} + g_{-\alpha_i} + g_0 + g_{\alpha_i} + g_{2\alpha_i}$ of g, because j is bounded by m_i. If φ is the canonical homomorphism $I(R) \to S$, this submodule contains $\varphi(R) = \varphi(R \otimes 1)$. Now the sum of all finite-dimensional g_i-submodules of S is readily seen to be stable under the action of each root-space (as in [Hu], §21.2) of g, and contains the generating set $\varphi(R)$, so is equal to S. It follows that each $y \in Y_i$, as well as the corresponding $x \in g_{\alpha_i}$, acts locally nilpotently on S, and thus that

$$A_i = \exp(Y)\exp(x)\exp(y)$$

is a well-defined linear automorphism of S.

Because the adjoint action of t_0 on $U(g)$ is F-diagonalizable, S is a sum of t_0-weight spaces, and S is spanned by elements $\varphi(u \otimes f_1 \dots f_s)$, $f_j \in \sum_{i=1}^{\ell} g_{-\alpha_i}$, so that all t_0-weights of S have the form $\lambda - \Sigma k_i \alpha_i$, k_i non-negative integers, and weight spaces have finite dimension. Let μ be a weight, S_μ the corresponding weight space, and fix i. Then S_μ is contained in a finite g_i-submodule, stable under A_i, and conjugation by A_i maps S_μ to $S_{\mu^{\sigma_i}}$, where σ_i is the Weyl reflection associated with α_i. Likewise A_i maps $S_{\mu^{\sigma_i}}$ to S, and A_i is invertible, so that μ^{σ_i} is a t_0-weight if and only if μ is, and the two weight spaces have the same dimension. It now follows that this is the case when σ_i is replaced by any element of the Weyl group W of our system of roots.

By minimizing Σk_i for the weights $\mu = \lambda - \Sigma k_i \alpha_i$ in the W-orbit of a given weight we find a weight μ in the W-orbit such that $\mu(T_i) \ge 0$ for all i. Now λ is a rational linear combination of $\alpha_1, \dots, \alpha_\ell$, as is μ, and the condition $\mu(T_i) \ge 0$ for all i implies that all coefficients are non-negative, by an argument as in [Hu], §21, making appeal to a positive definite rational inner product $(\mu|\nu)$ on the rational span of the $\{\alpha_j\}$ such that $\alpha_j(T_i)$ is a positive multiple of $(\alpha_i \mid \alpha_j)$. But these coefficients are of the form $j_i - k_i$, where $\lambda = \Sigma j_i \alpha_i$ and the k_i are non-negative integers. It follows that there are only finitely many W-orbits of weights of S. Because W is finite, the finiteness of dimension of S is proved.

Because R is g_0-irreducible and generates $I(R)$, each proper g-submodule of $I(R)$ meets R only in zero, and is contained in the sum of the weight-spaces $I(R)_\mu$ for μ (strictly) less than λ. Thus the sum of all proper g-submodules is proper, and is the unique maximal g-submodule of $I(R)$. In particular, if $R = M^+$ as before, the $\mathcal{g}$-module M is uniquely determined by the $\mathcal{g}_0$-module structure of R, as the unique irreducible $\mathcal{g}$-quotient of $I(R)$.

Lemma 2.3. Let the irreducible $\mathcal{g}_0$-module R be as at the beginning of this section, i.e., annihilated by all $\pi(z)$ as defined there. Then $I(R)^- \cap R = \{0\}$.

Proof. Because each generator $u \otimes f_1 \dots f_{m_i+1}$ for $I(R)^-$ is a t_0-weight vector and R is the subspace $I(R)_\lambda$, to have $I(R)^- \cap R \neq \{0\}$ would entail having a non-zero element of R of the form $u \otimes f_1 \dots f_{m_i+1}\, x_1 \dots x_s$ for some i, where all $f_j \in \mathcal{g}_{-\alpha_i}$ and $x_1, \dots, x_s \in \mathcal{g}_0 \cup \bigcup_{i=1}^{\ell} (\mathcal{g}_{\alpha_j} \cup \mathcal{g}_{-\alpha_j})$. We may assume s is minimal for all such elements. Then because R is a $\mathcal{g}_0$-module and ad $\mathcal{g}_0$ stabilizes all root-spaces, no x_j is in $\mathcal{g}_0$. By comparing weights, some x_k is in $\sum_{j=1}^{\ell} \mathcal{g}_{\alpha_j}$. Letting k be minimal with this property, we may assume by rearranging commuting factors among the x's that either $k = 1$ or $x_{k-1} \in \mathcal{g}_{-\alpha_j}$, where $x_k \in \mathcal{g}_{\alpha_j}$. But then $x_{k-1}x_k = x_k x_{k-1} + [x_{k-1}, x_k]$, the second term on the right being in $\mathcal{g}_0$, and the minimality of s gives $u \otimes f_1 \dots x_s = u \otimes f_1 \dots x_k x_{k-1} \dots x_s$. That is, x_1 may be taken to lie in $\bigcup_j \mathcal{g}_{\alpha_j}$, in which case necessarily $x_1 = e_1 \in \mathcal{g}_{\alpha_i}$.

Now suppose $1 \le t \le m_i$ and that we have shown there is such a shortest element of the form $u \otimes f_1 \dots f_{m_i+1} e_1 \dots e_t x_{t+1} \dots x_s$, where $e_1, \dots, e_t \in \mathcal{g}_{\alpha_i}$. Again some $x_k \in \bigcup_j \mathcal{g}_{\alpha_j}$ because $t \le m_i$, and again, we may assume $k = t + 1$. If $x_{t+1} \in \mathcal{g}_{\alpha_j}$ for $j \neq 1$, then

$$0 \neq u \otimes f_1 \dots f_{m_i+1} e_1 \dots e_t x_{t+1} \quad \text{belongs to the weight}$$

$\mu = \lambda - r\,\alpha_i + \alpha_j$ for $r > 0$, and this is impossible because $\mu \not\leq \lambda$. Hence $x_{t+1} = e_{t+1} \in \mathcal{g}_{\alpha_i}$, and we have a non-zero element of R of the form

$$u \otimes f_1 \dots f_{m_i+1} e_1 \dots e_b x_{b+1} \dots x_s$$

for all $b \le m_i+1$. But then $u \otimes f_1 \dots f_{m_i+1} e_1 \dots e_{m_i+1} = u \otimes \pi(f_1 \dots e_{m_i+1}) = u\, \pi(f_1 \dots e_{m_i+1}) = 0$, and we have a contradiction.

Proposition 2.3. Let R be an irreducible $\mathcal{g}_0$-module of t_0-weight λ as before, annihilated by all $\pi(z) \in U(\mathcal{g}_0)$, for $z = f_1 \dots f_{m_i+1}\, e_1 \dots e_{m_i+1}$, all i, with f_j running over $\mathcal{g}_{-\alpha_i}$, e_j over $\mathcal{g}_{\alpha_i}$. Then $M = I(R)/I^-(R)$ is the finite-dimensional irreducible $\mathcal{g}$-module with the $\mathcal{g}_0$-module R as highest t_0-weight space, i.e., $R = M^+$.

Proof. By Lemmas 2 and 3, M is non-zero and finite-dimensional, and R is its weight space of weight λ. Moreover, R generates M and $R \subseteq M^+$. The complete reducibility of M and of the action of t_0 in M implies that g_0 acts completely reducibly in M [J2], therefore in M^+, and if $M = M_1 \oplus \dots \oplus M_d$, where the M_j are g-irreducible, then at least one of the M_j^+ must be g_0-isomorphic to R. In particular, M_j^+ has t_0-weight λ, so $M_j^+ \subseteq R$, so $M_j^+ = R$. Because R generates M, $M = M_j$ is irreducible and $R = M^+$. The proof of the proposition is complete.

Proposition 3 gives a prescription for giving presentations for all irreducible g-modules of highest t_0-weight λ satisfying our conditions of dominant integrality. They are in one-one correspondence with the irreducible g_0-modules R of the proposition. It is a relatively easy matter to show that there are only finitely many such R when t_0 is a maximal split toral subalgebra, but we shall get more precise information, from which this will follow, in individual cases. (See [Se3], Prop. I.10, and the cases of reduced relative root systems handled individually in that work.)

§3. Generalities Concerning (finite) Splitting.

Let h be a Cartan subalgebra of g containing t_0, and let K be a finite extension field of F splitting the action of h. Each root-space of $K \otimes g$ relative to $K \otimes h$ lies in $K \otimes g_0$ or in some $K \otimes g_\alpha$, and has K-dimension one. Thus all roots of $K \otimes g$ relative to $K \otimes h$ have rational values at $T_1, \dots, T_\ell$, and we may supplement $T_1, \dots, T_\ell$ by $h_{\ell+1}, \dots, h_r$, elements of $K \otimes h$ where all roots take rational values, to obtain a K-basis for $K \otimes h$. We may then order the roots γ of $K \otimes g$ relative to $K \otimes h$ with reference to our given order on the "roots of g relative to t_0" by setting $\gamma_1 > \gamma_2$ if $\gamma_1|_{t_0} > \gamma_2|_{t_0}$ (in any ordering of the roots of t_0, together with $0 \in t_0^*$, in which "positive roots" are positive), or if $\gamma_1|_{t_0} = \gamma_2|_{t_0}$ and the first of $h_{\ell+1}, \dots, h_r$ where the values of γ_1 and γ_2 differ, say h_j, has $\gamma_1(h_j) > \gamma_2(h_j)$.

If M is an irreducible $K \otimes g$-module (of finite dimension), we may replace t_0 by $K \otimes t_0$, n by $K \otimes n$ to have the same conditions 1)-5) as before (as Y_i, Z_i, X_i we may take the same subsets of $g_{-\alpha_i}$, $g_{-2\alpha_i}$, $g_{-2\alpha_i}$, respectively). Thus $M^+ = \{u \in M | u\, n^+ = 0\}$ is a $K \otimes t_0$-weight space and an irreducible $K \otimes g_0$-module. Because of the chosen ordering, $n^+ \subset K \otimes n^+ \subseteq \sum_{\gamma>0} (K \otimes g)_\gamma$, the root-spaces $(K \otimes g)_\gamma$ being taken with respect to $K \otimes h$. Here M is absolutely irreducible with a one-dimensional highest $(K \otimes h)$-weight space V ([J3], Chapter VII), and $V \subset M^+$. If $V = M_\mu$, $\mu \in (K \otimes h)^*$, then $\mu|_{K \otimes t_0}$ is the highest $K \otimes t_0$-weight of M.

Because [K:F] is finite, we may "restrict scalars" and regard M as finite-dimensional g-module. Let R be an irreducible g-submodule of M. Each non-zero element of K transforms R by scalar multiplication into a g-module isomorphic to R, and the sum of these is stable under both K and g, so must be M. That is, <u>all irreducible g-submodules of M are isomorphic</u>. With μ as above, their highest

$\mathfrak{t}_0$-weight is $\mu|_{\mathfrak{t}_0}$.

Conversely, let S be an irreducible $\mathfrak{g}$-module, so that $K \otimes S = M$ is a $K \otimes \mathfrak{g}$-module, therefore a direct sum $M_1 \oplus M_2 \oplus \ldots \oplus M_s$ of irreducible $K \otimes \mathfrak{g}$-modules. As $\mathfrak{g}$-module, M is a sum of $[K:F]$ copies of S, so that every irreducible $\mathfrak{g}$-submodule of every M_i is isomorphic to S. In particular, there is an irreducible $K \otimes \mathfrak{g}$-module, all of whose irreducible $\mathfrak{g}$-submodules are isomorphic to S.

A useful test for completeness of certain families of irreducible $\mathfrak{g}$-modules will be the following: Let S, M and $M_1, \ldots, M_s$ be as in the last paragraph. Suppose that S' is an irreducible $\mathfrak{g}$-module such that some M_j is $K \otimes \mathfrak{g}$-isomorphic to a submodule of $K \otimes S'$. Then S is $\mathfrak{g}$-isomorphic to S'.

For each M_j is $\mathfrak{g}$-isomorphic to a sum of copies of S and all the irreducible $\mathfrak{g}$-submodules of the distinguished M_j are isomorphic to S'.

In a more awkward form, this last is to be found in Chapter II of [Se3]. There one also finds a version of the following principle, for the case where $\mathfrak{t}_0$ is a maximal split torus. We restate it for our context:

Theorem 2.1. Let $\mathfrak{g}$, $\mathfrak{t}_0$, K and other notations be as above. Let $\pi_1, \pi_2, \ldots, \pi_r$ be the fundamental dominant integral functions associated with $K \otimes \mathfrak{g}$ and $K \otimes \mathfrak{h}$, and to the simple roots associated with our ordering of the roots of $K \otimes \mathfrak{h}$ in $K \otimes \mathfrak{g}$, and let M_i be the irreducible $K \otimes \mathfrak{g}$-module of highest $K \otimes \mathfrak{h}$-weight π_i. Let S_i be an irreducible $\mathfrak{g}$-submodule of M_i, so that S_i^+ is an irreducible $\mathfrak{g}_0$-module of $\mathfrak{t}_0$-weight $\pi_i|_{\mathfrak{t}_0}$.

Then if S is an irreducible $\mathfrak{g}$-module of highest $\mathfrak{t}_0$-weight λ, one has $\lambda = \sum_{i=1}^{r} n_i \cdot \pi_i|_{\mathfrak{t}_0}$ for some non-negative integers n_i. For some such choice of $n_1, \ldots, n_r$, the irreducible $\mathfrak{g}_0$-module S^+ is a submodule of the completely reducible $\mathfrak{g}_0$-module $\bigotimes_{i=1}^{r} (\otimes^{n_i} S_i^+)$. Every irreducible $\mathfrak{g}_0$-submodule of such a module is the module S^+ for an irreducible $\mathfrak{g}$-module of highest $\mathfrak{t}_0$-weight λ.

Proof. With n_i as indicated, the $K \otimes \mathfrak{g}$-submodule of $\bigotimes_{i=1}^{r} (\otimes^{n_i} M_i)$, the tensor products being taken over K, generated by the tensor product of the highest $K \otimes \mathfrak{h}$-weight vectors, is the irreducible $K \otimes \mathfrak{g}$-module of highest $K \otimes \mathfrak{h}$-weight $\sum_{i=1}^{r} n_i \pi_i$, and the highest $\mathfrak{t}_0$-weight space of any irreducible $\mathfrak{g}$-submodule S is a $\mathfrak{g}_0$-module S^+, with S as in the first sentence of the second paragraph of the theorem. With S_i and S_i^+ as in the statement, $\bigotimes_{i=1}^{r} (\otimes^{n_i} S_i)$ is a completely reducible $\mathfrak{g}$-module, all of whose $\mathfrak{t}_0$-weights μ satisfy $\mu \leq \lambda$, and whose λ-weight space is $\bigotimes_{i=1}^{r} (\otimes^{n_i} S_i^+)$, a completely reducible $\mathfrak{g}_0$-module.

Now $\bigotimes_{i=1}^{r} (\otimes^{n_i} M_i)$, as $\mathfrak{g}$-module, is a homomorphic image of $\bigotimes_{F,\, i=1}^{r} (\otimes_F^{n_i} M_i)$, a direct sum of copies of $\bigotimes_{i=1}^{r} (\otimes^{n_i} S_i)$, so all the $\mathfrak{g}$-irreducible submodules of the first of these are submodules of the last. Now S is a $\mathfrak{g}$-submodule of the first, with highest $\mathfrak{t}_0$-weight λ, and the only elements of the last of $\mathfrak{t}_0$-weight λ are those of $\bigotimes_{i=1}^{r} (\otimes^{n_i} S_i^+)$. Thus S^+ is a $\mathfrak{g}_0$-submodule of the tensor product as claimed. The final statement is seen by noting that if R is a $\mathfrak{g}_0$-irreducible submodule of the tensor product, then R has $\mathfrak{t}_0$-weight λ and is the $\mathfrak{t}_0$-subspace of weight λ in the $\mathfrak{g}$-submodule S of $\bigotimes_i (\otimes^{n_i} S_i)$ generated by R, a submodule that is necessarily irreducible because all other $\mathfrak{t}_0$-weights are strictly less than λ. Thus $R = S^+$, with S as claimed.

CHAPTER 3: INVOLUTORIAL ALGEBRAS AND MODULES FOR THEIR SKEW ELEMENTS

§1. Structure and Conventions.

Let D be a central division algebra over a field Z of characteristic zero, with $[D:Z] = d^2 < \infty$. We assume D has an involution $a \to a^*$, which necessarily stabilizes Z. If Z is not fixed by the involution, Z is a quadratic extension of the fixed subfield F of Z, and the involution * is <u>of second kind</u>. If Z is fixed by the involution, we set F = Z and say that * is <u>of first kind</u>. Upon passage to a splitting field, we find in this case that the *-fixed elements of D have F-dimension either $\binom{d+1}{2}$ or $\binom{d}{2}$. In the former case, * is said to be <u>of orthogonal type</u>; in the latter, <u>of symplectic type</u>.

Let U be a left D-vector space carrying a non-degenerate hermitian ($\varepsilon = 1$) or anti-hermitian ($\varepsilon = -1$) D-valued form, and let V be a nonsingular subspace such that, if (u,v) is the form in U, the orthogonal complement of V in U has D-basis $u_1,\ldots,u_\ell,u_{\ell+m+1},\ldots,u_{m+2\ell}$, where m is the dimension of V, with $(u_i,u_j) = 0$ unless $i+j = m + 2\ell + 1$, and with $(u_i, u_{m+2\ell+1-i}) = 1$, $1 \leq i \leq \ell$. The general assumptions on the form are the identities

$$(u_1 + u_2,v) = (u_1,v) + (u_2,v);\ (au,v) = a(u,v);\ (v,u) = \varepsilon(u,v)^*;\quad \text{thus}$$

$$(u,av) = (u,v)a^*.$$

The F-Lie algebra g is taken to be the derived Lie algebra of the Lie algebra of all D-endomorphisms S of U (written on the right) such that

$$(uS, v) + (u, vS) = 0 \qquad (1)$$

for all $u,v \in U$. In general, if the involution * is of first kind, the derived Lie algebra is equal to the original algebra. The only exceptions arise when $m = 0$, $\ell = 1$ and $D = F$, with $\varepsilon = 1$; or when $m = 1$, $\ell = 0$ and D is a quaternion algebra over F, with either $\varepsilon = 1$ and * of orthogonal type or $\varepsilon = -1$ and * of symplectic type; or when $m = 2$, $\ell = 0$, $D = F$ and $\varepsilon = 1$. The Lie algebra g is central simple over F whenever $\ell > 0$, except when $[g:F] = 6$, a case we shall exclude. We also assume $V \neq \{0\}$. When $D = F$ and $\varepsilon = -1$, we are in the symplectic case, and both g and the Lie algebra L of skew transformations of V are split. Accordingly, their finite-dimensional representations are known, and <u>we exclude the case</u> $D = F$, $\varepsilon = -1$ <u>from further consideration</u>.

Every element a of D may be uniquely written, as in Chapter 1, as $a = \alpha 1 + a_0$, where $\alpha \in Z$, $a_0 \in [DD]$. We write $\alpha = t(a)$, so that t is an F-linear map $D \to Z$ with kernel $[DD]$. If $w_1,\ldots,w_n$ is an arbitrary D-basis for U, then $\mathfrak{g}$ may be described as the set of those S satisfying (1) and such that $\sum_{i=1}^{n} t(s_{ii}) = 0$, where $w_i S = \sum_{j=1}^{n} s_{ij} w_j$ for each i.

Let $\mathfrak{t}_0$ be the subspace of $\mathfrak{g}$ annihilating V and stabilizing each Fu_j, $1 \le i \le \ell$, $1 \le m+2\ell+1-i \le \ell$ - we always assume $\ell > 0$. Then $T \in \mathfrak{t}_0$ means $u_i T = r_i u_i$, $1 \le i \le \ell$, where $r_i \in F$, and $u_{m+2\ell+1-i} T = -r_i u_{m+2\ell+1-i}$ for such i. An F-basis for $\mathfrak{t}_0$ thus consists of the matrix-unit combinations

$$X_i = E_{ii} - E_{m+2\ell+1-i,m+2\ell+1-i}, \quad 1 \le i \le \ell,$$

and $\mathfrak{t}_0$ is an F-split toral subalgebra of $\mathfrak{g}$. We show that the conditions 1)-5) of §1 of Chapter 2 are satisfied.

The centralizer $\mathfrak{g}_0$ of $\mathfrak{t}_0$ in $\mathfrak{g}$ is the stabilizer in $\mathfrak{g}$ of V and of every Du_i. For fixed i, $1 \le i \le \ell$, the elements of $\mathfrak{g}$ mapping u_i into V and V into $Du_{\ell+2m+1-i}$, while annihilating $u_{m+2\ell+1-i}$ and all other u_j, belong to a $\mathfrak{t}_0$-weight whose value at X_i is -2, and at all other X_j is zero, while those mapping V into Du_i and $u_{m+2\ell+1-i}$ into V, while annihilating u_i and all other u_j, belong to the negative of this $\mathfrak{t}_0$-weight. We let $\alpha_\ell \in \mathfrak{t}_0^*$ be the weight of the elements of $\mathfrak{g}$ mapping u_ℓ into V and V into $Du_{\ell+m+1}$, $T_\ell = 2(E_{m+\ell+1,m+\ell+1} - E_{\ell\ell}) \in \mathfrak{t}_0$.

For $1 \le i < j \le \ell$, the elements of $\mathfrak{g}$ annihilating V and all u_k, $k \ne i$, $m+2\ell+1-j$, and mapping u_i into Du_j, $u_{m+2\ell+1-j}$ into $Du_{m+2\ell+1-i}$ form a $\mathfrak{t}_0$-weight space, as do those similarly mapping u_j into Du_i and $u_{m+2\ell+1-i}$ into $Du_{m+2\ell+1-j}$. The two $\mathfrak{t}_0$-weights thus associated with the pair i,j are negatives of one another. For $1 \le i < \ell$, with $j = i+1$, we let $\alpha_i \in \mathfrak{t}_0^*$ be the weight of the former, $T_i = X_{i+1} - X_i$. Then the spaces associated with a general pair $i < j$, $1 \le i,j \le \ell$, have weights

$\pm(\alpha_i + \ldots + \alpha_{j-1})$, and those of the preceding paragraph have weights

$\pm(\alpha_i + \ldots + \alpha_\ell)$, $1 \le i \le \ell$. It is readily verified that $\alpha_1,\ldots,\alpha_\ell$ form a basis for $\mathfrak{t}_0^*$.

Those S in $\mathfrak{g}$ annihilating V and all u_k for $k \ne i,j$ as above, while mapping u_i into $Du_{m+2\ell+1-j}$, u_j into $Du_{m+2\ell+1-i}$, form a $\mathfrak{t}_0$-weight space of weight

$$\alpha_i + \ldots + \alpha_{j-1} + 2\alpha_j + \ldots + 2\alpha_\ell,$$

and the negative of this is the weight obtained by interchanging the roles of u_i and $u_{m+2\ell+1-i}$, u_j and $u_{m+2\ell+1-j}$. Finally, those S in $\mathfrak{g}$ annihilating V and all u_k, $k \ne i$, while mapping u_i into $Du_{m+2\ell+1-i}$ form a non-zero subspace except when $\varepsilon = 1$ and $D = F$, of $\mathfrak{t}_0$-weight $2(\alpha_i + \ldots + \alpha_\ell)$, and $-2(\alpha_i + \ldots + \alpha_\ell)$ is the weight of the corresponding space with u_i and $u_{m+2\ell+1-i}$ interchanged. The algebra g is the sum of g_0 and these weight-spaces.

From the listings of weights, we see that the t_0-weights form an indecomposable system of roots of non-reduced type BC_ℓ, with base $\alpha_1,\dots,\alpha_\ell$, except when $D = F$, $\varepsilon = 1$, when the type is B_ℓ, with the same base. The root-space g_{α_ℓ} consists of all $S_{v,u_{m+2\ell+1}}$, $v \in V$, where $S_{v,w}$ sends u to $\varepsilon(u,v)w - \varepsilon(u,w)v$, and $g_{-\alpha_\ell}$ consists of all $S_{u_\ell,v}$. For $1 \le i < \ell$, g_{α_i} consists of all $S_{u_{m+2\ell+1-i},au_{i+1}}$, $a \in D$, and $g_{-\alpha_i}$ of all $S_{u_i,au_{m+2\ell-i}}$. When $2\alpha_\ell$ is a root, $g_{2\alpha_\ell}$ consists of all $S_{u_{\ell+m+1},bu_{\ell+m+1}}$, $b^* = -\varepsilon\, b$, and $g_{-2\alpha}$ of all such S_{u_ℓ,bu_ℓ}. In general, one has (in <u>right</u> action on U)

$$[S_{v,w}, S_{x,y}] = S_{vS_{x,y},w} + S_{v,wS_{x,y}}, \tag{2}$$

and the $S_{w,x}$ generate the F-vector space of all $S \in \mathrm{End}_D(U)$ satisfying (1).

From (2) applied to the data above it is readily seen that 5) of Chap. 2, §1 is satisfied, and one has, for $1 \le i < \ell$, $a \ne 0$ in D,

$$[S_{u_{m+2\ell+1-i},au_{i+1}}, S_{u_i,cu_{m+2\ell-i}}]$$

$$= \varepsilon\, S_{au_{i+1},cu_{m+2\ell-i}} - \varepsilon\, S_{u_{m+2\ell-i},\, a\, c^* u_i},$$

which for c invertible, $a = \varepsilon\, c^{*-1}$, is T_i. If $v \in V$, $(v,v) \ne 0$ - the set of such v is non-empty, and therefore Zariski-dense, by the exclusion of the case $\varepsilon = -1$, $D = F$ - let $w = 2(v,v)^{-1}v$. Then

$$[S_{u_\ell,v}, S_{u_{m+2\ell+1},w}] = T_\ell.$$

Other than in the case $\varepsilon = 1$, $D = F$, let $b \in D$, $b^* = -\varepsilon\, b \ne 0$ and let $c = -\frac{1}{2} b^{*-1}$. Then $[S_{u_\ell,bu_\ell}, S_{u_{\ell+m+1},\, cu_{\ell+m+1}}] = T_\ell$, and the condition 2) of Chapter 2, §1 is satisfied. Moreover, for $v \in V$,

$$[S_{u_\ell,bu_\ell}, S_{u_{m+\ell+1},v}] = S_{v,bu_\ell} + S_{u_\ell,bv}$$

$$= -\varepsilon S_{bu_\ell,v} + S_{u_\ell,bv} = -\varepsilon S_{u_\ell,b^*v} + S_{u_\ell,bv} = 2\, S_{u_\ell,bv},$$

and 2bv runs over V as v does. This demonstrates 3), and 4) follows from

$$[S_{u_\ell,v}, S_{u_\ell,w}] = -\varepsilon S_{u_\ell,(v,w)u_\ell}$$

$$= S_{(v,w)u_\ell,u_\ell} = S_{u_\ell,(v,w)^*u_\ell}$$

$$= \frac{1}{2} S_{u_\ell,((v,w)^*-\varepsilon(v,w))u_\ell} \quad \text{for all } v,w \in V.$$

§2. Even Clifford Algebras and Certain Admissible Modules.

The results of §1 above show that all the considerations of Chapter 2 apply in the setting of this chapter. We use them in particular to investigate irreducible $\mathfrak{g}$-modules M whose highest $\mathfrak{t}_0$-weight has the form $\lambda = k\lambda_\ell$, where $\lambda_1,\ldots,\lambda_\ell$ is the basis for $\mathfrak{t}_0^*$ dual to $T_1,\ldots,T_\ell$, and where k is a non-negative integer. By §2 of Chapter 2, the highest weight $\mathfrak{g}_0$-modules for such λ (the "λ-admissible" $\mathfrak{g}_0$-modules, as in Chapter 1) are the irreducible $\mathfrak{g}_0$-modules annihilated by all T_i for $i < \ell$, on which T_ℓ acts as the scalar k, and annihilated by

$$\pi(S_{u_i, au_{m+2\ell-i}} \; S_{u_{i+1}, bu_{m+2\ell+1-i}}) \in U(\mathfrak{g}_0) \tag{3}$$

for all $i < \ell$, all $a,b \in D$, as well as by all

$$\pi(S_{u_\ell, v_1} \cdots S_{u_\ell, v_{k+1}} \; S_{u_{m+2\ell+1-i}, w_1} \cdots S_{u_{m+2\ell+1-i}, w_{k+1}}) \tag{4}$$

for $v_1,\ldots,v_{k+1}$: $w_1,\ldots,w_{k+1} \in V$.

Now (3) is equal to

$$[S_{u_i, au_{m+2\ell-i}}, \; S_{u_{i+1}, bu_{m+2\ell+1-i}}]$$

$$= -\varepsilon\, S_{b^*u_{i+1}, au_{m+2\ell-i}} + \varepsilon\, S_{u_i, abu_{m+2\ell+1-i}}$$

$$= S_{au_{m+2\ell-i}, b^*u_{i+1}} - S_{abu_{m+2\ell+1-i}, u_i},$$

with matrix $-b^*a^*E_{ii} + a^*b^*E_{i+1,i+1} - baE_{m+2\ell-i,m+2\ell-i}$

$$+ ab\, E_{m+2\ell+1-i,\; m+2\ell+1-i}. \tag{5}$$

When $b = 1$, all elements (5) of $\mathfrak{g}_0$ annihilate M^+, for arbitrary a. Replacing a by another element a' of D and forming commutators, we see that M^+ is annihilated by all elements

$$[a,a']E_{m+2\ell+1-i,\; m+2\ell+1-i} + [aa']\, E_{m+2\ell-i,\; m+2\ell-i} \tag{6}$$

$$- [a,a']^*E_{i+1,i+1} - [aa']^*E_{i,i}.$$

With $b = 1$, a replaced by $[aa']$ in (5), we add the result to (6), and likewise subtract, to conclude that M^+ is annihilated by all

$$cE_{ii} - c^*E_{m+2\ell+1-i,\; m+2\ell+1-i}, \quad 1 \le i \le \ell, \tag{7}$$

$c \in [DD]$. Thus if c is general in D, the action of (7) on M^+ is the same as that of $t(c)X_i$. The general element of $\mathfrak{g}_0$ has the form

$$X = \sum_{i=1}^{\ell} (a_iE_{ii^*} - a_i^*E_{m+2\ell+1-i,\; m+2\ell+1-i}) + S, \quad \text{where} \tag{8}$$

S is a skew transformation $V \to V$, and its action is the same as that of $\sum_{i=1}^{\ell} t(a_i)X_i + S$, or of $(\sum_{i=1}^{\ell} t(a_i))X_\ell + S$, because all $X_{i+1} - X_i$ annihilate M^+. From the definition of T_ℓ and its action, we conclude that X acts by the action of S, plus the scalar from Z

$$-\frac{k}{2}\sum_{i=1}^{\ell} t(a_i).$$

Thus the structure of the g_0-module M^+ is essentially determined by its structure as module for the Lie algebra L of skew transformations $S:V \to V$.

In case $L = [LL]$, the trivial extension of such S, to U, annihilating all u_j, is in g, so L may be assumed to annihilate all u_j. Otherwise, if $S \in L$, and if $v_1,\ldots,v_m$ is a D-basis for V, $v_iS = \sum_j s_{ij}v_j$, $t(\Sigma s_{ii}) \neq 0$, we extend S to act as the Z-scalar $\frac{1}{2\ell}t(\Sigma s_{ii})$ on each u_j. Except when the involution in D is of second kind, the quantity $t(\Sigma s_{ii})$ is equal to zero; in the exceptional case, $L = F\zeta I_V \oplus [LL]$, where I_V is the identity mapping $V \to V$, $\zeta \in Z$ has $\zeta^* = -\zeta \neq 0$, and $t(\Sigma s_{ii})$ is an F-multiple of ζ. Thus the extension to U is in g.

For this extension C of ζI_V to U, we may write every element of g_0 uniquely in the form

$$X = \sum_{i=1}^{\ell} S_{u_i,a_iu_{m+1\ell+1-i}} + \beta C + S, \text{ where } \beta \in F,\ S \in g \tag{9}$$

stabilizes V and annihilates all u_j, and where the $a_i \in D$ satisfy $\sum_i t(a_i) \in F$. For if $t(a) = t^+(a) + t^-(a)\zeta$, where $t^+(a), t^-(a) \in F$, the condition that an element X as in (9) be in g is that $\sum_{i=1}^{\ell}(t(a_i) - t(a_i^*)) + \sum_j t(s_{jj}) = 0$. Now $t(a)^* = t(a^*)$ for all $a \in D$, so the above relation is

$$\sum_{i=1}^{\ell} 2t^-(a_i)\zeta + \sum_j t(s_{jj}) = 0.$$

In particular, $\sum_j t^+(s_{jj}) = 0$, and $2\sum_{i=1}^{\ell} t^-(a_i) + \sum_j t^-(s_{jj}) = 0$. Thus $X - \beta C$, for $\beta = \frac{1}{m}\Sigma t^-(s_{jj})$, when written in the form (9), has "S" and the "a_i" as required in our refined expression (8). In all other cases, we have (9) without the term "βC".

For brevity, write $u = u_\ell$, $y = u_{m+\ell+1}$. Then for $v,w \in V$,

$$\pi(S_{u,v}S_{y,w}) = [S_{u,v}S_{y,w}] = S_{w,v} - \varepsilon S_{u,(v,w)y}$$

acts as a transformation $-\varepsilon\psi_1(v;w)$ of M^+. The pairing $\psi_1(v;w)$ is evidently F-bilinear in v and w, and $\psi_1(v;w) + \varepsilon\psi_1(w;v)$ acts as does

$$\varepsilon S_{u,((w,v)+\varepsilon(v,w))y} = \varepsilon S_{u,((w,v)+(w,v)^*)y}$$

$$= -((w,v)+(w,v)^*)E_{\ell\ell} + ((w,v)+(w,v)^*)E_{m+\ell+1,m+\ell+1},$$

or as $k\ t\frac{((w,v)+(w,v)^*)}{2} = \frac{k}{2}t((w,v)+(w,v)^*) \in F$. Moreover, we have that

$[\psi_1(v;w), \psi_1(x;z)]$ is the action of the commutator

$$[S_{w,v} - \varepsilon S_{u,(v,w)y}, S_{x,z} - \varepsilon S_{u,(x,z)y}]$$

$$= S_{wS_{x,z},v} + S_{w,vS_{x,z}} + [S_{u,(v,w)y}, S_{u,(x,z)y}]$$

$$= -\varepsilon S_{v,wS_{x,z}} - \varepsilon S_{vS_{x,z},w} - \varepsilon S_{(x,z)^*u,(v,w)y}$$

$$+ \varepsilon S_{u,(v,w)(x,z)y}$$

$$= -\varepsilon S_{v,wS_{x,z}} - \varepsilon v S_{vS_{x,z},w} + \varepsilon S_{u,[(v,w),(x,z)]y},$$

and we have seen that the last term acts as zero. Combining the above with the fact that $S_{x,z}$ acts skewly on V, we have

$$[\psi_1(v;w), \psi_1(x;z)] = \psi_1(vS_{x,z};w) + \psi_1(v;wS_{x,z}).$$

For $a \in D$, $S_{aw,v} = S_{w,a^*v}$, and

$$S_{u,(v,aw)y} - S_{u,(a^*v,w)y} = S_{u,[(v,w),a^*]y}$$

annihilates M^+. Thus

$$\psi_1(v;aw) = \psi_1(a^*v;w) .$$

Let $\psi: V \otimes_F V \to \mathrm{End}_F(M^+)$ be the F-linear mapping such that $\psi(v \otimes w) = \psi_1(v;w)$. What we have shown is

<u>Lemma 3.1</u>. a) $\psi(av \otimes w) = \psi(v \otimes a^*w)$;

b) $[\psi(v \otimes w), \psi(x \otimes z)] = \psi(vS_{x,z} \otimes w) + \psi(v \otimes wS_{x,z})$;

c) $\psi(v \otimes w) + \varepsilon\, \psi(w \otimes v) = \frac{k}{2}t((w,v) + (w,v)^*)$,

<u>for all</u> $v,w,x,z \in V$ <u>and all</u> $a \in D$.

Next we show by induction the following

<u>Lemma 3.2</u>. <u>For</u> $v,w,x \in V$, <u>let</u> $[v,w,x] = vS_{w,x} - \varepsilon(w,x)v \in V$. <u>Then the action on</u> M^+ <u>of</u>

$$\pi(S_{u,v_1} \cdots S_{u,v_t} S_{y,w_1} \cdots S_{y,w_t}) \in U(g_0) \text{ is}$$

<u>given, for each positive integer</u> t, <u>by</u> $(-\varepsilon)^t\, \psi_t(v_1,\dots,v_t;\, w_1,\dots,w_t)$, ψ_t <u>defined inductively as</u>

$$\sum_{i=1}^{t} \psi_1(v_i;w_1)\,\psi_{t-1}(v_1,\ldots,\hat{v}_i,\ldots,v_t;\,w_2,\ldots,w_t)$$

$$+ \sum_{i<j} \psi_{t-1}(v_1,\ldots,[v_i,v_j,w_1],\ldots,\hat{v}_j,\ldots,v_t;w_2,\ldots,w_t),$$

with $[v_i,v_j,w_1]$ *being the* i-th *argument in the second sum.*

Proof. For $t = 1$, the assertion was the definition of ψ_1. Assuming it for lower values of t, consider the element

$B = S_{u,v_1}\ldots S_{u,v_t}S_{y,w_1}\ldots S_{y,w_t}$ of $U(g)$. If $X,Y \in U(g)$, write $X \sim Y$ if $\pi(X) = \pi(Y)$. Then

$$B \sim \sum_{j=1}^{t} S_{u,v_1}\ldots[S_{u,v_j}, S_{y,w_1}]\ldots S_{u,v_t}S_{y,w_2}\ldots S_{y,w_t}$$

$$= \sum_{j=1}^{t} [S_{u,v_j}S_{y,w_1}]S_{u,v_1}\ldots\hat{S}_{u,v_j}\ldots S_{u,v_t}S_{y,w_2}\ldots S_{y,w_t}$$

$$+ \sum_{i<j} S_{u,v_1}\ldots[S_{u,v_i}[S_{u,v_j}S_{y,w_1}]]\ldots\hat{S}_{u,v_j}\ldots S_{u,v_t}S_{y,w_2}\ldots S_{y,w_t}.$$

Now $[S_{u,v_j}\, S_{y,w_1}] = S_{w_1,v_j} - \varepsilon\, S_{u,(v_j,w_1)y}$ acts on M^+ by $-\varepsilon\,\psi_1(v_j;w_1)$, and $[S_{u,v_i}[S_{u,v_j}S_{y,w_1}]] = -\,\varepsilon\, S_{u,[v_i,v_j,w_1]}$. By induction, $\pi(B)$ acts on M^+ as stated in the lemma.

Thus $\psi:V \otimes_F V \to End_F(M^+)$ satisfies

d) $\psi_{k+1} = 0$, where $\psi_t:V^{2t} \to End_F(M^+)$ is defined as in Lemma 3.2 with $\psi_1(v;w) = \psi(v \otimes w)$.

The pair consisting of an associative F-algebra $A_k = A_k(V)$ with unit, an F-linear mapping $\varphi:V \otimes_F V \to A_k$ satisfying the identities a) – d) as given for ψ, and universal with these properties, generalizes the even Clifford algebra, and will be called *the* k-th *even Clifford algebra of* V (the form (v,w) being understood). (The definition by these conditions also applies when $D = F$, $\varepsilon = 1$.) A number of its properties have been developed in [Se4], [Se5], of which the second reference is more relevant here. In particular, $A_k(V)$ is finite-dimensional and there is a morphism of F-Lie algebras

$$\gamma:L \oplus FI_V \to A_k(V) \text{ with image } \varphi(V \otimes V) + F1,$$

a generating set for the associative algebra $A_k(V)$, such that $\gamma(I_V) = 1$,

$$\gamma(S_{v,w}) = \varphi(v \otimes w) - \frac{k}{4}\, t((w,v) + (w,v)^*)1.$$

From the universality of (A_k, φ), there is a unique structure of right A_k-module on M^+ with $x\varphi(v \otimes w) = x\,\psi(v \otimes w)$ for all $x \in M^+$, v and $w \in V$. Assuming that M is an irreducible g-module, the fact that the action of any element of g_0 on the irreducible g_0-module M^+ differs only by a scalar in F from an element of

$\psi(V \otimes V)$ shows that M^+ is an irreducible right A_k-module. When the involution in D is of first kind, one has for $x \in M^+$, v and $w \in V$,

$$x\,\gamma(S_{v,w}) = x\varphi(v \otimes w) - \frac{k}{4}\, t((w,v) + (w,v)^*)x$$

$$= x\,\psi(v \otimes w) - \frac{k}{4}\, t((w,v) + (w,v)^*)x$$

$$= -\varepsilon\, x(S_{w,v} - \varepsilon\, S_{u,(v,w)y}) - \frac{k}{4}\, t((w,v) + (w,v)^*)x$$

$$= x\, S_{v,w} + x(-\varepsilon(v,w)^* E_{\ell\ell} - \varepsilon(v,w)E_{m+\ell+1,m+\ell+1})$$

$$- \frac{k}{4}\, t((w,v) + (v,w)^*)x.$$

As we have seen before, the second term is $\frac{k}{2}\, t(\varepsilon(v,w)^*)x = \frac{k}{2}\, t((w,v))x$ $= \frac{k}{4}\, t((w,v) + (w,v)^*)x$. That is, the action of L on M^+ is compatible with the structure of A_k-module on M^+ via the mapping $\gamma : L \to A_k$.

The same conclusion holds when * is of second kind. The proof of this assertion requires some care in handling the element C. For $\varepsilon = -1$ it is given in [Se5] (§5); there is no gain in insight to be had by repeating that proof here for $\varepsilon = \pm 1$, and we omit it. As a conclusion, we have

Theorem 3.1. There is a one-one correspondence between $k\lambda_\ell$-admissible $\mathfrak{g}_0$-modules and irreducible right modules for $A_k(V)$, the underlying vector spaces being the same. In our embedding of L in $\mathfrak{g}_0 \subset \mathfrak{g}$, the action of $S \in L$ on the module agrees with that of $\gamma(S) \in A_k(V)$, and the element $\sum_{i=1}^{\ell} S_{u_i, a_i u_{m+2\ell+1-i}}$, $t^-(\Sigma a_i) = 0$, acts as the scalar $\frac{k}{2}\,\varepsilon\, t(\Sigma a_i) \in F$. Thus the irreducible $\mathfrak{g}$-modules of highest $\mathfrak{t}_0$-weights $k\lambda_\ell$ are in one-one correspondence with the irreducible $A_k(V)$-modules, by the considerations of Chapter 2.

Proof. One needs only remark that an irreducible $A_k(V)$-module, when made into a $\mathfrak{g}_0$-module as above, is annihilated by the ideal in $U(\mathfrak{g}_0)$ generated by the $\pi(z)$ because of the defining identities of $A_k(V)$, the action of $\mathfrak{t}_0$, and the prescriptions concerning the action of elements $\sum_i S_{u_i, a_i u_{m+2\ell+1-i}}$ as above. Except for the case of second kind [Se5], all these facts are contained in reversible steps above.

In [Se5], it is shown, at least for $\varepsilon = -1$, that A_k is a semi-simple associative algebra whenever A_k is non-zero. In the case of involution of first kind, where V and D are large enough so that L is simple (or even semi-simple) that is clear, because every A_k-module is an L-module via γ, so is L-semisimple, so is A_k-semisimple. Thus minimal right ideals in A_k afford all $k\lambda_\ell$-admissible $\mathfrak{g}_0$-modules.

§3. Splitting Data; Completeness of the Clifford Modules.

We now develop splitting information in detail, in order to invoke the considerations of Chapter 2, §3. This information will yield a basic set of t_0-highest weights $\pi_i|t_0$, as in Theorem 2.1, and in particular, which highest weights of representations of $K \otimes g$ have restriction $k\lambda_\ell$ to t_0. This last will enable us, in principle, to determine the dimension of A_k and of its center, as well as its irreducible modules. Furthermore, we shall see that every irreducible $[LL]$-module is an irreducible A_k-module for suitable k, and thereby complete the constructive representation theory of the central simple algebras $[LL]$. The algebras $A_k(V)$ play a role here completely analogous to that of the symmetric powers $S^k(A)$ of Chapter 1 in the representations of [AA]. In the spirit of that chapter, our considerations construct all irreducible modules of all central simple Lie algebras except for forms of the split exceptional algebras $G_2 - E_8$ and exceptional forms of D_4. (See [J3], Chapter 10, for the classification.)

We consider $End_D(U)$ with the involution σ such that $(uT,v) = (u,vT^\sigma)$ for all $u,v \in U$, $T \in End_D(U)$, so that g is the derived algebra of the F-Lie algebra of those T such that $T^\sigma = -T$. Except when $D = F$ and $\varepsilon = -1$, V has a D-basis $v_{\ell+1},\ldots,v_{\ell+m}$ with $(v_i,v_j) = \delta_{ij}\gamma_i$, $0 \neq \gamma_i \in D$, $\gamma_i^* = \varepsilon\gamma_i$. In the exceptional case, $m = 2r$ is even and V has an F-basis $v_{\ell+1},\ldots,v_{\ell+m}$ with $(v_i,v_j) = \delta_{i+j,2\ell+m+1}\,\mathrm{sgn}(j-i)$. Here g is split simple, with splitting Cartan subalgebra h having F-basis the elements $S_{u_i,u_{2\ell+m+1-i}}$, $1 \le i \le \ell$, of t_0, along with the $S_{v_i,v_{2\ell+m+1-i}}$, $\ell < i \le \ell+r$. The "simple corrots" of h may be taken to be the $S_{u_{i+1},u_{2\ell+m-i}} - S_{u_i,u_{2\ell+m+1-i}}$, $1 \le i < \ell$, the element $-S_{u_\ell,u_{\ell+m+1}} + S_{v_{\ell+1},v_{\ell+m}}$, the $S_{v_{i+1},v_{2\ell+m-i}} - S_{v_i,v_{2\ell+m+1-i}}$, $\ell+1 \le i < \ell+r$, and $-S_{v_{\ell+r},v_{\ell+r+1}}$, associated with a simple system of roots $\gamma_1,\ldots,\gamma_{\ell+r}$ (in that order) of type $C_{\ell+r}$:

Here $g_0 = t_0 + K(V)$, where $K(V)$ is the Lie algebra of skew transformations of V, with Cartan subalgebra h_0 with basis the coroots of $\gamma_{\ell+1},\ldots,\gamma_{\ell+r}$, and the restrictions to h_0 of these roots as simple roots. The fundamental weights π_j of g relative to h are $\pi_j = \gamma_1 + 2\gamma_2 + \ldots + j(\gamma_j + \ldots + \gamma_{\ell+r-1}) + \frac{j}{2}\gamma_{\ell+r}$, $j < \ell+r$,

$$\pi_{\ell+r} = \gamma_1 + 2\gamma_2 + \ldots + (\ell+r-1)\gamma_{\ell+r-1} + \frac{\ell+r}{2}\gamma_{\ell+r}.$$

The restriction of γ_i to t_0 is α_i, $1 \le i \le \ell$, while those of the remaining γ_j are all zero. If we let β_j be the restriction to h_0 of $\gamma_{\ell+j}$, $1 \le j \le r$, then $\gamma_i|h_0 = 0$ for $i < \ell$, while $\gamma_\ell|h_0 = -\pi_{\ell+1}|h_0 = -\beta_1 - \beta_2 - \ldots - \beta_{r-1} - \frac{1}{2}\beta_r$. If $\lambda_1,\ldots,\lambda_\ell$ are the fundamental weights for t_0, we thus have $\pi_j|t_0 = \lambda_j$, $1 \le j \le \ell$;

$\pi_\ell|_{t_0} = 2\lambda_{\ell_{\ell-1}} = \pi_j|_{t_0}$ for all $j \geq \ell$. Thus $\pi = \sum_{i=1}^{\ell+r} n_i \pi_i$ has restriction to t_0 equal to $\sum_{i=1} n_i \lambda_i + (2 \sum_{i\geq\ell} n_i)\lambda_\ell$.

To have $\pi|_{t_0} = k\lambda_\ell$ we must have $n_i = 0$ for $i < \ell$ and $k = 2 \sum_{i\geq\ell} n_i \equiv 0$ (mod 2)-always assuming $r > 0$. By the remarks of §2.1, it follows that g has no irreducible modules of highest t_0-weight $k\lambda_\ell$ unless k is even.

When M is such a module, the highest weight of the irreducible $K(V)$-module M^+ is the restriction to h_0 of $\pi = \sum_{i\geq\ell} n_i \pi_i$, or

$$\sum_{i>\ell} n_i \pi_i + n_\ell \pi_\ell|h_0$$

$$= \sum_{i>\ell} n_i \omega_{i-\ell} + n_\ell \ell(\gamma_\ell + \ldots + \gamma_{\ell+r-1} + \frac{1}{2}\gamma_{\ell+r})\Big|_{h_0} = \sum_{i>\ell} n_i \omega_{i-\ell},$$

where $\omega_1,\ldots,\omega_r$ are the fundamental weights associated with the base $\beta_1,\ldots,\beta_r$ for the roots of $K(V)$ relative to h_0. Here $2(\sum_{i>\ell} n_i) = k - 2n_\ell$.

Now let $\omega = \sum_{i>\ell} n_i \omega_{i-\ell}$ be any dominant integral function on h_0 (n_i non-negative integers), and let k be an even integer, $k \geq 2 \sum_{i>\ell} n_i$. Set $2n_\ell = k - \sum_{i>\ell} 2n_i$ and let M be the irreducible g-module of highest h-weight $\sum_{j\geq\ell} n_j \pi_j = \pi$. Then $\pi|_{t_0}$ is the highest t_0-weight, or $k\lambda_\ell$, and the highest h_0-weight of the action of $K(V_0)$ on M^+ is ω. Thus M^+, the irreducible $K(V)$-module of highest weight ω, is the highest t_0-weight space of an irreducible g-module of highest t_0-weight $k\lambda_\ell$, whenever k is even and $k \geq 2 \sum_{i>\ell} n_i$. In our case this is equivalent with having M^+ an irreducible module for $A_k(V)$. We have, for the case $D = F$, $\varepsilon = -1$, with notations as above:

Theorem 3.2. $A_k(V) = \{0\}$ unless k is even. When k is even, the irreducible right $A_k(V)$-modules are the irreducible right $K(V)$-modules whose highest weights $\sum_{i=1}^{r} m_i\omega_i$ satisfy $2 \sum m_i \leq k$.

The point of the theorem is that the finite-dimensional semisimple associative algebras $A_0(V)$, $A_2(V)$, $A_4(V)$,..., have as their irreducible right modules (up to equivalence) an increasing family $S_0 \subset S_2 \subset S_4 \subset \ldots$ of finite sets of irreducible right $K(V)$-modules, whose union consists of all finite-dimensional irreducible $K(V)$-modules.

In the sequel we have a D-basis for V as when the case $D = F$, $\varepsilon = -1$ is excluded, and we proceed from there. Let E_i be a maximal *-stable subfield of D containing $(v_i,v_i) = \gamma_i \neq 0$, $\ell < i \leq \ell+m$, and let E_0 be an arbitrary maximal *-stable subfield of D. Then E_0 and each E_i is a maximal subfield of D, of dimension d over Z, where $[D:Z] = d^2$, and a finite field extension K of F may be chosen so that $K \otimes_F E_0$ and each $K \otimes_F E_i$ is a product of d or $2d$ copies of K, according as $Z = F$ or $[Z:F] = 2$. We consider the K-involutorial algebra

$(K \otimes_F \mathrm{End}_D(U),\ 1 \otimes \sigma)$, which as algebra is $K \otimes_F M_{m+2\ell}(D) \approx M_{m+2\ell}(K \otimes_F D)$. If $t = [E_j:F]$, let $e_1^{(j)},\ldots,e_t^{(j)}$ be a complete family of primitive idempotents in E_j, which in case $Z \neq F$ may be so chosen that $e = \sum_{i=1}^{d} e_i^{(j)}$ and $f = \sum_{i=1}^{d} e_{d+i}^{(j)}$ are in $K \otimes_F Z$, with $e^* = f$, $f^* = e$ (see [Se5], p. 435).

When $Z = F$ we have for each j that $e_1^{(j)},\ldots,e_d^{(j)}$ are a complete set of primitive idempotents in $K \otimes_F D \approx M_d(K)$, and the $*$-stability of E_j implies that these $e_i^{(j)}$ are permuted by (the K-linear extension of) $*$. For fixed j, k we evidently have $K \otimes_F D = \sum_{\mu,\nu=1}^{d} e_\mu^{(j)} (K \otimes_F D) e_\nu^{(k)}$; the sum is direct, and the simplicity of $K \otimes_F D$ implies that no term is zero. By dimensions, we see that each term has dimension <u>one</u>.

If $\varepsilon = -1$, then no E_j $(j > 0)$ is fixed by $*$, and the permutation p of the corresponding $e_\mu^{(j)}$ is non-trivial: $e_\mu^{(j)*} = e_{p(\mu)}^{(j)}$. Here we fix E_0 as one of the E_j for $j > 0$ so that the same applies when $j = 0$. Then γ_j is a unit in $K \otimes_F E_j$, so $\gamma_j = \sum_\mu \varkappa_{j\mu} e_\mu^{(j)}$, all $\varkappa_{j\mu} \neq 0$ in K, $\gamma_j^* = -\gamma_j = \sum_\mu \varkappa_{j\mu} e_{p(\mu)}^{(j)}$, and it follows that $\varkappa_{j,p(\mu)} = -\varkappa_{j\mu}$ for each μ, and that p <u>has no fixed points</u>. Thus the idempotents $e_\mu^{(j)}$ may be labeled $e_1^{(j)},\ldots,e_{2s}^{(j)}$ with $e_\mu^{(j)*} = e_{2s+1-\mu}^{(j)}$. If $\varepsilon = 1$ and $D = F$, we take K to be a finite extension field of F such that $K \otimes_F V$ is a <u>split</u> quadratic space over K, <u>i.e.</u>, there is a K-basis $w_1,\ldots,w_m$ such that the bilinear extension of the original symmetric form satisfies $(w_i,w_j) = \delta_{i+j,m+1}$. If $\varepsilon = 1$ and $D \neq F$, the Lie algebra $K(V)$ and the involutorial algebra $(\mathrm{End}_D(V),\ \sigma)$ are unchanged if we take $\delta \in D$, $\delta^* = -\delta \neq 0$, replace $*$ by the involution $a \to \delta^{-1} a^* \delta$ and the form (v,w) on V by $\langle v,w \rangle = (v,w)\delta$, <u>antihermitian</u> with respect to the new involution. Accordingly <u>we shall assume</u> $\varepsilon = -1$ <u>except when</u> $D = F$.

The case $Z \neq F$ thus allows us to assume $\varepsilon = -1$. Here the fact that $e^* = f$ as above again allows a labeling of the $e_\mu^{(j)}$, $1 \le \mu \le 2d$, so that $e_\mu^{(j)*} = e_{2d+1-\mu}^{(j)}$. The fact that e, f are central idempotents means that for fixed j,k,

$$K \otimes_F D = \sum_{\mu,\nu=1}^{d} e_\mu^{(j)} (K \otimes_F D)\, e_\nu^{(k)} \oplus \sum_{\mu,\nu=1}^{d} e_{d+\mu}^{(j)} (K \otimes_F D)\, e_{d+\nu}^{(k)} .$$

The two summands are, respectively, $(K \otimes_F D)e$ and $(K \otimes_F D)f$, each a central simple algebra isomorphic to $M_d(K)$. Again each term in each sum is a space of dimension one.

The involution σ in $\mathrm{End}_D(U)$ sends cE_{ii} to $c^* E_{m+2\ell+1-i,m+2\ell+1-i}$, $1 \le i \le \ell$, and $c\,E_{\ell+i,\ell+i}$ to $(\gamma_{\ell+1}\, c^*\, \gamma_{\ell+i}^{-1})\, E_{\ell+i,\ell+i}$, $1 \le i \le m$, where $c\,E_{ij}$ sends the i-th basis vector to c times the j-th vector. Passing to our splitting, with $\varepsilon = -1$, we have, for $1 \le i \le \ell$, $(e_\mu^{(0)} E_{ii})^\sigma = e_{t+1-\mu}^{(0)} E_{m+2\ell+1-i,m+2\ell+1-i}$, $1 \le \mu \le t$, and $(e^{(j)} E_{\ell+j,\ell+j})^\sigma = e_{t+1-\mu}^{(j)} E_{\ell+j,\ell+j}$, $1 \le j \le m$, $1 \le \mu \le t$. All the elements here are primitive idempotents in $K \otimes_F (\mathrm{End}_D(U))$ and the sum of the distinct ones is the identity.

Still with $\varepsilon = -1$ (and $D \neq F$), and with $Z = F$, we arrange the idempotents above as follows: Let $f_{(i-1)t+j} = e_j^{(0)} E_{ii}$, $1 \le j \le t$, $1 \le i \le \ell$, and $f_{t(m+2\ell)+1-((i-1)t+j)} = e^{(0)}_{t+1-j} E_{m+2\ell+1-i,\, m+2\ell+1-i}$, thereby defining $f_1,\dots,f_{\ell t}$ and $f_{mt+\ell t+1},\dots,f_{t(m+2\ell)}$. We have seen that $t = 2s$ is even, so we set $f_{\ell t+(i-1)s+j} = e_j^{(i)} E_{\ell+i,\ell+i}$, $1 \le i \le m$, $1 \le j \le s$, and $f_{t(m+2\ell)+1-(\ell t+(i-1)s+j)} = e^{(i)}_{t+1-j} E_{\ell+i,\ell+i}$, thereby defining $f_{\ell t+1},\dots,f_{\ell t+ms}$, $f_{\ell t+ms+1},\dots,f_{mt+\ell t}$. The involution σ then sends f_μ to $f_{t(m+2\ell)+1-\mu}$ for each μ. Writing g for $s(m+2\ell)$, $A = K \otimes \mathrm{End}_D(U) \approx M_{2g}(K)$, with $f_1,\dots,f_{2g}$ a complete set of primitive idempotents. Then A is the direct sum of the one-dimensional spaces $f_i A f_j$. If $i+j \ne 2g+1$ and $f_i a f_j \ne 0$, then $f_i a f_j - f_{2g+1-j} a^\sigma f_{2g+1-i}$ is a nonzero element of $K \otimes_F \mathfrak{g}$, while $f_i a f_j + f_{2g+1-j} a^\sigma f_{2g+1-i}$ is hermitian with respect to the involution in A. We also have

$$(f_i a f_{2g+1-i})^\sigma = f_i a^\sigma f_{lg+1-i},$$

so that for each i one and only one of the possibilities holds:

(a) There is $a \in A$, $a^\sigma = a$, such that $f_i a f_{2g+1-i} \ne 0$;

(b) There is $a \in A$, $a^\sigma = -a$, such that $f_i a f_{2g+1-i} \ne 0$.

If h is the number of indices i for which (b) holds, then $\dim \mathfrak{g} = \dim_K(K \otimes_F \mathfrak{g}) = \binom{2g}{2} + h$.

Now $K \otimes_F \mathfrak{g}$ is the Lie algebra of skew elements of $A \approx M_{2g}(K)$ with respect to an involution, and such involutions are associated with non-degenerate symmetric or alternate forms in $2g$ variables. Accordingly the dimension of $K \otimes_F \mathfrak{g}$ is either $\binom{2g}{2}$ or $\binom{2g+1}{2}$, so that $h = 0$ or $2g$. That is, either (a) holds for all i, or (b) holds for all i. In the former case, $K \otimes_F \mathfrak{g}$ is a (split) simple K-Lie algebra of type D_g, in the latter of type C_g. (A splitting is displayed below.) Continuing with $\varepsilon = -1$, $Z = F$, $D \ne F$, let $\mathfrak{h}$ be the set of elements of $\mathfrak{g}$ mapping each u_i, $1 \le i \le \ell$, $\ell + m < i \le m + 2\ell$, to an E_0-multiple of itself, and each $v_{\ell+j}$, $1 \le j \le m$, to an E_j-multiple of itself. Then $\mathfrak{h}$ is a commutative subalgebra containing $\mathfrak{t}_0$, and is readily seen to be a Cartan subalgebra of $\mathfrak{g}$. The Cartan subalgebra $K \otimes \mathfrak{h}$ of $K \otimes \mathfrak{g}$ then has as K-basis all $f_i - f_i^\sigma = f_i - f_{2g+1-i}$, $1 \le i \le g$, and is a splitting Cartan subalgebra in the sense of [J3], Chapter 4. The roots may be ordered so that the positive root-spaces are those of the form $\{f_i a f_j - f_{2g+1-j} a^\sigma f_{2g+1-i} \mid a \in A,\ i < j,\ i+j \ne 2g+1\}$ and, if $h = 2g$, $\{f_i a f_{2g+1-i} \mid a^\sigma = -a,\ 1 \le i \le g\}$. Corresponding to simple roots $\varepsilon_1,\dots,\varepsilon_g$ we have $\{f_1 a f_2 - f_{2g-1} a^\sigma f_{2g}\},\dots,\{f_{g-1} a^\sigma f_g - f_{g+1} a^\sigma f_{g+2}\}$, and

$$\left\{\begin{array}{ll} \{f_g a f_{g+1} \mid a^\sigma = -a\}, & \text{if } h = 2g; \\ \{f_g a f_{g+2} - f_{g-1} a^\sigma f_{g+1}\}, & \text{if } h = 0 \end{array}\right\}, \text{ respectively,}$$

giving Dynkin diagrams

C_g: 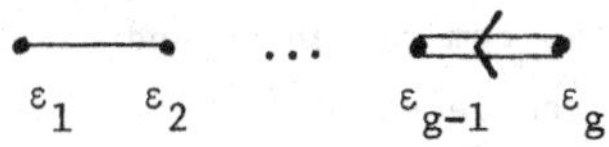

, $h = 2g$, or

D_g: 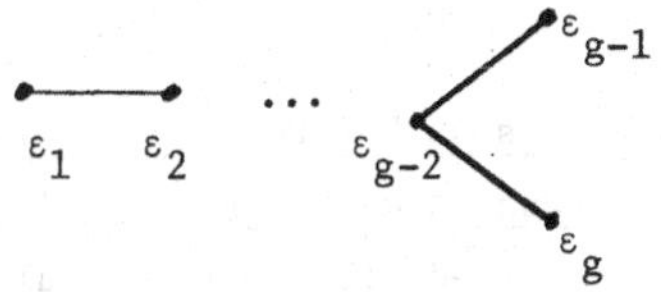

, if $h = 0$.

We have $\varepsilon_i|_{t_0} = 0$ except when $i \le \ell t = \ell d$ and $d|i$. In those cases, $\varepsilon_{id}|_{t_0} = \alpha_i$. The simple roots of $K \otimes_F K(V)$ relative to the intersection h_V with $K \otimes_F h$ of this simple subalgebra are the restrictions of $\varepsilon_{\ell d+1}, \ldots, \varepsilon_g$, a system of type $C_{g-\ell d}$ resp. $D_{g-\ell d}$. In case $h = 2g$, a fundamental weight

$$\pi_i = \sum_{j=1}^{i-1} j\, \varepsilon_j + i(\varepsilon_i + \ldots + \varepsilon_{g-1} + \frac{\varepsilon_g}{2}),$$

as when $D = F$, $\varepsilon = -1$, can only have restriction to t_0 of the form $k\lambda_\ell$ if $i \ge d\ell$, in which case the restriction is $2d\lambda_\ell$. The restriction of such a π_i for $i > d\ell$ to h_V is the $(i - d\ell)$-th fundamental weight $\omega_{i-d\ell}$ with respect to the base of roots with $\beta_{i-d\ell} = \varepsilon_i|_{h_V}$, while $\pi_{d\ell}|_{h_V} = 0$.

By the results of Chapter 2, the only irreducible g-modules M of highest t_0-weight $k\lambda_\ell$ are those for which the irreducible constituents over K have highest weights $\sum_{i=\ell d}^{g} m_i \pi_i$ with $2d \sum_{i=\ell d}^{g} m_i = k$. Over K, M^+ is the sum of irreducible $K \otimes g_0$-modules, which are irreducible $K \otimes K(V)$-modules in a way that determines their structure as $K \otimes g_0$-modules, and their highest weights must be of the form $\sum_{i=1}^{g-\ell d} m_i \omega_i$ with $\sum m_i \le \frac{k}{2d}$, an integer. As when $D = F$, $\varepsilon = -1$, each irreducible $K \otimes K(V)$-module with such a highest weight has the property that its irreducible $K(V)$-submodules are highest t_0-weight spaces of irreducible g-modules of highest t_0-weight $k\lambda_\ell$, therefore are modules for $A_k(V)$. We have shown

<u>Theorem 3.3. Suppose $\varepsilon = -1$, $D \neq F$, $F = Z$, and $K(V)$ is an F-form of the split simple algebra of type C_n. Then $A_k(V) = \{0\}$ unless $2d|k$, where $[D:F] = d$. When $k = 2dm$, the irreducible $A_k(V)$-modules are the irreducible $K(V)$-modules whose constituents upon extension to a splitting field K as above have highest weights $\sum_{i=1}^{n} m_i \omega_i$ satisfying $\sum m_i \le m$. All irreducible finite-dimensional $K(V)$-modules are obtained by this procedure, for suitable k. If S_k is the family of equivalence classes of irreducible $K(V)$-modules that are realized as $A_k(V)$-modules, we have</u>

$$S_0 \subset S_{2d} \subset S_{4d} \quad \ldots \quad .$$

The theorem is obtained by embedding V in a suitably larger space U as above.

The completeness of the set of S_k's follows from the remarks above and §2.3. If $2d \nmid k$, $A_k(V)$ has no irreducible modules, so must reduce to $\{0\}$.

By analogous reasoning, we obtain

Theorem 3.4. *Suppose $\varepsilon = -1$, $Z = F \neq D$, and $K(V)$ is an F-form of the split simple algebra of type D_n $(n \geq 3)$. Then $A_k(V) = \{0\}$ unless $d|k$, where $[D:F] = d^2$. When $k = dm$, the irreducible $A_k(V)$-modules are the irreducible $K(V)$-modules whose constituents upon extension to the splitting field K have highest weights $\omega = \sum_{i=1}^{n} m_i \omega_i$ satisfying $s(\omega) = 2 \sum_{i=1}^{n-2} m_i + m_{n-1} + m_n \leq m$, $s(\omega) \equiv m \pmod 2$. All irreducible $K(V)$-modules occur as irreducible $A_k(V)$-modules for suitable k. With S_k as in Theorem 3.3, we have*

$$S_0 \subset S_{2d} \subset S_{4d} \subset \dots ,$$

$$S_d \subset S_{3d} \subset \dots ,$$

the unions of the two chains being disjoint.

One must here distinguish between

$$\pi_i = \sum_{j<i} j\, \varepsilon_j + i(\varepsilon_i + \dots + \varepsilon_{g-2} + \tfrac{1}{2}\varepsilon_{g-1} + \tfrac{1}{2}\varepsilon_g), \quad i \leq n-2,$$

and the pair

$$\pi_{g-1} = \tfrac{1}{2}\Big(\sum_{j<g-1} j\, \varepsilon_j + \tfrac{g}{2}\varepsilon_{g-1} + \tfrac{g-2}{2}\varepsilon_g\Big),$$

$$\pi_g = \tfrac{1}{2}\Big(\sum_{j<g-1} j\, \varepsilon_j + \tfrac{g-2}{2}\;{}_{g-1} + \tfrac{g}{2}\varepsilon_g\Big).$$

Those π_i with $\ell d \leq i < g-1$ have restriction $2d\lambda_\ell$ to t_0, while the restriction of each of the last two is $d\lambda_\ell$. The assertion about disjointness of the unions of the two chains needs a little more argument, especially when $n = 4$; such an argument may be found in [Se5], pp. 455ff.

The case $Z \neq F$, $\varepsilon = -1$ is worked out in full in [Se5]. In this case the derived algebra $K'(V)$ is a central simple Lie algebra of (outer) type A. A splitting of $K'(U)$ as above yields a diagram

$\varepsilon_1 \quad \varepsilon_2 \qquad\qquad \varepsilon_{g-2} \quad \varepsilon_{g-1}$

where each of the two central simple K-algebras whose sum is $K \otimes_F \mathrm{End}_D(U)$ is isomorphic to $M_g(K)$. For $i \leq [\frac{g}{2}]$, the restriction to t_0 of ε_i is zero unless $d \mid i$, in which case it is α_i. Moreover, we have $\varepsilon_{g-i}\big|_{t_0} = \varepsilon_i\big|_{t_0}$ for all i. For $i \leq [\frac{g}{2}]$,

$$\pi_i = \tfrac{1}{g}\Big[(g-i)\sum_{i=1}^{i} j\, \varepsilon_j + i \sum_{j=i+1}^{g-1} (g-j)\, \varepsilon_j\Big]$$

has restriction $k\lambda_\ell$ to t_0 if and only if $i \geq \ell d$, and then the restriction is $2d\lambda_\ell$. Moreover, $\pi_{g-i}|_{t_0} = \pi_i|_{t_0}$. The corresponding diagram for $K'(V)$ is

$\beta_1 \quad \beta_2 \quad \cdots \quad \beta_{n-1}$,

where $\beta_i = \varepsilon_{i-\ell d}|_{h_V}$, $\ell d + 1 \leq i \leq g-\ell d-1$, so $n = g - 2\ell d$.

The corresponding results are expressed in:

Theorem 3.5. *Suppose* $\varepsilon = -1$, $Z \neq F$, *and* $K'(V)$ *is an F-form of the split simple Lie algebra of type* A_n $(n \geq 1)$. *Then* $A_k(V) = \{0\}$ *unless* $2d|k$, *where* $[D:Z] = d^2$. *When* $k = 2dm$, *each irreducible* $A_k(V)$-*module is an irreducible* $K(V)$-*module with at most two irreducible* $K'(V)$-*constituents. Upon extension to the splitting field* K, *these irreducible* $K'(V)$-*modules have constituents all of whose highest weights* $\omega = \sum_{i=1}^{q-1} m_i \omega_i$ *satisfy* $s(\omega) = \sum m_i \leq \frac{k}{2d}$. *All irreducible* $K'(V)$-*modules occur in irreducible* $A_k(V)$-*modules for suitable* k, *and if* S_k *is the family of those occurring in* $A_k(V)$-*modules, we have*

$$S_0 \subset S_{2d} \subset S_{4d} \subset \cdots .$$

The distinction between $K(V)$ and $K'(V)$ is essential in the above. If $\zeta \in Z$, $\zeta^* = -\zeta \neq 0$, so that ζI_V spans the center of $K(V)$, one sees in [Se5] (p. 404) that in any irreducible $A_k(V)$-module the transformation T representing ζI_V satisfies $T^2 = r^2\zeta^2$, where r is a rational number, limited by k to a finite set of values. When $r = 0$, one has $T = 0$, and the irreducible $A_k(V)$-module is an irreducible $K'(V)$-module. When $r \neq 0$, T satisfies an irreducible quadratic equation and any irreducible $K'(V)$-submodule of our irreducible $A_k(V)$-module is mapped by T onto an isomorphic $K'(V)$-submodule, the sum of the two $K'(V)$-submodules being the full module. (The last assertion of Theorem 5.3 of [Se5] is incorrect, and is corrected in the statement above.)

When $D = Z \neq F$, an explicit determination of the structure of $A_2(V)$ (with $\varepsilon = -1$) is given in [Se4] (Theorem 5). For the most part, the irreducible $K(V)$-modules occurring as $A_2(V)$-modules are the exterior powers $\Lambda^j(V)$ (formed over Z).

As the final case of this chapter, we return to $D = F$, $\varepsilon = 1$. Upon passage to a finite extension K, we may assume $K \otimes_F V$ has a K-basis $u_{\ell+1},\ldots,u_{\ell+m}$ with $(u_i,u_j) = \delta_{i+j,2\ell+m+1}$. Thus $K \otimes_F g$ is split by the Cartan subalgebra of transformations diagonal with respect to the basis $u_1,\ldots,u_{2\ell+m}$ for $K \otimes U$, which Cartan subalgebra has the form $K \otimes h$, h a Cartan subalgebra of g, upon suitable formation of the $u_{\ell+i}$ from the $v_{\ell+i}$. The Dynkin diagram for $K \otimes g$ relative to $K \otimes h$ and canonical ordering of the roots is

B_n:  $\varepsilon_1 \quad \varepsilon_2 \quad \cdots \quad \varepsilon_{n-1} \quad \varepsilon_n$, if m is odd, $n = \frac{2\ell+m-1}{2}$;

or D_n: 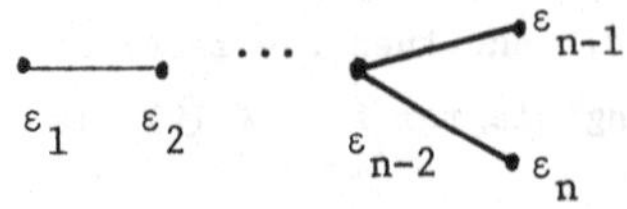

, if m is even, $n = \frac{2\ell+m}{2}$. One has $\varepsilon_i|_{\mathfrak{t}_0} = \alpha_i$, $1 \le i \le \ell$, with $\varepsilon_i|_{\mathfrak{t}_0} = 0$ if $i > \ell$. Moreover, $m \ne 1,2$, or 4, because $K(V)$ is to be central simple. The subdiagram corresponding to those ε_i with $i > \ell$ is the Dynkin diagram of $K \otimes K(V)$ relative to $\mathfrak{h}_V$.

When m is even, we see as above that the only π_i with restriction to $\mathfrak{t}_0$ equal to $k\lambda_\ell$ are those with $i \ge \ell$, and that $\pi|_{\mathfrak{t}_0} = \sum_{i \ge \ell} m_i \pi_i|_{\mathfrak{t}_0} = (2\sum_{i=\ell}^{n-2} m_i + m_{n-1} + m_n)\lambda_\ell$. When m is odd, $\pi_i = \sum_{j<i} j\varepsilon_j + i(\sum_{j \ge i} \varepsilon_j)$ for $i < n$, $\pi_n = \frac{1}{2}\sum_{j=1}^{n} j\varepsilon_j$. Again the restriction of $\pi = \sum m_i \pi_i$ to $\mathfrak{t}_0$ is of the form $k\lambda_\ell$ if and only if $m_i = 0$ for all $i < \ell$, with $k = 2\sum_{i=}^{n-1} m_i + m_n$. In the respective cases, we have $\omega = \pi|_{\mathfrak{h}_V} = \sum_{i=\ell+1}^{n} m_i \omega_{i-\ell} = \sum_{i=1}^{n-\ell} s_i \omega_i$, and we set $s(\omega) = 2\sum_{i<n-\ell-1} s_i + s_i + s_{n-\ell-1} + s_{n-\ell}$ resp. $2\sum_{i<n-\ell} s_i + s_{n-\ell}$. The conclusions, drawn as before, are then:

Theorem 3.6. *Let V be an F-vector space of dimension other than 1,2,4, carrying a nondegenerate symmetric bilinear form. Then $A_k(V) \ne \{0\}$ for every k. The irreducible $A_k(V)$-modules are the irreducible $K(V)$-modules whose constituents upon extension to the splitting field K have highest weights ω satisfying $s(\omega) \le k$, $s(\omega) \equiv k \pmod 2$. All irreducible $K(V)$-modules occur as irreducible $A_k(V)$-modules for suitable k. With S_k as in Theorem 5, we have*

$$S_0 \subset S_2 \subset S_4 \subset \dots ,$$

$$S_1 \subset S_3 \subset S_5 \subset \dots ,$$

the unions of the two chains being disjoint.

In [Se4] it is shown that S_2 consists essentially of the $K(V)$-modules $\Lambda^j(V)$. If $\dim V = 2t$ is even, it is possible that $\Lambda^t(V)$ is irreducible, but in other cases (such as the split one) $\Lambda^t(V)$ decomposes into two non-isomorphic absolutely irreducible modules, of the same dimension. All the other $\Lambda^j(V)$ are absolutely irreducible. S_1 consists of a single irreducible spin module if dim V is odd or if dim V is even and the center of the even Clifford algebra $C\ell^+(V)$ ($\simeq A_1(V)$) is a field. In the remaining case, S_1 consists of two "half-spin" modules.

Theorems 1.1, 3.2-3.6 may be claimed to give a constructive description of all irreducible finite-dimensional modules for all classical central simple Lie algebras, in the same sense as Theorem 1.1 gives such a description for the algebras $s\ell(n,D)$.

CHAPTER 4: CONSTRUCTION OF MODULES WITH PRESCRIBED RELATIVE HIGHEST WEIGHTS, FOR THE ISOTROPIC ALGEBRAS OF CHAPTER 3

§1. The Fundamental Relative Weights.

Let the involutorial division algebra D be as in Chapter 3, and let the D-vector space U be as there. We study the process of constructing the F-irreducible right modules for the Lie algebra $\mathfrak{g} = K'(U)$ from the fundamental ones, under the assumption that the D-combinations of $u_1,\ldots,u_\ell$ form a (nonzero) maximal totally isotropic subspace of U. When $D = F$ and $\varepsilon = -1$, the case is the (split) symplectic one, and - in contrast to the approach of Chapter 3 - nothing is to be added this way to the classical theory. When $D = F$ and $\varepsilon = 1$, the program aimed at here is carried out in Chapter VII of [Se3]. Accordingly, we may assume $D \neq F$, and therefore we may also assume $\varepsilon = -1$, as in §3.3.

The effect of the assumption that $u_1,\ldots,u_\ell$ span a maximal totally isotropic subspace is to make our $\mathfrak{t}_0$ a <u>maximal</u> split toral subalgebra of $\mathfrak{g}$. When $V = \{0\}$, the system of roots of $\mathfrak{g}$ relative to $\mathfrak{t}_0$ is reduced, and our program is carried out in Chapters V and VI of [Se3]. We therefore assume $V \neq \{0\}$. Then V is an anisotropic space.

By what we have seen in Chapter 3, or by the general considerations of [Se1], Chapter 1, the hypotheses (1) - (5) of Chapter 2, §1 are satisfied by $\mathfrak{g}$, $\mathfrak{t}_0$, and the $\mathfrak{t}_0$-roots of $\mathfrak{g}$. We retain all the notation of that chapter. In particular, if m is the D-dimension of V, and if $u,v \in V$, then for $Z = F(\zeta) \neq F$, $\zeta^* = -\zeta$,

$$S_{u,v} - \frac{2}{m}\, t^-((u,v))\zeta\, I_V$$

is in $K'(V)$.

In Chapter 3, §3, splitting information has been developed to display a Cartan subalgebra $\mathfrak{h}$ of $\mathfrak{g}$ containing $\mathfrak{t}_0$, $K \otimes_F \mathfrak{h}$ splitting $K \otimes_F \mathfrak{g}$, with base $\varepsilon_1,\ldots,\varepsilon_n$ for the roots of $K \otimes \mathfrak{g}$ relative to $K \otimes \mathfrak{h}$ such that the restriction of each ε_i to $\mathfrak{t}_0$ is either zero or a member of the chosen base $\{\alpha_j\}$ for the roots of $\mathfrak{g}$ relative to $\mathfrak{t}_0$. As there, let $\pi_1,\ldots,\pi_n$ be the fundamental weights relative to $K \otimes \mathfrak{h}$, associated with $\varepsilon_1,\ldots,\varepsilon_n$, and let $\lambda_1,\ldots,\lambda_\ell$ be the fundamental weights of $\mathfrak{g}$ relative to $\mathfrak{t}_0$ and the base $\alpha_1,\ldots,\alpha_\ell$. Then we have seen that, if either $Z \neq F$ or if the involution * in D is "of orthogonal type", the restriction to $\mathfrak{t}_0$ of π_i is $2d\lambda_\ell$, where $d^2 = [D:Z]$, with exceptions (that are easily computed on the basis of §3.3) as below:

When $Z = F$:

$$\pi_i\big|_{t_0} = i\,\lambda_1,\ 1 \le i \le d;$$

$$\pi_{id+k}\big|_{t_0} = (d-k)\lambda_i + k\,\lambda_{i+1},\ 1 \le i < \ell-1,\ 1 \le k \le d;$$

$$\pi_{(\ell-1)d+k}\big|_{t_0} = (d-k)\lambda_{\ell-1} + 2k\lambda_\ell,\ 1 \le k < d.$$

Thus each of the restrictions on the right is obtained exactly once as the restriction of a fundamental weight π_i, while $2d\lambda_\ell$ may be obtained many times, depending on the size of [D:F] and that of [V:D].

With $n = g-1$ as in §3.3, we find for $Z \neq F$ that

$$\pi_i\big|_{t_0} = i\,\lambda_1 = \pi_{g-i}\big|_{t_0},\ 1 \le i \le d;$$

$$\pi_{id+k}\big|_{t_0} = (d-k)\lambda_i + k\,\lambda_{i+1} = \pi_{g-(id+k)}\big|_{t_0},\ 1 \le i < \ell-1,\ 1 \le k \le d;$$

$$\pi_{(\ell-1)d+k}\big|_{t_0} = (d-k)\lambda_{\ell-1} + 2k\lambda_\ell = \pi_{g-(\ell-1)d-k}\big|_{t_0},\ 1 \le k < d;$$

are the exceptions to the "rule" $\pi_i\big|_{t_0} = 2d\lambda_\ell$. Each of these restriction-values is obtained exactly <u>twice</u>.

The case $Z = F$, with the involution * in D of symplectic type has $\pi_{n-1}\big|_{t_0} = d\lambda_\ell = \pi_n\big|_{t_0}$, and otherwise $\pi_i\big|_{t_0} = 2d\lambda_\ell$, with exceptions for π_j, $j < \ell d$, exactly as in the other case above with $Z = F$.

The program of Chapter 2, based on Cartan multiplication, reduces the rational representation theory of g to the construction of all irreducible g-modules of these highest t_0-weights, and then, by §2.3, to the construction of all λ-admissible (irreducible) g_0-modules for these values of λ. We subdivide these λ's into groups according to the role played by λ_ℓ. The first group is that already treated in Chapter 3.

§2. Weights of the Form $\lambda = k\lambda_\ell$.

We have seen in Chapter 3 that the λ-admissible g_0-modules are exactly the irreducible right modules for $A_k(V)$. To make this explicit, the action on the λ-admissible g_0-module W of the generator $\varphi(u \otimes v)$ for $A_k(V)$ $(u,v \in V)$ is that of $[S_{u_\ell,u}, S_{u_{\ell+m+1},v}] \in g_0$.

In the reverse direction, if ψ is the embedding of $K(V) + FI_V$ in A_k, an irreducible right A_k-module W becomes a λ-admissible g_0-module by writing the general element H of g_0 in the form

$$H = \sum_{i=1}^{\ell} S_{u_i,a_i u_{2\ell+m+1-i}} + R, \tag{1}$$

where $a_i \in D$, $\sum_i t^-(a_i) = 0$, and where $R = T + M$, with $T \in K'(V)$ (annihilating the u_i). Here M is defined by extending a transformation $M_0 = \mu\zeta I_V$, $\mu \in F$, to be equal to $-\frac{m}{2}\mu\zeta I_{V\perp}$ on $V^\perp$, the span of $u_1,\ldots,u_\ell$, $u_{\ell+m+1},\ldots,u_{2\ell+m}$. Then the action on $w \in W$ of this H is taken to be

$$-\frac{k}{2}\sum_{i-1}^{\ell} t^+(a_i)w + w\,\psi(T + M_0)\ . \tag{2}$$

(If $Z = F$, $t^- = 0$ and the elements M_0, M do not appear.)

From the results of Chapter 3, we now have the solutions in the relevant cases:

Theorem 4.1. When the involution in D is of symplectic type, the $d\lambda_\ell$-admissible g_0-modules are the irreducible modules for $A_d(V)$, with g_0-structure given by (2). There are at most two nonisomorphic $d\lambda_\ell$-admissible g_0-modules. In all cases the $2d\lambda_\ell$-admissible g_0-modules are the irreducible $A_{2d}(V)$-modules, with g_0-structure given by (2). Bounds on their number are as follows:

$Z = F$, $*$ of orthogonal type: $m\frac{d}{2} + 1$.

$Z \neq F$: $md + 1$.

$Z = F$, $*$ of symplectic type: $m\frac{d}{2} + 2$.

When $*$ is of symplectic type, all $2d\lambda_\ell$-admissible g_0-modules are contained in tensor products of two $d\lambda_\ell$-admissible modules.

The bounds given are simply the number of dominant integral functions on $K \otimes h$ whose restrictions to t_0 are equal to $d\lambda_\ell$ resp. $2d\lambda_\ell$.

In connection with Theorem 1, it should be noted that the assumption $\ell \geq 2$ was used at points in Chapter 3, in particular in reducing the action of elements like $S_{u_\ell, au_{\ell+m+1}}$ to the case where a is scalar. However, the formula (2) above applies even when $\ell = 1$ to make each irreducible $A_d(V)$ - resp. $A_{2d}(V)$-module into a $d\lambda_\ell$ - resp. $2d\lambda_\ell$-admissible g_0-module for each positive value of ℓ. Combining the information on highest weights relative to $K \otimes h$ with these restrictions to t_0 (also applicable for all ℓ) with the observations of Chapters 2 and 3, showing that the number of nonisomorphic $d\lambda_\ell$-resp. $2d\lambda_\ell$-admissible g_0-modules, upon splitting, is given precisely by the cited bound (as computed for $\ell \geq 2$, in application to $A_k(V)$), we see that the conclusions of Theorem 1 apply equally well when $\ell = 1$.

On a more trivial level, it is clear as above that $A_0(V) = F$ enables us to construct the only 0-admissible g_0-module by the formula (2) with $k = 0$. Of course the associated g-module is then 1-dimensional.

§3. Weights Not Involving λ_ℓ.

Here the results are identical with those of Chapters V and VI of [Se3], for the corresponding cases when $m = 0$. To have weights $(\neq 0)$ not involving λ_ℓ we must have $\ell > 1$. The highest weight g_0-module M^+ must be annihilated by $[g_{-\alpha_\ell}, g_{\alpha_\ell}]$, that is, by all

$$[S_{u_\ell,v}, S_{u_{\ell+m+1},w}] \in g_0, \tag{3}$$

for all $v,w \in V$, or by all

$$S_{w,v} + S_{u_\ell,(v,w)u_{\ell+m+1}}. \tag{4}$$

Interchanging v and w in (4) and subtracting (here $\varepsilon = -1$), we see that M^+ must be annihilated by all $S_{u_\ell,a\,u_{\ell+m+1}}$, $a^* = a$, and thus by all $S_{u_k,c\,u_{\ell+m+1}}$, where c is a sum of commutators of $*$-fixed elements of D. When $Z = F$, it follows (cf. [Se3], V. 2) that all $S_{u_\ell,c\,u_{\ell+m+1}}$, $c \in D$, annihilate M^+, except perhaps when $d = 2$ and the involution $*$ is of symplectic type. (This is the case where the division algebra D is quaternionic and $*$ is the standard involution in a quaternion algebra.)

With this exception (still assuming $Z = F$), it now follows that all $S_{v,w}$ annihilate M^+, and then that the action on M^+ of the general element

$$H = \sum_{i=1}^{\ell} S_{u_i,a_iu_{2\ell+m+1-i}} + T, \quad T \in K(V), \tag{5}$$

is given, for $\lambda = k\lambda_j$, by that of

$$S_{u_j,t(a_1+\ldots+a_j)u_{2\ell+m+1-j}},$$

which is scalar multiplication by $-kt(a_1 +\ldots+ a_j)$, provided $j > 1$. In this case we see from our list of restrictions that we must have $d|k$, and that there is a unique (1-dimensional) $k\lambda_j$-admissible g_0-module for each such k.

Still with $Z = F$, and excluding the case $d = 2$ with standard quaternionic involution, we consider the case $\lambda = k\lambda_1$, where H of (5) acts on M^+ as does $S_{u_1,a\,u_{2\ell+m}}$. We also consider the cases $\lambda = (d-k)\,\lambda_{j-1} + k\lambda_j$, $0 < j < d$, $j < \ell$, where H acts as does

$$t(a_1 +\ldots+ a_{j-1})S_{u_{j-1},u_{2\ell+m+2-j}} + S_{u_j,a_ju_{2\ell+m+1-j}}$$

for $j > 2$, and as does $S_{u_1,a_1u_{2\ell+m}} + S_{u_2,a_2u_{2\ell+m-1}}$ for $j = 2$. In the first case, it follows by [Se3], §V.3, that M^+ is an irreducible (right) module for the k-th symmetric power $S^k(D)$, with H acting on $w \in M^+$ to send w to $-w\rho_k(a_1)$, where $\rho_k:D \to S^k(D)$ is the canonical map: $\rho_k(a) = a \otimes 1 \otimes\ldots\otimes 1 +\ldots+ 1 \otimes\ldots\otimes 1 \otimes a$. Conversely, each irreducible right $S^k(D)$-module is the highest weight-space for an irreducible $\mathfrak{g}$-module of highest $\mathfrak{t}_0$-weight $k\lambda_1$. As in §V.7 of [Se3], all $k > 1$ are redundant with respect to generation of irreducible $\mathfrak{g}$-modules by Cartan multiplication. These observations will be incorporated in the comprehensive theorem (Theorem 2) to be stated in §5.

By the considerations of [Se3], Chapter V, all these observations apply equally well in the quaternionic case excluded thus far, provided the index $j < \ell-1$. Likewise, the considerations there on weights $\lambda = (d-k)\lambda_{j-1} + k\lambda_j$ apply to this quaternionic case if $j < \ell-1$. From [Se3], §V.4, we see that here there is a unique irreducible $S^k(D)$-module $\mathcal{W}$ that is simultaneously a module for $S^{d-k}(D^{op})$, such that for $a \in D$, $w \in \mathcal{W}$,

$$w\,\rho_k(a) + w\,\rho_{d-k}(a) = dt(a)w.$$

This $\mathcal{W}$ is a $\mathfrak{g}_0$-module, with

$$wH = -w\,\rho_k(a_j) - dt(a_1 + \ldots + a_{j-1})w,$$

and is the unique λ-admissible $\mathfrak{g}_0$-module.

Now let $d = 2$, $*$ be of symplectic type. We still must consider $\lambda = 2\lambda_{\ell-1}$ if $\ell > 2$; $\lambda = \lambda_1, 2\lambda_1$ if $\ell = 2$; and $\lambda = \lambda_{\ell-2} + \lambda_{\ell-1}$. In the last case, let $\mathcal{W}$ be the unique irreducible module for $D = S^1(D)$ as in the last paragraph: that is, $\mathcal{W} \cong D$, with

$$w\,H = -w\,a_{\ell-1} - 2\,t(a_1 + \ldots + a_{\ell-2})w.$$

(Here "2t" is just the ordinary quaternionic trace.)

When $\lambda = 2\lambda_{\ell-1}$, $\ell > 2$, let $\mathcal{W}$ be a one-dimensional F-space, with $wH = -2t(a_1 + \ldots + a_{\ell-1})w$. When $\ell = 2$, $\lambda = \lambda_1$, let $\mathcal{W} = D$, with $wH = -wa_1$, and when $\lambda = 2\lambda_1$, let $\mathcal{W}$ be an irreducible right $S^2(D)$-module (necessarily of dimension 1 or 3, by the Appendix to Chapter 1), setting

$$w\,H = -w\,\rho_2(a_1).$$

In the last case it should be noted that there are exactly two dominant integral functions π on $K \otimes \mathfrak{h}$ with $\pi|_{\mathfrak{t}_0} = 2\lambda_1$, <u>viz</u>. $2\pi_1$ and π_2. Combined with the uniqueness of such π for the other values of λ treated here, this yields as in [Se3], Chapter V, that the λ-admissible $\mathfrak{g}_0$-modules above are exhaustive for these values of λ.

When $*$ is of second kind, we write $t(a) = t^+(a) + t^-(a)\zeta$, where $t^+(a)$, $t^-(a) \in F$. The general element H of $\mathfrak{g}_0$, as before, has the form

$$H = \sum_{i=1}^{\ell} S_{u_i, a_i u_{2\ell+m+1-i}} + T, \quad T \in K(V), \tag{6}$$

where, if $v_1, \ldots, v_m$ is any D-basis for V, $v_i T = \sum_{j=1}^{m} a_{ij} v_j$, $t(T) = \sum_i t(a_{ii}) =$

$(\sum_i t^-(a_{ii}))\zeta$ is independent of basis, so that H satisfies the condition

$$-2 \sum_{i=1}^{\ell} t^-(a_i) + \zeta^{-1} t(T) = 0.$$

Now, as in §VI.2 of [Se3], each $k\lambda_j$-admissible g_0-module, $1 < j < \ell$, $k > 0$, is an irreducible right $S^k(D)$-module, the enveloping algebra of this action being commutative. It then follows that $d|k$; for $k = d$, the module is a one-dimensional Z-vector space W, the action of H as in (6) above being multiplication by

$$-dt(a_1 + \ldots + a_j).$$

For $\lambda = k\lambda_1$, $k > 0$, it again follows as in [Se3], §VI.3, that a λ-admissible g_0-module W is an irreducible right module for $S^k(D)$ (symmetric power over F), the action on $w \in W$ of H sending w to $-w\,\rho_k(a_1)$. All such W for $k > 1$ occur in the k-fold tensor power of that for $k = 1$, and this last is (a one-dimensional right vector space over) D.

When $\lambda = (d-k)\lambda_{j-1} + k\lambda_j$, $0 < k < d$, $j < \ell$, let W be the unique irreducible right $S^k(D)$-module which is simultaneously a right $S^{d-k}(D^{op})$-module, the images $\rho_k(a)$, $\rho_{d-k}(a)$ of $a \in D$ being related in their action on $w \in W$ by

$$w\,\rho_k(a) + w\,\rho_{d-k}(a) = d\,t(a)w.$$

(W is a Z-vector space and all mappings are Z-mappings.) This W is the unique λ-admissible g_0-module, with

$$w\,H = -2\,\rho_k(a_j) + d\,t(\sum_{i=1}^{j-1} a_i)w.$$

§4. Weights Involving Both $\lambda_{\ell-1}$ and λ_ℓ.

Here it is necessary to go to a synthesis of the constructions involving symmetric powers and those involving higher even Clifford algebras. The weights are those of the form $\lambda = (d-k)\lambda_{\ell-1} + 2k\lambda_\ell$, $0 < k < d$. Assuming $\ell > 2$ to begin with, we see as before (cf. [Se3], §V.2) that the action on a λ-admissible g_0-module of the element H as in (5) or (6) agrees with that of

$$S_{u_{\ell-1},t(a_1 + \ldots + a_{\ell-1})u_{\ell+m+2}} + S_{u_\ell,a_\ell u_{\ell+m+1}} + T.$$

For $a \in D$, define $\rho(a)$ to be the action on such a module W of

$$-S_{u_\ell,au_{\ell+m+1}} + t^-(a)\,S_{u_{\ell-1},\zeta\,u_{\ell+m+2}},$$

the second term omitted if the involution is of first kind. For $v,w \in V$, define $g(v,w)$ to be the action on W of the element of g_0,

$$S_{u_\ell,(v,w)u_{\ell+m+1}} + S_{v,w} = [S_{u_\ell,v}, S_{u_{\ell+m+1},w}].$$

The fact that $\mathcal{W}$ is a $\mathfrak{g}_0$-module, in which $H_{\ell-1} = T_\ell - T_{\ell-1} = S_{u_\ell,u_{\ell+m+1}} - S_{u_{\ell-1},u_{\ell+m+2}}$ is represented by the scalar d-k and $H_\ell = -2T_\ell$ by 2k, yields the following identities:

$$g(av,w) - g(v,a^*w) = \rho([(v,w),a]) = [\rho(a), g(v,w)]; \tag{7}$$

$$g(v,w) - g(w,v) = \rho((w,v) - (v,w)); \tag{8}$$

$$\rho([a,b]) = [\rho(a),\rho(b)]; \tag{9}$$

$$\rho(1) = k1; \tag{10}$$

$$[g(v,w),g(x,y)] = g(vS_{x,y},w) + g(v,wS_{x,y}) + \rho([(v,w),(x,y)]), \tag{11}$$

for all $v,w,x,y \in V$, all $a,b \in D$.

With π the projection of $\mathcal{U}(\mathfrak{g})$ on $\mathcal{U}(\mathfrak{g}_0)$ as in Chapter 2, the condition of λ-admissibility entails that $\mathcal{W}$ must be annihilated by the following elements of $\mathcal{U}(\mathfrak{g}_0)$:

$$\pi(S_{u_{\ell-1},a_1u_{\ell+m+1}} \cdots S_{u_{\ell-1},a_{d-k+1}u_{\ell+m+1}} S_{u_\ell,b_1u_{\ell+m+2}} \cdots$$
$$\cdots S_{u_\ell,b_{d-k+1}u_{\ell+m+2}}), \tag{12}$$

for all $a_i,b_i \in D$;

$$\pi(S_{u_\ell,a_1u_\ell} \cdots S_{u_\ell,a_{k+1}u_\ell} S_{u_{\ell+m+1},b_1u_{\ell+m+1}} \cdots$$
$$\cdots S_{u_{\ell+m+1},b_{k+1}u_{\ell+m+1}}), \tag{13}$$

for all $a_i,b_i \in D$, with $a_i^* = a_i$, $b_i^* = b_i$; and

$$\pi(S_{u_\ell,v_1} \cdots S_{u_\ell,v_{2k+1}} S_{u_{\ell+m+1},w_1} \cdots S_{u_{\ell+m+1},w_{2k+1}}), \tag{14}$$

for all $v_i,w_i \in V$. In terms of the previous chapter, these have the respective interpretations:

$\sigma = dt - \rho$ satisfies the d-k+1-th symmetric identity (of Chapter 1, §3) on all elements of D; (12)'

ρ satisfies the k+1-th symmetric identity on all products of (two) *-fixed elements of D (cf. [Se3], Ch. V); (13)'

The map $\varphi: V \otimes_F V \to \mathrm{End}_F(\mathcal{W})$ defined by $\varphi(u \otimes v) = g(u,v)$ satisfies the (2k+1)-th identity of Chapter 3, §2. (14)'

Conversely, if $\mathcal{W}$ is a finite-dimensional F-space and if $\rho: D \to \mathrm{End}_F(\mathcal{W})$,

$\varphi : V \otimes_F V \to \mathrm{End}_F(\mathcal{W})$ are linear mappings satisfying (7)-(11) and (12)'-(14)'; if $\mathcal{W}$ is also irreducible for the combined action; then $\mathcal{W}$ is a λ-admissible $\mathfrak{g}_0$-module. For this one needs only define an action of $\mathfrak{g}_0$ so that $\mathcal{W}$ is annihilated by all

$$[S_{u_{i-1}, a\, u_{m+2\ell+1-i}}, S_{u_i, b\, u_{m+2\ell+2-i}}], \quad 1 \le i < \ell,$$

$a, b \in D$, with the action of

$$t^-(a) S_{u_{\ell-1}, \zeta u_{\ell+m+2}} \quad \text{being} \quad \rho(a),$$

and with the action of

$$S_{u_\ell, (v,w) u_{m+\ell+1}} + S_{v,w} \quad \text{for} \quad v, w \in V$$

being that of $\varphi(v \otimes w)$. The identities (7)-(11) guarantee that there is such a well-defined action, as below:

Consider the F-vector space $D \oplus (V \otimes_F V)$, and form the quotient $\mathcal{B}$ of its tensor algebra $\mathcal{T}$ by the ideal generated by the elements

$$au \otimes v - u \otimes a^* v - [(u,v), a]; \tag{7)'$$

$$au \otimes v - u \otimes a^* v - (a \otimes (u \otimes v) - (u \otimes v) \otimes a); \tag{7)"$$

$$u \otimes v - v \otimes u - ((v,u) - (u,v)); \tag{8)'$$

$$a \otimes b - b \otimes a - [a,b]; \tag{9)'$$

$$1_D - k\, 1_{\mathcal{T}}; \tag{10)'$$

$$(u \otimes v) \otimes (x \otimes y) - (x \otimes y) \otimes (u \otimes v) - u\, S_{x,y} \otimes v \tag{11)'$$

$$- u \otimes v\, S_{x,y} - [(u,v),(x,y)],$$

for all $a, b \in D$; all $u, v, x, y \in V$. Let θ be the canonical mapping from $\mathcal{T}$ onto $\mathcal{B}$.

There is a unique homomorphism of associative algebras with unit from $\mathcal{T}$ to $\mathrm{End}_D(U)$ sending $a \in D$ to

$$t^-(a) S_{u_{\ell-1}, \zeta\, u_{\ell+m+2}} - S_{u_\ell, a\, u_{\ell+m+1}} + t^+(a) S_{u_\ell, u_{\ell+m+1}} + k t^+(a)\, I_U,$$

and, for $v, w \in V$, $v \in w$ to

$$S_{u_\ell, (v,w) u_{\ell+m+1}} + S_{v,w}\,.$$

From (7)-(11), all of (7)', (7)", (8)', (9)', (11)' lie in the kernel of this map, and by definition $1_D \in D$ is sent to kI_U, so (10)' also lies in the kernel. Thus the Lie subalgebra of $\mathcal{B}$ generated by $\theta(D \oplus (V \otimes V))$ is represented on U. The presence in the kernel of the elements (7)', (7)", (8)', (11)' shows that the image in End(U) of this subalgebra is just the image of the space $\theta(D \oplus (V \otimes V))$. The subspace $F1_{\mathcal{B}} + \theta(D \oplus (V \otimes V))$ is a Lie subalgebra of $\mathcal{B}$, mapped into $FI_U + g_0 \subset \mathrm{End}_D(U)$. If $*$ is of first kind, the image of this last subalgebra is

$$\{\mu I_U + S_{u_\ell, au_{\ell+m+1}} + T \mid \mu \in F,\ a \in D,\ T \in K(V)\},$$

a subalgebra of $FI_U + g_0$ of dimension

$$1 + \dim D + \binom{m}{2} \dim D + ms, \tag{15}$$

where s is the F-dimension of $\{a \in D \mid a^* = a\}$.

From (8)' and (9)' one sees, as in 2) of 2 of [Se5], that $FI_U + \theta(D \oplus (V \otimes V))$ has F-dimension at most equal to (15). Thus the last map above is an isomorphism of Lie algebras, and the algebra A obtained by adjoining the relations (12)'-(14)' to those defining $\mathcal{B}$ has a Lie subalgebra that is a homomorphic image of the above subalgebra of $FI_U + g_0$, in such a way that this Lie subalgebra generates A. In fact the subalgebra in question of $FI_U + g_0$ is an ideal and an ideal direct summand, a complementary ideal being the set of all elements

$$\sum_{i=1}^{\ell-1} S_{u_i, a_i u_{2\ell+m+1-i}}, \quad a_i \in D.$$

Extending our homomorphism into A to be zero on this complement, we have a Lie morphism of $FI_U + g_0$ into A, sending $S_{u_\ell, au_{\ell+m+1}}$ to the image of $-a \in D$, I_U to 1_A and $S_{v,w}$ to the sum of the image of $v \otimes w$ and that of $(v,w) \in D$ for $v,w \in V$. That is, each irreducible right A-module becomes a λ-admissible g_0-module.

When $*$ is of second kind, one modifies the above as follows: Consider the set of elements of g_0 that stabilize each of the sets V, Du_ℓ, $Du_{\ell+m+1}$, $Fu_{\ell-1}$, $Fu_{\ell+m+2}$, and that annihilate all other u_i. These form an ideal in $FI_U + g_0$, a complementary ideal J being the set of all

$$\sum_{i=1}^{\ell-1} S_{u_i, a_i u_{2\ell+m+1-i}} \quad \text{with} \quad \sum_i t^-(a_i) = 0.$$

Then, as in the case where $*$ is of first kind, there is a morphism of Lie algebras $FI_U + g_0 \to A$ mapping J to zero, I_U to 1_A,

$$S_{u_\ell, au_{\ell+m+1}} - t^-(a) S_{u_{\ell-1}, \zeta u_{\ell+m+2}}$$

to the image in A of $-a \in D$, and

$$S_{v,w} + S_{u_\ell, (v,w)u_{\ell+m+1}} \in g_0$$

to the image of $v \otimes w$. Again each irreducible right A-module is a λ-admissible g_0-module.

If η is the canonical homomorphism of T onto A, one sees from (12)', (13)',

(14)', combined with the characterizations of $S^{d-k}(D^{op})$ in [Se3], Chapter III, that the image $\eta(\mathcal{T}(D))$ in A of $\mathcal{T}(D) \subset \mathcal{T}$ is finite-dimensional, and from (7)" we see that A is generated as left $\eta(\mathcal{T}(D))$-module by monomials (noncommutative) in the $v_{ij} = \eta(v_i \otimes v_j)$, where $\{v_i\}$ is an F-basis for V. From (8)' and (11)', these monomial generators can be taken in a form where each v_{ij} occurs at most once (with a suitable exponent), i.e., as monomials where the v_{ij} occur in a fixed order. Finally, applying (14)' with all v's equal to v_i and all w's equal to v_j shows, as in §2 of [Se5], that v_{ij}^{2k+1} can be reduced to an $\eta(\mathcal{T}(D))$-combination of such monomials, each of degree $< 2k+1$, and finally that A has finite dimension, at most

$$(d^2 m)^{4k+2} \dim S^{d-k}(D^{op}).$$

(A better value will be given later.)

For further properties of A, we extend F to a finite splitting extension K, following the conventions of Chapter 3. When * is of first kind, the fact that there is only one π relative to $K \otimes h$ with restriction λ to t_0 tells us that if W is an irreducible right A-module, so $W = M^+$ for an irreducible g-module M of highest t_0 weight λ by the above, then all the irreducible constituents of $K \otimes M$, as $K \otimes g$-module, have highest weight π. It follows by Theorem 2.1 that g has only one irreducible module of highest t_0-weight λ, so that there is only one λ-admissible g_0-module, and only one irreducible A-module.

When * is of first kind, A-modules are modules for $\mathcal{S} = F \oplus D \oplus K(V)$, whose center is $F \oplus F1_D$ except when $m = 1$, D is quaternionic and * of symplectic type, when $K(V)$ has F-dimension one. From the identities defining A, F and 1_D are represented by scalars on each A-module. Excluding the exceptional case, it follows that each A-module is a completely reducible $\mathcal{S}$-module, hence completely reducible as A-module, and that A is (finite-dimensional) semisimple. We can say more:

Proposition 4.1. When * is of first kind, except when $m = 1$, D is quaternionic and * is symplectic, A is central simple.

Proof. Once A is known to be nonzero, the proposition will follow from semisimplicity and the (absolute irreducibility and) isomorphism of all irreducible $K \otimes A$-modules, which is a consequence of the fact that π is the only dominant integral function on $K \otimes h$ with restriction λ to t_0. To see that $A \neq \{0\}$, let R be an irreducible right $K \otimes g$-module of this highest weight π, and let R^+ be the subspace annihilated by $K \otimes N^+$ (or by N^+). From the considerations of Chapter 2 (now using, for example, non-zero F-rational points of the root spaces as Zariski-dense sets), R^+ is a weight space relative to t_0 and an irreducible $K \otimes g_0$-module, containing the highest $K \otimes h$-weight space of R. Thus the t_0-weight of R^+ is $\pi|_{t_0} = \lambda$. Now it follows as in Chapter 3 that R^+ must be an irreducible $K \otimes A$-module, so that $K \otimes A \neq \{0\}$.

The remaining exceptional case where $*$ is of first kind is treated as in the proof of Prop. 7.3 of [Se5]. Namely, one shows that a basis for $\mathcal{K}(V)$ may be taken to consist of an element whose image in the (field-extended) generalized even Clifford algebra $K \otimes (\mathcal{A}_j(V))$ satisfies a semisimple polynomial for each even integer $j \geq 0$. The element in question, denoted $S_{e_1u,\varkappa^{-1}e_2u}$ in the reference, corresponds to the image in our $K \otimes \mathcal{A}$ of

$$(e_1u, \varkappa^{-1} e_2u) + (e_1u \otimes \varkappa^{-1} e_2u),$$

the first term lying in $K \otimes D$ and the second in $K \otimes (V \otimes_F V)$. The first term is equal to e_1, an idempotent in $K \otimes D$, and from (11)' it follows that the images in $K \otimes \mathcal{A}$ of these two terms commute. The canonical image of e_1 in any symmetric power, $S^j(K \otimes D)$, is a sum of commuting idempotents, so is semisimple, and the image in $K \otimes \mathcal{A}$ of $e_1u \otimes \varkappa^{-1}e_2u$ is seen, using (14)' and the fact that $k = 1 = d-k$, to satisfy the semisimple polynomial $X(X+1)(X+2)$. Thus $S_{e_1u,\varkappa^{-1}e_2u}$ is represented by a semisimple transformation in each $K \otimes \mathcal{A}$-module, and it now follows that $\mathcal{A}$ is a semisimple algebra in this case as well. The rest of the proof of Proposition 4.1 goes as before, and we have strengthened the result to

<u>Proposition 4.1'</u>. <u>The algebra</u> $\mathcal{A}$ <u>is central simple whenever</u> $*$ <u>is of first kind.</u>

Now let $*$ be of second kind. To show $\mathcal{A}$ is semisimple (possibly zero, for now) is equivalent to showing that the image in $\mathcal{A}$ of the center of the reductive Lie algebra $F1_{\mathcal{A}} + \theta(D + (V \otimes V))$ is represented by semisimple transformations on any $K \otimes \mathcal{A}$-module. Recalling that $F1_{\mathcal{A}} + \theta(D + (V \otimes V))$ identifies with a Lie subalgebra of $\mathrm{End}_D(U)$ having basis for its center I_U,

$$S_{u_{\ell-1},\zeta u_{\ell+m+2}} - S_{u_\ell,\zeta u_{\ell+m+1}}, \quad \sum_{i=1}^{m} S_{v_{\ell+1},\theta_i^{-1} v_{\ell+i}} + \frac{m}{2} S_{u,\zeta u_{\ell+m+1}},$$

where $2(v_{\ell+i},v_{\ell+i}) = \zeta\,\theta_i$, $1 \leq i \leq m$, we see that the last two of these elements must be shown to have semisimple representants.

The former of these last two has image in $\mathcal{A}$ lying in $\eta(\mathcal{T}(D))$, an element which is the image in $\mathcal{A}$ of the canonical image in $S^{d-k}(D^{op})$ of $\zeta \in D$, via the mapping of $S^{d-k}(D^{op})$ onto $\eta(\mathcal{T}(D))$. In [Se3], §IV.4, it is shown that the image of ζ in $S^{d-k}(D^{op})$ is semisimple.

The remaining element is the sum of commuting elements

$$S_{v_{\ell+i},\theta_i^{-1} v_{\ell+i}} + \frac{1}{2} S_{u_\ell, u_{\ell+m+1}}, \quad 1 \leq i \leq m, \tag{16}$$

so that it suffices to show that each term (16) has a semisimple representant. The image in $\mathcal{A}$ of (16) is that of $v_{\ell+i} \otimes \theta_i^{-1} v_{\ell+i}$, and the arguments used to prove Proposition 7.1 of [Se5] apply to show that its image in $K \otimes \mathcal{A}$ may be written as a sum of commuting terms, each of which has a nonzero K-multiple satisfying a polynomial

$\sum_{j=0}^{2k} (X + j)$, so is semisimple. Thus A <u>is semisimple</u>.

Here $K \otimes g$ has two inequivalent irreducible modules whose highest weights π have restriction λ to t_0. As before, each irreducible right A-module W has the property that each irreducible $K \otimes g_0$-constituent of $K \otimes W$ is isomorphic to the highest t_0-weight space of one of these two $K \otimes g$-modules, and that these two are nonisomorphic absolutely irreducible $K \otimes A$-modules. That is, <u>$K \otimes A$ is the sum of two minimal</u> (2-sided) <u>ideals, each of which is a central simple</u> K-<u>algebra</u>.

Concerning A itself, it follows that the center of A has F-dimension two. Either this center is a field extension, over which A is central simple, or it splits. In the latter case, A is the sum of two minimal ideals, each a central simple F-algebra. These alternatives are considered more closely below.

Consider now the irreducible $K \otimes g_0$-modules occurring as $K \otimes A$-modules. For the associated $K \otimes g$-modules, the possibilities are as follows: Z = F, * orthogonal type:

$, \quad \pi = \pi_{(\ell-1)d+k}$.

Z = F, * symplectic type:

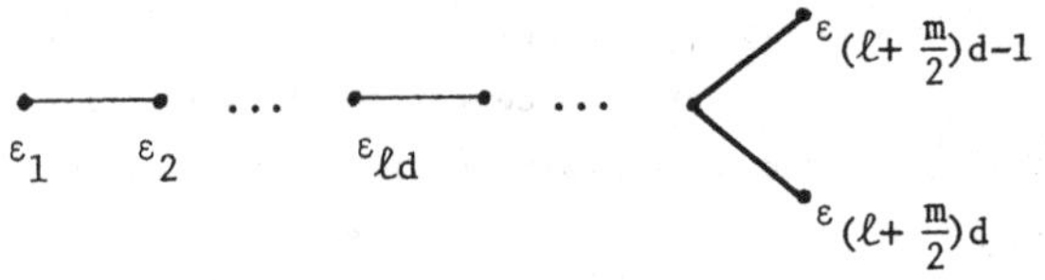

$, \quad \pi = \pi_{(\ell-1)d+k}$.

When d = 2, this becomes

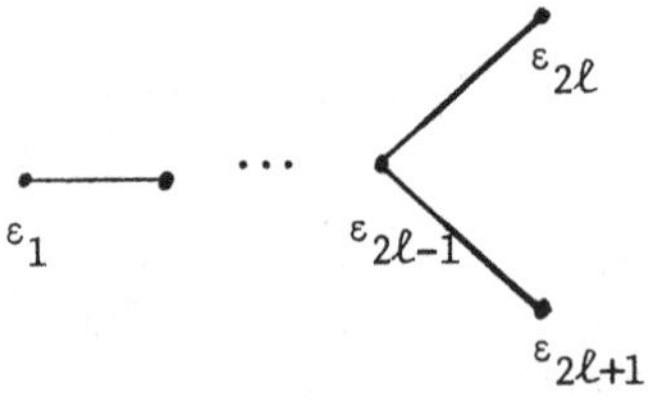

$, \quad \pi = \pi_{2\ell-1}, \quad \varepsilon_{2\ell}|_{t_0} = \alpha_\ell = \varepsilon_{2\ell+1}|_{t_0}$

(In all above cases d is even, indeed a power of 2.)

Z ≠ F:

$, \quad \pi = \pi_{(\ell-1)d+k}$ or $\pi_{(\ell+m+1)d-k}$.

The highest t_0-weight spaces of these modules are those obtained by operating successively on a highest weight vector by root-vectors corresponding to the negatives of fundamental roots, constrained to the connected components containing ε_j of the diagram obtained by deleting all ε_i whose restriction to t_0 is non-zero, when $\pi = \pi_j$. In each case this space is a module for the simple subalgebra of $K \otimes g_0$

generated by root-vectors for these roots and their negatives. In each case, the system has type A. In the first two cases, it is

in the third,

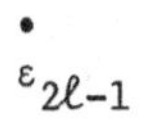

and in the last,

As a representation of a split algebra of type A, the pair (algebra, module) is isomorphic to a pair $(\mathrm{End}_K(X)', \Lambda^k(X))$, where X is a vector space over K, of dimension d except for the case $d = 2$, $m = 1$, $*$ symplectic, where X has dimension 2. Apart from this case, we have $K \otimes A \simeq M_{\binom{d}{k}}(K)$ for $*$ of first kind, $K \otimes A \simeq M_{\binom{d}{k}}(K) \times M_{\binom{d}{k}}(K)$ for $*$ of second kind, while $K \otimes A \simeq M_2(K)$ in the exceptional case. Thus A <u>has</u> F-<u>dimension</u> $\binom{d}{k}^2$, $2\binom{d}{k}^2$, 4, <u>respectively</u>.

From [Se3], Chapter IV, we note that when $*$ is of first kind the dimension above is the same as that of the minimal ideal in $S^{d-k}(D^{op})$ which "is" also one in $S^k(D)$, such that the relation $\sigma(a) + \rho(a) = dt(a)$ is satisfied for the mappings σ,ρ of D into these ideals obtained by composing their projectors with the canonical maps ρ_{d-k}, ρ_k of D into $S^{d-k}(D^{op})$ resp. $S^k(D)$. When $*$ is of orthogonal type, the identities characterize the ideals in question (cf. [Se3], pp. 84-85). When $*$ is of symplectic type, the considerations of §V.6 of [Se3] show that the identities above only involving elements of D characterize this same ideal in $S^{d-k}(D^{op})$ - the exceptional case "$\lambda = \lambda_{r-1} + (\frac{d}{2} - 1)\lambda_r$" has no counterpart here because $m > 0$. In these cases, this ideal, of dimension $\binom{d}{k}^2$, is mapped faithfully into A (being itself simple), and therefore its image coincides with A.

It is not clear how to make an explicit identification of the images in A of elements $v \otimes w$ in $V \otimes V$ with elements of $S^{d-k}(D^{op})$ distinguished in the last paragraph. We therefore choose, for our "constructive" approach, to retain <u>all</u> the generators and all the relations (7)'-(14)' in the definition of A, so as to be able to give explicitly the action of $\mathfrak{g}_0$ on an irreducible A-module.

At the beginning of this section, we assumed $\ell > 2$. As at the end of §2, this restriction may now be dropped, for analogous reasons. Namely, with A as above, the same definitions make an irreducible A-module into a λ-admissible $\mathfrak{g}_0$-module, and the structure of A is independent of ℓ, so is as for large ℓ. Thus $K \otimes A$ has one or two (absolutely) irreducible modules, according as $*$ is of first or second kind, while $K \otimes \mathfrak{g}$ has at the same time one or two irreducible modules whose highest weight has restriction λ to $\mathfrak{t}_0$. Now it follows by Th. 2.1 that the irreducible A-modules constitute all the λ-admissible $\mathfrak{g}_0$-modules.

§5. Summary. Generation of all Irreducible $\mathfrak{g}$-Modules of Given Highest $\mathfrak{t}_0$-Weight.

With $\mathfrak{g} = K'(U)$ above, we have determined a set Λ of "fundamental weights" λ and have shown, for each of these, how to construct all λ-admissible $\mathfrak{g}_0$-modules. If μ is a given dominant integral function on $\mathfrak{t}_0$, then all μ-admissible $\mathfrak{g}_0$-modules can be constructed by forming all tensor products of admissible $\mathfrak{g}_0$-modules of $\mathfrak{t}_0$-weights in Λ, such that these weights form a partition of μ, and decomposing the tensor product into its $\mathfrak{g}_0$-irreducible summands. Each of these summands will be μ-admissible. (Of course, there will be redundancy in the procedure.) To obtain all irreducible $\mathfrak{g}$-modules of highest weight μ, one lets $\mathfrak{p} = \mathfrak{p}^+$ be as in Chapter 2, and for each μ-admissible $\mathfrak{g}_0$-module W, one factors the right $U(\mathfrak{g})$-module $W \otimes_{U(\mathfrak{p})} U(\mathfrak{g})$ by the submodule generated by all

$$w \otimes f_1^{(i)} \cdots f_{k_i+1}^{(i)}, \quad 1 \leq i \leq \ell; \quad w \in W; \; f_j^{(i)} \in \mathfrak{g}_{-\alpha_i}, \quad k_i = \mu(H_i).$$

Eliminating redundancies, the fundamental sets Λ are as follows:

<u>When $*$ is of symplectic type</u>:

$$\lambda_1; \; d\lambda_j (1 < j \leq \ell); \; (d-k)\lambda_{j-1} + k\lambda_j \quad (1 < j < \ell, \; 0 < k < d);$$

$$(d-k)\,\lambda_{\ell-1} + 2k\lambda_\ell \quad (0 < k < d). \quad \text{If } \ell = 1, \; \Lambda = \{d\lambda_\ell\}.$$

<u>Otherwise</u>:

$$\lambda_1; \; d\lambda_j \; (1 < j < \ell); \; 2d\lambda_\ell \, ; \; (d-k)\lambda_{j-1} + k\lambda_j \quad (1 < j < \ell, \; 0 < k < d);$$

$$(d-k)\,\lambda_{\ell-1} + 2k\,\lambda_\ell (0 < k < d). \quad \text{If } \ell = 1, \Lambda = \{2d\,\lambda_\ell\}.$$

In §3, we have seen why weights $k\lambda_1$, $k > 1$, are redundant if $\ell > 1$; the omission of $2d\lambda_\ell$ in the symplectic case is possible because all irreducible modules for the even Clifford algebra $A_{2d} = A_{2d}(V)$ occur in tensor products of (two) irreducible A_d-modules (cf. [Se5]).

The conclusions of §§2-4 are summarized here:

<u>Theorem 4.2. For the various $\lambda \in \Lambda$ the λ-admissible $\mathfrak{g}_0$-modules are as follows:</u>

$\lambda = \lambda_1$ $(\ell > 1)$: <u>The unique irreducible module for $S^1(D^{op}) \simeq D$, which we identify with D. The action on D of the general element H of $\mathfrak{g}_0$ is given in §3.</u>

$\lambda = d\lambda_j$, $1 < j < \ell$: <u>A unique irreducible $\mathfrak{g}_0$-module which is a one-dimensional Z-module; the action of the general element H is given in §3.</u>

$\lambda = (d-k)\lambda_{j-1} + k\lambda_j$, $1 < j < \ell$, $0 < k < d$: A unique irreducible $\mathfrak{g}_0$-module, which may be identified with a certain minimal right ideal in $S^{d-k}(D)$, if * is of first kind; in $S^{d-k}(D(Z))$, if * is of second kind (cf. [Se3], Proposition VI.1). The action of the general element H is given in §3.

$\lambda = d\lambda_\ell$ (* of first kind, symplectic type): A minimal right ideal in the even Clifford algebra $A_d = A_d(V)$, the action of the general element H as specified in §2. There are one or two λ-admissible modules.

$\lambda = 2d\lambda_\ell$ (* of other types): A minimal right ideal in the even Clifford algebra $A_{2d} = A_{2d}(V)$, the action of H as specified in §2. The number of inequivalent λ-admissible modules is at most $\frac{1}{2}(md + 2)$ resp. $md + 1$, as * is of orthogonal type resp. second kind.

$\lambda = (d-k)\lambda_{\ell-1} + 2k\lambda_\ell$, $0 < k < d$: A minimal right ideal in the semisimple associative algebra A constructed in §4, the action of the general element H as specified there. When * is of first kind, A is central simple and the module is unique; when * is of second kind, the number of irreducible modules is one or two.

Finally, it may be of interest to compare the above with the approach of Chapter 3, by considering which irreducible modules for even Clifford algebras $A_k(U)$ correspond to our fundamental weights. From the splitting data it is clear that the $\mathfrak{g} = K'(U)$-modules of highest weight $d\lambda_\ell$, where * is of symplectic type, are the irreducible $A_d(U)$-modules. Except when $d = 1$ (so $D = Z \neq F$, by our assumptions), these $\mathfrak{g}$-modules do not exhaust the $A_{2d}(U)$-modules: for example, there is an $A_{2d}(U)$-module which is a $\mathfrak{g}$-module of highest $\mathfrak{t}_0$-weight $2\lambda_1$. Thus the two modes of generating all irreducible $K'(U)$-modules do not run identical courses.

CHAPTER 5: CONSTRUCTION OF EXCEPTIONAL ALGEBRAS FROM QUADRATIC FORMS

In [Sel], the isotropic exceptional simple Lie algebras with non-reduced root-systems divided into two classes, according as the root-space of a "longest" (i.e. doubled) root has dimension greater than one or equal to one. Those of the former type are associated with quadratic forms satisfying certain necessary conditions. The sufficiency of these conditions and their use in constructing the Lie algebra were treated rather casually in [Sel]. In some cases they are not new (see, e.g., [He]), and other constructions may be given based on pairs of composition algebras (see [A3], the cases of p. 147). Because generalizations of Clifford algebras have been central to earlier chapters, and because the whole construction can be based on a quadratic form, we give the construction in our terms here with some thoroughness. This form will play a prominent part in the constructions of irreducible modules in Chapter 6.

§1. The Fundamental Quadratic Spaces.

Let F be a field as before, V a vector space over F, of odd dimension $n = 2\ell + 1$. Assume that V carries a non-degenerate symmetric bilinear form (u,v), whose Clifford algebra we denote by $C\ell(V)$. To fix conventions, we stipulate that $C\ell(V)$ is the algebra generated by the subspace V with relations $uv + vu = 2(u,v)1$ for all $u,v \in V$. Then

$$C\ell(V) = C\ell^+(V) \oplus C\ell^-(V),$$

where $C\ell^+(V)$ is the subspace of linear combinations of products of even numbers of elements of V, and $C\ell^-(V)$ likewise for odd products. Then $C\ell^+(V)$ is a subalgebra, the even Clifford algebra, and $C\ell^-(V)$ is left and right $C\ell^+(V)$-module. There is in $C\ell^-(V)$ an element $z \neq 0$ central in $C\ell(V)$; z is unique to within scalar multiples, and may be taken to be the product of all elements in any orthogonal basis for V. Thus z is invertible and $C\ell^-(V) = z\, C\ell^+(V)$. We distinguish two cases:

(i) If $z^2 \in F^2$, we may assume $z^2 = 1$. Then if M is a simple right $C\ell(V)$-module, either $wz = w$ for all $w \in M$ or $wz = -w$ for all w, and M is a simple $C\ell^+(V)$-module. (ii) If $z^2 \notin F^2$, then $F(z)$ is a central quadratic subfield of $C\ell(V)$, $C\ell(V) \cong F(z) \otimes_F C\ell^+(V)$, a central simple algebra over $F(z)$. Our irreducible $C\ell(V)$-module M is a vector space over $F(z)$, and is a simple module for $F(z) \otimes_F C\ell^+(V)$.

Let σ be the involution in $C\ell(V)$ that fixes V, τ the involution that is -1 on V. Both induce the same involution in $C\ell^+(V)$; products of $0,4,8, \ldots$ elements of an orthogonal basis are fixed, while products of $2,6,10, \ldots$ are sent to their negatives. The space of fixed elements of $C\ell^+(V)$ has dimension $\binom{n}{0} + \binom{n}{4} + \binom{n}{8} + \ldots$, which is equal to $2^{\ell-1}(2^{\ell}-1)$ if $\ell \equiv 1,2 \pmod 4$,

$2^{\ell-1}(2^{\ell}+1)$ if $\ell \equiv 0,3 \pmod 4$.

(For instance, expand $(1 \pm i)^n$.) We have $z^{\sigma} = (-1)^{\ell} z$, $z^{\tau} = (-1)^{\ell+1} z$.

Some low-dimensional cases will concern us:

1) $\ell = 1$, case i): $C\ell^+(V)$ is a (generalized) quaternion algebra, and σ sends each element of quaternionic trace zero to its negative. An irreducible right $C\ell^+(V)$-module M admits two structures of irreducible right $C\ell(V)$-module extending its structure of $C\ell^+(V)$-module, distinguished by whether z acts as 1 or -1. If $C\ell^+(V)$ is split, M has dimension 2 and there is a nondegenerate alternate bilinear form q_1 on M such that for all $w,w' \in M$, $a \in C\ell^+(V)$,

$$q_1(wa, w') = q_1(w, w'a^{\sigma}).$$

From $(zv)^{\sigma} = -zv$, all $v \in V$, it follows (cf. 2) below) that

$$q_1(wv, w') = -q_1(w, w'v).$$

2) $\ell = 2$, case i): $C\ell^+(V)$ is a central simple F-algebra of dimension 2^4, which we further assume to be <u>split</u>. Then $C\ell^+(V) \approx \mathrm{End}_F(M)$, where the 4-dimensional F-vector space M is the unique irreducible right $C\ell^+(V)$-module. As in 1), M admits two structures of irreducible right $C\ell(V)$-module. The involution σ in $C\ell^+(V)$ is associated as in 1) with a non-degenerate alternate bilinear form on M, say q_2. If z acts as $\varepsilon (= \pm 1)$ on M in the structure of M as $C\ell(V)$-module, then for each $v \in V$, zv is a linear combination of products of 4 orthogonal basis elements of V, so is σ-fixed in $C\ell^+(V)$. Thus

$$q_2(wv,w') = \varepsilon\, q_2(wzv,w') = \varepsilon\, q_2(w,w'zv) = q_2(w,w'v)$$

for all $w,w' \in M$; all $v \in V$.

3) $\ell = 2$, case ii): Here we further assume the central simple $F(z)$-algebra $C\ell(V) = F(z) \otimes_F C\ell^+(V)$ to be split, hence isomorphic to $\mathrm{End}_{F(z)}(M)$, where M is a 4-dimensional $F(z)$-vector space. Here σ fixes $F(z)$, so is associated as in 2) with a non-degenerate alternate $F(z)$-bilinear form on M. The involution τ maps z to $-z$; we consider two extensions of τ to involutions in $C\ell(V)$:

a) the involution τ of $C\ell(V)$;

b) the "conjugate-transpose" map on $M_4(F(z)) \approx \mathrm{End}_{F(z)}(M)$, where τ induces the conjugation in $F(z)$.

Combining a) and b), we conclude as in the usual theory of involutions of second kind in matrix algebras that there is an $F(z)$-hermitian or antihermitian inner product h on M such that for all $w,w' \in M$, $a \in C\ell(V)$,

$$h(wa, w') = h(w, w'a^{\tau}).$$

Thus $h(\xi w,w') = \xi h(w,w')$ for $\xi \in F(z)$, $h(w',w) = \varepsilon h(w,w')^{\tau}$ where $\varepsilon = \pm 1$. Replacing h by zh if necessary, we may assume $\varepsilon = -1$. Now write $h(w,w') = h^{+}(w,w') + h^{-}(w,w')$, where $h^{+}(w,w') \in F$, $h^{-}(w,w') \in Fz$. Both h^{+} and h^{-} are F-bilinear, h^{+} is alternate, h^{-} symmetric, and $h^{+}(zw,w') = zh^{-}(w,w')$. Moreover, h^{+} is non-degenerate, and if $v \in V$, $h(wv,w') = -h(w,w'v)$.

4) $\ell = 3$, case ii): Again we add the assumption that the central simple $F(z)$-algebra $C\ell(V) = F(z) \otimes_F C\ell^{+}(V)$ is split, thus isomorphic to $\mathrm{End}_{F(z)}(M)$, M an 8-dimensional $F(z)$-space. The involution σ maps z to $-z$, and as in 3) there is an $F(z)$-antihermitian inner product h on M such that for all $w,w' \in M$, $a \in C\ell(V)$, $h(wa,w') = h(w,w'a^{\sigma})$; in particular, $h(wv,w') = h(w,w'v)$ for all $v \in V$. Again h has a decomposition $h = h^{+} + h^{-}$, as in 3), with $h^{+}: M \times M \to F$ alternate, F-bilinear and non-degenerate.

5) $\ell = 3$, case i): Here we make the added assumption that the central simple F-algebra $C\ell^{+}(V)$, of dimension 2^6, is of the form $\mathrm{End}_{\mathcal{Q}}(M)$, where $\mathcal{Q}$ is a quaternion algebra (possibly split) over F, and M is a free left $\mathcal{Q}$-module of rank 4. If $\mathcal{Q}$ is a division algebra, M is an irreducible right $C\ell^{+}(V)$-module, of F-dimension 16. If $\mathcal{Q}$ is split, with orthogonal primitive idempotents e_1, e_2, $e_1 + e_2 = 1$, we have $M = e_1M + e_2M$, the e_iM being isomorphic irreducible $C\ell^{+}(V)$-modules, of F-dimension 8, and $C\ell^{+}(V) \cong \mathrm{End}_F(e_iM)$, $i = 1,2$.

Again M and each e_iM (if present) admits an action of z (as ± 1), hence of $C\ell(V)$. If e_1, e_2 are present, we choose the action on M so that z acts as either 1 on both e_iM, or as -1 on both. With $b \to \bar{b}$ the canonical involution in $\mathcal{Q}$, we see by comparing the effect of σ on $C\ell^{+}(V)$ with the conjugate-transpose in $M_4(\mathcal{Q}) \cong C\ell^{+}(V)$ that there is a non-degenerate $\mathcal{Q}$-valued hermitian or antihermitian form $h(w,w')$ on M such that for all $a \in C\ell^{+}(V)$,

$$h(wa,w') = h(w,w'a^{\sigma}).$$

The space of σ-fixed elements of $C\ell^{+}(V)$ has dimension $1 + \binom{7}{4} > \frac{1}{2} \cdot 2^6$, from which it follows that h must be antihermitian. As above, $h(wv,w') = -h(w,w'v)$ for all $v \in V$, and $h = h^{+} + h^{-}$, where h^{+} has values in F, h^{-} in $\mathcal{Q}_0 = \{b \in \mathcal{Q} \mid \bar{b} = -b\}$. The form h^{+} is alternate and non-degenerate, while h^{-} is symmetric.

6) $\ell = 4$, case i): Again we assume a structure on $C\ell^{+}(V)$ as in 5): $C\ell^{+}(V) \cong \mathrm{End}_{\mathcal{Q}}(M)$, M a free left $\mathcal{Q}$-module of rank 8. Again there is a $\mathcal{Q}$-valued

antihermitian form $h = h^+ + h^-$, with $h(wa,w') = h(w,w'a^\sigma)$. In this case, $(zv)^\sigma = zv$, so $h(wv,w') = h(w,w'v)$ for all $v \in V$.

7) $\ell = 5$, case i): The added assumption here is that $C\ell^+(V)$ is split; thus $C\ell^+(V) \cong \mathrm{End}_F(M)$, $[M:F] = 2^5$. As in 2), there is a non-degenerate alternate bilinear form q_5 on M, such that $q_5(wa,w') = q_5(w,w'a^\sigma)$ for all $a \in C\ell^+(V)$. The element z acts on M as ± 1, and M is a $C\ell(V)$-module. From $(zv)^\sigma = -zv$ for $v \in V$, we have $q_5(wv,w') = -q_5(w,w'v)$.

8) $\ell = 6$, case i): Again we assume $C\ell^+(V)$ is split, so is as in 7) with $[M:F] = 2^6$. There is a non-degenerate alternate bilinear form q_6 on M, with $q_6(wa,w') = q_6(w,w'a^\sigma)$, $q_6(wv,w') = q_6(w,w'v)$, the notions as before.

9) We also consider one even-dimensional case, with $\dim V = 6$. Here $C\ell(V)$ is a 64-dimensional central simple algebra, which we assume to be of the form $\mathrm{End}_F(M)$, $[M:F] = 8$. The space of σ-fixed elements of $C\ell(V)$ has dimension 28, so there is a non-degenerate alternate bilinear form q on M such that for all $w,w' \in M$, $a \in C\ell(V)$, $q(wa,w') = q(w,w'a^\sigma)$.

§2. The [VV]-Module Morphisms.

The involution σ (or τ) maps each element x of the Lie subalgebra $[VV] \subset C\ell^+(V)$ to its negative. Regarding F as trivial right [VV]-module, we note that in the cases of §1, $w \otimes w' \to q_1(w,w')$, $q_2(w,w')$, $h^+(w,w')$, $q_5(w,w')$, $q_6(w,w')$, $q(w,w')$ all define morphisms of Lie [VV]-modules $M \otimes M \to F$.

In cases 3)-6), Fz resp. $\mathcal{Q}_0$ may be identified with a space of F-endomorphisms of M centralizing the action of [VV], so as a trivial right [VV]-module. Then $w \otimes w' \to h^-(w,w')$ defines a morphism $M \otimes M \to Fz$ resp. $\mathcal{Q}_0$ of [VV]-modules.

Let Q be the alternate mapping q_1, q_2, h^+, q_5, q_6 or q, according to the case of §1. Fixing $w,w' \in M$, consider the linear map $v \to Q(wv,w')$ of V into F. There is a unique $P(w,w') \in V$ with $Q(wv,w') = (v,P(w,w'))$, and $(v,P(w',w)) = Q(w'v,w) = -Q(w,w'v) =$

$$Q(wv,w') = (v,P(w,w')) \quad \text{in 1), 3), 5), 7);}$$

$$-Q(wv,w') = -(v,P(w,w')) \quad \text{in the other cases.}$$

Thus $P(w',w) = \pm P(w,w')$ according to these cases.

For all $v' \in V$, $(v',P(wv,w') - P(w'v,w)) = Q(wvv',w') - Q(w'vv',w) = Q(w(vv' + v'v),w') = 2(v',v)\,Q(w,w')$. Thus

$$P(wv,w') - P(w'v,w) = 2Q(w,w')v. \tag{1}$$

Still fixing $w,w' \in M$, consider the mapping $\varphi: V \to V$ defined by

$v\varphi = P(wv,w') + P(w'v,w)$. Expansion using the definition of P yields

$$(v\varphi, v') + (v,v'\varphi) = 2(v,v')(Q(w,w') + Q(w',w)) = 0.$$

Thus $\varphi:V \to V$ is linear and skew with respect to the inner product in V, so there is a unique $S(w,w') \in [VV]$ such that, for all v,

$$[v, S(w,w')] = P(wv,w') + P(w'v,w). \tag{2}$$

Clearly S is symmetric in w and w'.

Now one verifies that $P:M \otimes M \to V$ is a morphism of $[VV]$-modules, or that:

$$P(wx,w') + P(w,w'x) = [P(w,w'),x] \tag{3}$$

for all $w,w' \in M$, all $x \in [VV]$, and likewise

$$S(wx,w') + S(w,w'x) = [S(w,w'),x], \tag{4}$$

expressing the fact that $S:M \otimes M \to [VV]$ is a morphism of (right) $[VV]$-modules.

Substituting $v'' = P(w,w')$ in the fundamental relation $[v[v'v'']] = 4(v,v')v'' - 4(v,v'')v$ gives

$$[v[v',P(w,w')]] = 2P(w(vv' + v'v),w') - 4(v,P(w,w'))v'$$

$$= P(w(vv' + v'v),w') + P(w,w'(vv' + v'v))$$

$$- 2(v,P(w,w'))v' - 2\theta(v,P(w',w))v', \quad \text{where}$$

$\theta = \pm 1$ according as P is symmetric or skew. We write this as

$$[v[v',P(w,w')]] = P(wv'v,w') + P(wvv',w') - 2Q(wv,w')v'$$

$$+ \theta P(w'v'v,w) + \theta\{P(w'vv',w) - 2Q(w'v,w)v'\},$$

and use (1) to replace the terms in braces, obtaining

$$[v, S(wv',w') + \theta S(w'v',w)].$$

That is,

$$[v,P(w,w')] = S(wv,w') + \theta S(w'v,w) \tag{5}$$

where $P(w',w) = \theta P(w,w')$.

§3. Decomposition of Tensor Products.

We concentrate on the cases where V has odd dimension $2\ell + 1$, beginning with those where the center of $C\ell(V)$ is split, and where $C\ell^+(V)$ is split; e.g., 1), 2), 7), 8) of §1. Thus M is a $C\ell(V)$-module and $C\ell^+(V) \approx \mathrm{End}_F(M)$. The decomposition of the $[VV]$-module $M \otimes M$ has been known at least since the classic work of Brauer and Weyl [BW]. It will be useful to recall it, in terms of the notations of §2.

Let $0 \le j \le \ell$, and $q_\ell : M \times M \to F$ be a non-degenerate alternate or symmetric bilinear form associated with the involution σ in $C\ell^+(V)$ as in 1), 2), 7), 8): $q_\ell(wa,w') = q_\ell(w,w'a^\sigma)$. From dimensions of spaces of fixed elements and their degrees, we find in general

$$q_\ell(w',w) = (-1)^{\binom{\ell+1}{2}} q_\ell(w,w'),$$

$$q_\ell(wv,w') = (-1)^\ell q_\ell(w,w'v),$$

for all $w,w' \in M$, and all $v \in V$. Here q_ℓ defines a morphism of right $[VV]$-modules $M \otimes M \to F$. If $(M \otimes M)^+$ denotes the symmetric elements of $M \otimes M$, and $(M \otimes M)^-$ the skew elements, then q_ℓ defines a $[VV]$-morphism from $(M \otimes M)^+$ or $(M \otimes M)^-$ to $\Lambda^0(V) = F$, according as $\binom{\ell+1}{2}$ is even or odd.

As in §2, for each $w,w' \in M$ there is a unique $p_\ell(w,w') \in V$ satisfying $(v,p_\ell(w,w')) = q_\ell(wv,w')$ for all v. One has

$$p_\ell(wx, w') + p_\ell(w,w'x) = [p_\ell(w,w'),x] \tag{3)'}$$

for all $x \in [VV]$,

$$p_\ell(w',w) = (-1)^{\binom{\ell}{2}} p_\ell(w,w'),$$

$$p_\ell(wv,w') + (-1)^{\binom{\ell+1}{2}} p_\ell(w'v,w) = 2q_\ell(w,w')v. \tag{1)'}$$

For fixed w, w', the mapping $V \to V$ sending v to

$$p_\ell(wv,w') - (-1)^{\binom{\ell+1}{2}} p_\ell(w'v,w)$$

is skew with respect to the inner product in V, so there is a unique $s_\ell(w,w') \in [VV]$ with, for all v,

$$[v, s_\ell(w,w')] = p_\ell(wv,w') - (-1)^{\binom{\ell+1}{2}} p_\ell(w'v,w). \tag{2)'}$$

Here $s_\ell(w',w) = -(-1)^{\binom{\ell+1}{2}} s_\ell(w,w')$,

$$s_\ell(wx,w') + s_\ell(w,w'x) = [s_\ell(w,w'),x] \tag{4)'}$$

for all $x \in [VV]$, and, for all $v \in V$,

$$[v,p_\ell(w,w')] = s_\ell(wv,w') + (-1)^{\binom{\ell}{2}} s_\ell(w'v,w). \tag{5)'}$$

Associated with a fixed sequence $v_1,\ldots,v_j$ from V there is a map $M \otimes M \to F$ sending $w \otimes w'$ to

$$\sum_{\pi \in S_j} \operatorname{sgn}(\pi)\, q_\ell(wv_{\pi(1)}\cdots v_{\pi(j)},\, w').$$

Fixing w,w' and regarding the above as a map $X^jV \to F$, the map is alternating and multilinear, so defines an element of $(\Lambda^jV)^*$, a space which is isomorphic to Λ^jV by the inner product in V: if $y_1,\ldots,y_j \in V$, one associates with $y_1 \wedge\ldots\wedge y_j \in \Lambda^j(V)$ the alternating j-linear map

$$(v_1,\ldots,v_j) \to \sum_{\pi\in S_j} \operatorname{sgn}(\pi)\ (v_1,y_{\pi(1)}) \cdots (v_j,y_{\pi(j)}),$$

and the result is a $[VV]$-isomorphism $\Lambda^jV \to (\Lambda^jV)^*$.

For $x \in [VV]$ one has

$$q_\ell(w\, x\, v_1\cdots v_j,w') + q_\ell(w\, v_1\cdots v_j,w'x)$$
$$= -\sum_{i=1}^{j} q_\ell(w\, v_1\cdots[v_i x]\cdots v_j,w'),$$

so the composite mapping $\psi_j{:}M \otimes M \to \Lambda^jV$ is a $[VV]$-morphism. The parities already established show that ψ_j is symmetric in w,w' if $j\ell + \binom{j}{2} + \binom{\ell+1}{2}$ is even, skew if this number is odd. The parity of the quantity is the same as that of $\binom{\ell-j+1}{2}$; thus ψ_j is symmetric if $\ell - j \equiv 0,\ 3 \pmod 4$, and skew otherwise. Moreover, all the Λ^jV, $0 \le j \le \ell$, are nonisomorphic absolutely irreducible $[VV]$-modules, the sum of whose dimensions is $\frac{1}{2}\cdot 2^n = \dim(M \otimes M)$.

By the complete reducibility of $[VV]$-modules, the decomposition $M \otimes M \cong \oplus_{j=0}^{\ell} \Lambda^jV$ will follow once we know each ψ_j is nonzero. For this, we extend the base field to a splitting field for the form on V, so assume V has a basis $u_{-\ell},\ldots,u_{-1},u_0,u_1,\ldots,u_\ell$ with $(u_i,u_j) = \delta_{i,-j}$. Then for $\varepsilon = \pm 1$, we see as in Chapter III of [C] that the right ideal I in $C\ell(V)$ generated by $u_0 - \varepsilon, u_1,\ldots,u_\ell$ is maximal, and $C\ell(V)/I$ is a simple right $C\ell(V)$-module generated by an element annihilated by all of $u_1,\ldots,u_\ell,u_0 - \varepsilon$. Our M is $C\ell(V)$-isomorphic to one of these two modules.

Let w^+ generate M, $w^+u_i = 0$ for all $i > 0$, $w^+u_0 = \varepsilon w^+$. Then M is linearly generated by all $w^+u_{-J} = w^+u_{-j_t}u_{-j_{t-1}} \cdots u_{-j_1}$, where $J = \{j_1,\ldots,j_t\}$ is a subset of $L = \{1,\ldots,\ell\}$, with $j_1 <\ldots< j_t$ (if $t = 0$, $w^+u_{-\phi} = w^+$). If J, K are such subsets, consider $q_\ell(w^+u_{-J}, w^+u_{-K})$: when $J \cap K \neq \emptyset$, this is $\pm\, q_\ell(w^+u_{-J}u_{-K}, w^+) = 0$ because $u_{-J}u_{-K} = 0$. On the other hand, if $J \cap K = \emptyset$ but $J \cup K \neq L$, let $i \in L$, $i \notin J \cup K$. Then $w^+u_J = \pm\frac{1}{2}\, w^+u_{-(J\,\cup\,\{i\})}\, u_i$ and

$$q_\ell(w^+u_{-J}, w^+u_{-K}) = \pm \frac{1}{2} q_\ell(w^+u_{-(J \cup \{i\})} u_i, w^+u_{-K})$$

$$= \pm \frac{1}{2} q_\ell(w^+u_{-(J \cup \{i\})}, w^+u_{-K}u_i) = 0,$$

because $w^+u_{-K}u_i = \pm w^+u_iu_{-K} = 0$. Now the nondegeneracy of q_ℓ means that the basis $\{w^+u_{-J} | J \subseteq L\}$ for M has the property that $q_\ell(w^+u_{-J}, w^+u_{-K}) \neq 0$ exactly when J and K form a partition of L. All values of q_ℓ at pairs of basis elements are determined by $\gamma = q_\ell(w^+, w^+u_{-L})$, and we may normalize q_ℓ to assume $\gamma = 1$. We now claim that for each j, $0 \le j \le \ell$,

$$\psi_j(w^+u_{-j}u_{-(j-1)} \cdots u_{-1} \otimes w^+u_{-L}) \neq 0.$$

For $j = 0$, the value on the left is $q_\ell(w^+, w^+u_{-L}) = 1 \in F$. Otherwise, the displayed element, regarded as element of $(\Lambda^jV)^*$, sends $u_1 \wedge \ldots \wedge u_j$ to

$$\sum_{\pi \in S_j} \operatorname{sgn} \pi \; q_\ell(w^+u_{-j} \cdots u_{-1}u_{\pi(1)} \cdots u_{\pi(j)}, w^+u_{-L})$$

$$= \sum_{\pi \in S_j} q_\ell(w^+u_{-j} \cdots u_{-1}u_1 \cdots u_j, w^+u_{-L})$$

$$= 2^j \sum_{\pi \in S_j} q_\ell(w^+, w^+u_{-L}) = 2^j \, j! \neq 0.$$

There result the decompositions of $[VV]$-modules

$$(M \otimes M)^+ \cong \sum_{j \equiv 0,3 (\mathrm{mod}\ 4)} \Lambda^jV, \quad (M \otimes M)^- \cong \sum_{j \equiv 1,2 (\mathrm{mod}\ 4)} \Lambda^jV,$$

where j runs from 0 to ℓ. The maps "p_ℓ" are, to within scalar multiples, our morphisms $M \otimes M$ to V and the s_ℓ are essentially those $M \otimes M \to \Lambda^2V \cong [VV]$.

For our purposes, we shall determine the multiplicity of M in $(M \otimes M)^+ \otimes M$ in low-dimensional cases, but the considerations yield more general conclusions concerning the structure of these tensor products. Evidently $\Lambda^0V \otimes M = F \otimes M = M$. We show that in general, for $0 \le j \le \ell$, M occurs in $\Lambda^jV \otimes M$ with multiplicity <u>one</u>.

Namely, $(v_1, \ldots, v_j, w) \to \sum_{\pi \in S_j} \operatorname{sgn}(\pi)\, wv_{\pi(1)} \cdots v_{\pi(j)}$ defines a $[VV]$-morphism $(\Lambda^jV) \otimes M \to M$, which is at once seen to be non-zero from the split case. Thus M occurs in $\Lambda^jV \otimes M$ for each j.

To see that M occurs at most once, we return to the split case. Here a splitting Cartan subalgebra h for $[VV]$ has as basis the $[u_{-i}u_i]$, $1 \le i \le \ell$. Relative to h, a set of positive root-vectors is $\{[u_iu_j] | i < j, i + j > 0\}$, with $\{[u_iu_j] | i < j, i + j < 0\}$ a set of negative root-vectors. A highest weight vector in V is u_ℓ, belonging to weight $\mu_1 : \mu_1([u_{-i}u_i]) = 0$, $i < \ell$, $\mu_1([u_{-\ell}u_\ell]) = 4$. More generally, a highest weight vector for Λ^jV, $1 \le j \le \ell$, is

$u_{\ell-j+1} \wedge \ldots \wedge u_\ell$, belonging to the weight μ_j:

$\mu_j([u_{-i}u_i]) = 0$ for $i \le \ell-j$; $\mu_j([u_{-i}u_i]) = 4$ if $i > \ell-j$. A highest weight vector for M is w^+, belonging to the weight $\frac{1}{2}\mu_\ell$. With $\mu_0 = 0$, we prove by induction on j that

$$\Lambda^j V \otimes M \approx \sum_{i=0}^{j} M_{\mu_i + \frac{1}{2}\mu_\ell}, \tag{6}$$

where M_ω is the irreducible $[VV]$-module of highest weight ω. Thus $M_{\frac{1}{2}\mu_\ell} = M$.

Computation with Weyl's formula shows that the modules on both sides of (6) have the same dimension. No two summands on the right are isomorphic. Thus (6) will follow once it is known that each module on the right is a submodule, or homomorphic image, of $\Lambda^j V \otimes M$.

The positive roots of $[VV]$ relative to h have as base the roots of the following root-vectors:

$$[u_{-(\ell-1)}, u_\ell],\ ]u_{-(\ell-2)}, u_{\ell-1}],\ \ldots,\ [u_{-1}u_2]\ \cdot\ [u_0, u_1],$$

with respective roots

$$2\mu_1 - \mu_2,\ -\mu_1 + 2\mu_2 - \mu_3,\ \ldots,\ -\mu_{\ell-2} + 2\mu_{\ell-1} - \mu_\ell,\ \mu_\ell - \mu_{\ell-1},$$

which we abbreviate $\alpha_1, \ldots, \alpha_\ell$. For $j > 1$,

$$\mu_j - \mu_{j-1} = \alpha_j + \alpha_{j+1} + \ldots + \alpha_\ell, \quad \text{while} \quad \mu_1 - \mu_0 = \mu_1 = \alpha_1 + \ldots + \alpha_\ell.$$

It follows that $\mu_j + \frac{1}{2}\mu_\ell$ <u>is not a weight of any</u> $M_{\mu_i + \frac{1}{2}\mu_\ell}$ <u>for</u> $i < j$.

Now fix k, $0 < k \le \ell$, and suppose (6) holds for all $j < k$. There is a well-defined map $\Lambda^k V \otimes M \to \Lambda^{k-1} V \otimes M$ sending $(v_1 \wedge \ldots \wedge v_k) \otimes w$ to

$$\sum_{i=1}^{k} (-1)^{i-1} (v_1 \wedge \ldots \wedge \hat{v}_i \wedge \ldots \wedge v_k) \otimes w\, v_i, \tag{7}$$

and one easily sees this to be a $[VV]$-morphism. Iteration of the map yields a morphism $\Lambda^k V \otimes M \to \Lambda^j V \otimes M$ of $[VV]$-modules for each $j < k$, under which the image of

$$u_{\ell-k+1} \wedge \ldots \wedge u_\ell \otimes w^+ u_{-(\ell-j)} u_{-(\ell-j-1)} \cdots u_{-(\ell-k+1)}$$

is

$$\sum_{\pi \in S_{k-j}} \operatorname{sgn} \pi\ u_{\ell-j+1} \wedge \ldots \wedge u_\ell \otimes w^+ u_{-(\ell-j)} \cdots u_{-(\ell-k+1)} u_{\ell-k+\pi(1)} \cdots u_{\ell-k+\pi(k-j)}$$
$$= 2^{k-j}(k-j)!\ u_{\ell-j+1} \wedge \ldots \wedge u_\ell \otimes w^+,$$

a weight vector for the weight $\mu_j + \frac{1}{2}\mu_\ell$.

It follows that the image in $\Lambda^j V \otimes M$ of $\Lambda^k V \otimes M$ is not contained in $\Sigma_{i=0}^{j-1} M_{\mu_i + \frac{1}{2}\mu_\ell}$, and must contain $M_{\mu_j + \frac{1}{2}\mu_\ell}$. That is, each $M_{\mu_j + \frac{1}{2}\mu_\ell}$, $j < k$, is a submodule of an image of $\Lambda^k V \otimes M$, hence is a submodule of $\Lambda^k V \otimes M$. Finally, the submodule of $\Lambda^k V \otimes M$ generated by

$$(u_{\ell-k+1} \wedge \ldots \wedge u_\ell) \otimes w^+$$

is irreducible of highest weight $\mu_k + \frac{1}{2}\mu_\ell$. This completes the proof.

It will be noticed that the map $\Lambda^k V \otimes M \to \Lambda^{k-1} V \otimes M$ is onto in the split case, and is defined in all cases, so that $\Lambda^{k-1} V \otimes M$ is a summand of $\Lambda^k V \otimes M$. A complementary summand is irreducible in the split case, so is always irreducible. That is, whenever both the center of $C\ell(V)$ and the algebra $C\ell^+(V)$ are split, all irreducible [VV]-submodules of $\otimes^3 M$ are absolutely irreducible.

From the absolute irreducibility of M we see that if R is any finite-dimensional [VV]-module in which M occurs with multiplicity m, then $\dim(\mathrm{Hom}_{[VV]}(R,M)) = m$. The preceding calculations thus yield

$$\dim \mathrm{Hom}_{[VV]}((M \otimes M)^+ \otimes M, M) = \begin{cases} \frac{\ell+1}{2}, & \ell \text{ odd} \\ \frac{\ell}{2}, & \ell \equiv 2 \pmod 4 \\ \frac{\ell}{2} + 1, & \ell \equiv 0 \pmod 4 \end{cases}$$

$$\dim \mathrm{Hom}_{[VV]}((M \otimes M)^- \otimes M, M) = \begin{cases} \frac{\ell+1}{2}, & \ell \text{ odd} \\ \frac{\ell}{2} + 1, & \ell \equiv 2 \pmod 4 \\ \frac{\ell}{2}, & \ell \equiv 0 \pmod 4 \end{cases}$$

§4. Relations in $\mathrm{Hom}_{[VV]}(M \otimes M \otimes M, M)$.

The elements of $\mathrm{Hom}(M \otimes M \otimes M, M)$ in which the images of $w \otimes w' \otimes w''$ are, respectively, a), b), c) below, define elements of $\mathrm{Hom}_{[VV]}((M \otimes M)^+ \otimes M, M)$:

a) $q_\ell(w'',w)w' + q_\ell(w'',w')w$;

b) $w\, p_\ell(w',w'') + w' p_\ell(w,w'')$;

c) $w\, s_\ell(w',w'') + w' s_\ell(w,w'')$.

Similarly, each of the following defines an element of $\mathrm{Hom}_{[VV]}((M \otimes M)^- \otimes M, M)$:

d) $q_\ell(w'',w)w' - q_\ell(w'',w')w$;

e) $w\, p_\ell(w',w'') - w' p_\ell(w,w'')$;

f) $w\, s_\ell(w',w'') - w' s_\ell(w,w'')$.

The map

g) $w \otimes w' \otimes w'' \to q_\ell(w,w')w''$

defines an element of $\mathrm{Hom}_{[VV]}((M \otimes M)^+ \otimes M,M)$ for $\ell \equiv 0, 3 \pmod 4$, an element of $\mathrm{Hom}_{[VV]}((M \otimes M)^- \otimes M,M)$ otherwise. Likewise

h) $w \otimes w' \otimes w'' \to w''p_\ell(w,w')$ and

i) $w \otimes w' \otimes w'' \to w''s_\ell(w,w')$

define "+-maps" for $\ell \equiv 0, 1 \pmod 4$ in h), $\ell \equiv 1, 2 \pmod 4$ in i), and otherwise "- -maps".

The mappings above yield 5 elements of $\mathrm{Hom}_{[VV]}((M \otimes M)^+ \otimes M,M)$ if $\ell \equiv 0, 1 \pmod 4$ and 4 such elements otherwise, the numbers being reversed if "+" is replaced by " - ". From the dimensions at the end of §3 there must be, for each $\ell \leq 6$, a nontrivial relation (usually more than one) among these four or five mappings. The existence of such relations will be seen, from the point of view of this chapter, as a key to the existence of exceptional simple Lie algebras.

We shall extract bases from among a) - i) for each of the Hom-spaces in question, and we shall express the remaining maps as rational linear combinations of this basis. If the mappings $f_1,\ldots,f_d$, where d is the appropriate dimension, are claimed to be the basis, their linear independence will have been established by evaluating them at d elements $y_1,\ldots,y_d$ of $(M \otimes M \otimes M)_K$, where K is a splitting field, and showing them, as functions on the finite set $\{y_1,\ldots,y_d\}$, to be linearly independent (over K). Each of the remaining maps in the set of four/five must then be an F-linear combination of these d, with the coefficients determined by the values of f and the f_j on the test-set $\{y_1,\ldots,y_d\}$. These coefficients turn out to be rational. The y_j used will be of the form $w \otimes w' \otimes w''$, each $w^{(i)} \in M_K$; a triple (w,w',w'') giving one such y_j is referred to as a "test-triple". We shall not list test-triples, except in two illustrative cases, at the beginning and end of our list.

$\ell = 1$, "+": $d = 1$. Basis h), relations
a) = h), b) = -h), c) = h), i) = -h).

$\ell = 1$, "-": $d = 1$. Basis g), relations
d) = -g), e) = 3g), f) = -3g).

[Test-triple (w^+, w^+u_{-1}, w^+) - from M_K with notations as in §3.]

$\ell = 2$, "+": $d = 1$. Basis a), relations
b) = -a), c) = 2a), i) = -2a).

$\ell = 2$, "-": $d = 2$. Basis d), g), relations
e) = -3d) - 4g), f) = 2d) - 4g), h) = -2d) - g).

$\ell = 3$, "+": $d = 2$. Basis a), g), relations

b) = a) - 2g), c) = -3a) + 6g).

$\ell = 3$, "-": $d = 2$. Basis d), h), relations

e) = 3d) + 2h), f) = -d) + 2h), i) = 4d) + h).

$\ell = 4$, "+": $d = 3$. Basis a), g), h), relations

b) = a) + g) - h), c) = -4a) + 5g) + 3h).

$\ell = 4$, "-": $d = 2$. Basis d), e), relations

f) = -3d) + e), i) = 3d) - e).

$\ell = 5$, "+": $d = 3$. Basis a), b), h), relations

i) = -2a) - 2b) - h), c) = 2a) - 3b) - 4h).

$\ell = 5$, "-": $d = 3$. Basis d), e), g), relation

f) = 2d) + e) - 4g).

$\ell = 6$, "+": $d = 3$. Basis a), b), i), relation

c) = 3a) - 3b) + 2i).

$\ell = 6$, "-": $d = 4$. Basis d), e), g), h), relation

f) = d) + e) - 2g) - 2h).

[Test-triples" Let $L = \{1,2,\dots,6\}$:

$$(w,w',w'') = (w^+, w^+u_{-L}, w^+); \; (w^+, w^+u_{-6}, w^+u_{-L});$$

$$(w^+, w^+u_{-6}u_{-5}u_{-4}u_{-3}u_{-2}, w^+u_{-2}u_{-1});$$

$$(w^+, w^+u_{-L}, w^+u_{-6}u_{-4}u_{-5}u_{-3}).]$$

Finally, we derive similar information for the case 9) of §1, when V has dimension 6 and M is an absolutely irreducible $C\ell(V)$-module, $[M:F] = 8$. Here q is alternate, p is skew and s is symmetric. We shall be content with two relations: In "+":

$$c) = 3a) - 3b) + 2i). \tag{8}$$

In "-":

$$f) = d) + e) - g) - h). \tag{9}$$

It suffices to verify (8) and (9) when the form is split (as for $\ell = 3$ above, but with u_0 absent). A maximal right ideal in $C\ell(V)$ is generated by the u_i $(i > 0)$, and the quotient of $C\ell(V)$ by this ideal is an irreducible right $C\ell(V)$-module generated by an element w^+ with $w^+u_i = 0$ for all $i > 0$. The element

$$z = \sum_{J \subseteq \{1,2,3\}} (-1)^{|J|} u_{-J} u_J \quad \text{is central in} \quad C\ell^+(V),$$

with $vz = -zv$ for all $v \in V$, and $z^2 = 1$. A basis for M consists of all w^+u_{-J}, $J \subseteq L = \{1,2,3\}$, and we have $w^+u_{-J}u_J = 2^{|J|} w^+$. Thus $w^+z = -w^+$, and M splits as $C\ell^+(V)$-module into $M = M^+ \oplus M^-$, where $M^\varepsilon = \{w \in M | wz = \varepsilon w\}$. A basis for M^- consists of w^+ and the w^+u_{-J}, $|J| = 2$, while a basis for M^+ consists of those $w u_{-J}$, $|J|$ odd. To verify (8) and (9), it suffices to assume $w, w' w'' \in M^+ \cup M^-$.

As before, $q(w^+u_{-J}, w^+u_{-K}) = 0$ except when $J \cup K$ is a partition of L. It follows that $q(M^+,M^+) = 0 = q(M^-,M^-)$, so that q is a dual pairing of M^+ and M^-. We also have $M^+V = M^-$, $M^-V = M^+$, and thus $q(M^+V,M^-) = 0 = (V,p(M^+,M^-))$. That is, $p(M^+,M^-) = 0$. Also, $p(w^+,w^+u_{-3}u_{-2}) = u_1$, where $q(w^+, w^+u_{-L})$ is assumed to be 1, and $p(w^+u_{-3}, w^+u_{-2}) = u_1$ as well. Thus the restrictions of p to $M^+ \otimes M^+$ and to $M^- \otimes M^-$ are nonzero; each of $(M^+ \otimes M^+)^-$, $(M^- \otimes M^-)^-$ has dimension 6, so p yields an isomorphism of $[VV]$-modules $(M^\varepsilon \otimes M^\varepsilon)^- \to V$ for each value of ε. From the split setting $(M^\varepsilon \otimes M^\varepsilon)^+$ is irreducible and their (direct) sum for $\varepsilon = \pm 1$ is $\Lambda^3 V$. One sees as above that s vanishes on $M^\varepsilon \otimes M^\varepsilon$, so $s(M^+ \otimes M^-) \neq 0$. It is easy to see now that $M^+ \otimes M^- \cong [VV] \oplus F$.

Thus $(M^+ \otimes M^-) \otimes M^\varepsilon = ([VV] \otimes M^\varepsilon) \oplus M^\varepsilon$. Clearly M^ε is a homomorphic image (hence a submodule) of $[VV] \otimes M^\varepsilon$, and from split theory the multiplicity of M^ε in this tensor product is 1. That is,

$$\dim \mathrm{Hom}_{[VV]}((M^+ \otimes M^-) \otimes M^\varepsilon, M^\varepsilon) = 2.$$

Moreover $[VV] \otimes M^\varepsilon$ does not contain $M^{-\varepsilon}$ in the split case, so

$$\mathrm{Hom}_{[VV]}((M^+ \otimes M^-) \otimes M^\varepsilon, M^{-\varepsilon}) = 0.$$

Now $(M^\varepsilon \otimes M^\varepsilon)^- \otimes M^{\varepsilon'} = V \otimes M^{\varepsilon'}$ contains $M^{\varepsilon'}V = M^{-\varepsilon'}$ $(\varepsilon,\varepsilon' = \pm)$, the remaining 20-dimensional summand being absolutely irreducible. Thus

$$\mathrm{Hom}_{[VV]}((M^\varepsilon \otimes M^\varepsilon)^- \otimes M^{\varepsilon'}, M^{\varepsilon'}) = 0,$$

$$\dim \mathrm{Hom}_{[VV]}((M^\varepsilon \otimes M^\varepsilon)^- \otimes M^{\varepsilon'}, M^{-\varepsilon'}) = 1.$$

Finally, split theory shows $(M^\varepsilon \otimes M^\varepsilon)^+ \otimes M^\varepsilon$ to be the direct sum of two 20-dimensional absolutely irreducible modules, while $(M^\varepsilon \otimes M^\varepsilon)^+ \otimes M^{-\varepsilon}$ is the sum of M^ε and a 36-dimensional absolutely irreducible module:

$$\mathrm{Hom}_{[VV]}((M^\varepsilon \otimes M^\varepsilon)^+ \otimes M^\varepsilon, M^{\pm\varepsilon}) = 0 = \mathrm{Hom}_{[VV]}((M^\varepsilon \otimes M^\varepsilon)^+ \otimes M^{-\varepsilon}, M^{-\varepsilon});$$

$$\dim \mathrm{Hom}_{[VV]}((M^\varepsilon \otimes M^\varepsilon)^+ \otimes M^{-\varepsilon}, M^\varepsilon) = 1.$$

From these determinations, both members of each of (8), (9) are zero if all three of w, w', w'' are in the same M^{ε}, and it suffices to verify (8) and (9) when two arguments are in M^{ε}, one in $M^{-\varepsilon}$.

For $w, w' \in M^{\varepsilon}$, $w'' \in M^{-\varepsilon}$, (8) becomes

$$w\,s(w',w'') + w's(w,w'') = 3(q(w'',w)w' + q(w'',w')w), \tag{10}$$

and (9) is

$$ws(w',w'') - w's(w,w'') = -2\,w''p(w,w') + q(w'',w)w' - q(w'',w')w. \tag{11}$$

These correspond to relations in $\mathrm{Hom}_{[VV]}((M^{\varepsilon} \otimes M^{\varepsilon})^{\pm} \otimes M^{-\varepsilon}, M^{\varepsilon})$, in each case a space of dimension 1. Thus they may be verified by checking at one test-triple where some values are non-zero. For $\varepsilon = -1$, $(w^{+},w^{+}u_{-2}u_{-1}, w^{+}u_{-L})$ is such a test-triple (w,w',w''), and for $\varepsilon = 1$, $(w^{+}u_{-L}, w^{+}u_{-3}, w^{+})$ suffices.

We may therefore assume $w \in M^{\varepsilon}$, $w' \in M^{-\varepsilon}$. By the symmetry or skewness of (8), (9) in w, w', we may assume $w'' \in M^{\varepsilon}$. Then (8), (9) becomes

$$w\,s(w',w'') = 2w''s(w,w') + 3q(w'',w')w - 3w'(w,w'') \tag{12}$$

and

$$ws(w',w'') = -2q(w,w')w'' - q(w'',w')w - w'p(w,w''). \tag{13}$$

Here it suffices to display <u>two</u> test-triples on which $q(w'',w')w$ and $w'p(w,w'')$ give linearly independent maps and on which relations (12), (13) hold. For $\varepsilon = -1$, two such are $(w^{+},w^{+}u_{-L}, w^{+}u_{-2}u_{-1})$ and $(w^{+},w^{+}u_{-1}, w^{+}u_{-3}u_{-2})$; for $\varepsilon = 1$, $(w^{+}u_{-L},w^{+}, w^{+}u_{-3})$ and $(w^{+}u_{-L}, w^{+}u_{-3}u_{-2}, w^{+}u_{-1})$. The relations (8), (9) are thus established.

§5.A. Constructions From a 5-Dimensional Quadratic Space.

Let U be a 5-dimensional space satisfying the conditions imposed on "V" in 2) of §1. Replacing z by $-z$ if necessary, we may assume z acts as the identity on the irreducible $C\ell(V)$-module $M = M_U$, of dimension 4. We fix the alternate non-degenerate form q_2 on M_U, and derive from q_2 the pairings $p_2{:}M_U \times M_U \to U$ and $s_2{:}M_U \times M_U \to [UU]$ as in §2. (In this case, these are simply the "P,S" of §3, associated with "Q" = q_2.)

Take V as in 3), 5) or 7) of §1, $M = M_V$ as there. In cases 5), 7) we may assume as above that "z" acts as the identity on M_V. Let Q be the appropriate one of h^{+}, h^{+}, q_5 for 3), 5), 7), respectively, as in §2, P and S: $M_V \times M_V \to Fz$ resp. $\mathcal{Q}_0$ be as for 3) resp. 5), with $h^{-} = 0$ in 7). Then q_2 and p_2 are skew, while s_2 is symmetric; all of h^{-}, P, S are symmetric, while Q is skew. Let $\mathfrak{g}_2 = Fz$ in case 3), a commutative Lie algebra, $\mathfrak{g}_2 = \mathcal{Q}_0$ in case 5), a simple 3-dimensional Lie subalgebra of $\mathcal{Q}$, ($\mathfrak{g}_2 = \{0\}$ in case 7).

On the direct sum of vector spaces

$$\mathfrak{g} = \mathfrak{g}_2 \oplus [VV] \oplus [UU] \oplus (U \otimes V) \oplus (M_U \otimes M_V)$$

we define a skew-symmetric bilinear composition $[A,B]$ such that:

a) The Lie algebras $\mathfrak{g}_2$, $[UU]$, $[VV]$ are subalgebras, centralizing one another;

b) $[\mathfrak{g}_2, U \otimes V] = 0$;

c) For $u \in U$, $v \in V$, $x \in [UU]$, $[u \otimes v, x] = [ux] \otimes v \in U \otimes V$;

d) For $u \in U$, $v \in V$, $y \in [VV]$, $[u \otimes v, y] = u \otimes [vy] \in U \otimes V$;

e) For $m \in M_U$, $w \in M_V$, $a \in \mathfrak{g}_2$,

$[m \otimes w, a] = -m \otimes aw$;

f) $[m \otimes w, x] = mx \otimes w \in M_U \otimes M_V$, in notations above;

g) $[m \otimes w, y] = m \otimes wy$;

h) $[m \otimes w, u \otimes v] = mu \otimes wv \in M_U \otimes M_V$;

i) $[u \otimes v, u' \otimes v'] = (u,u')[v,v'] + (v,v')]u,u'] \in [VV] + [UU]$;

j) $[m \otimes w, m' \otimes w'] = \varkappa\, q_2(m,m')h^{-}[w,w'] + \lambda\, q_2(m,m')S(w,w')$

$$+ \lambda\, Q(w,w')s_2(m,m') - \lambda\, p_2(m,m') \otimes P(w,w'),$$

the terms being respectively in $\mathfrak{g}_2$, $[VV]$, $[UU]$, $U \otimes V$, where $\varkappa$, λ are constants whose relations to one another must be specified separately in the cases 3), 5), 7). We know that the product is skew when both factors are in the same summand; it is extended by skew-symmetry in the remaining cases. The choices of $\varkappa$, λ will be such that the Jacobi identity is satisfied.

Evidently $\mathfrak{g}_2 + [UU] + [VV] + U \otimes V$ is a subalgebra, the actions of $\mathfrak{g}_2$, $[UU]$, $[VV]$ on $U \otimes V$ make $U \otimes V$ a (right) module for each of these Lie algebras, and their actions centralize one another. To see that $\mathfrak{g}_2 + [UU] + [VV] + U \otimes V$ is a Lie algebra, it suffices to verify the Jacobi identity for triples of which two members come from $U \otimes V$, and the remaining member from each of the four summands. If the third member is from $\mathfrak{g}_2$, all terms in the Jacobi identity are zero.

Next note that

$$[[u \otimes v, u' \otimes v']x] + [[u' \otimes v', x]u \otimes v] + [[x, u \otimes v]u' \otimes v']$$

$$= (v,v')\,[[uu']x] + (v',v)[[u'x]u] - (v,v')[[ux]u'],$$

which is zero by the symmetry of (v,v') and by the Jacobi identity in $C\ell(U)$. A similar observation applies when $x \in [UU]$ is replaced by $y \in [VV]$.

Now

$$[[u \otimes v, u' \otimes v']u'' \otimes v''] = -(u,u')u'' \otimes [v''[vv']] - (v,v')[u''[uu']] \otimes v'',$$

and the sum over cyclic permutations is to be zero. This follows at once from $[u''[uu']] = 4(u'',u)u' - 4(u'',u')u$ and $[v''[[vv']] = 4(v'',v)v' - 4(v'',v')v$.

Next note that the right actions of g_2, $[UU]$, $[VV]$ in our product make $M_U \otimes M_W$ into a right module for each of these Lie algebras in such a way that their actions commute. For $[UU]$, $[VV]$, and for the commutativity of the action of g_2 with those of $[UU]$ and $[VV]$, this follows from e)-g) and the definition of g_2, using §1. For $a,a' \in g_2$,

$$[[m \otimes w, a]a'] - [[m \otimes w, a']a]$$
$$= m \otimes [a'(aw) - a(a'w)) = m \otimes [a'a]w$$
$$= - [m \otimes w, [a'a]] = [m \otimes w, [aa']],$$

and our condition is satisfied for two elements from g_2 as well.

Now

$$[[m \otimes w, u \otimes v]a] + [[u \otimes v, a]m \otimes w] + [[a, m \otimes w], u \otimes w]$$
$$= [mu \otimes wv, a] + [m \otimes aw, u \otimes v]$$
$$= -mu \otimes a(wv) + mu \otimes (aw)v = 0$$

because left multiplication by Fz resp. $\mathcal{Q}_0$ centralizes the action of $C\ell(V)$ on M_V. Also,

$$[[m \otimes w, u \otimes v]x] + [[u \otimes v, x]m \otimes w] + [[x, m \otimes w]u \otimes v]$$
$$= mux \otimes wv - m[ux] \otimes wv - (mx)u \otimes wv = 0,$$

and likewise with $x \in [UU]$ replaced by $y \in [VV]$.

One has

$$[[m \otimes w, u \otimes v]u' \otimes v'] + [[u \otimes v, u' \otimes v']m \otimes w] + [[u' \otimes v', m \otimes w]u \otimes v]$$
$$= muu' \otimes wvv' - (v,v')m[uu'] \otimes w - (uu')m \otimes w[vv']$$
$$- mu'u \otimes wv'v.$$

Substituting $\frac{1}{2}(uu' + u'u)$ for (u,u') and $\frac{1}{2}(vv' + v'v)$ for (v,v'), and writing $[uu'] = uu' - u'u$, $[vv'] = [vv'] = vv' - v'v$, it is immediate that the expression above is zero. The effect of this and the preceding paragraph is to show that $M_U \otimes M_V$ is a right module for the Lie algebra

$$g_2 + [UU] + [VV] + U \otimes V.$$

To consider the Jacobi identity for two factors from $M_U \otimes M_V$, it suffices to take these in the form $m \otimes w$, and the remaining factor from g_2, $[UU]$, $[VV]$, of the form $u \otimes v$, or of the form $m \otimes w$. Except for the last of these, where the relation between $\varkappa$ and λ must be considered, the argument is uniform. We treat those separate cases in subsequent sections, but the uniform ones here.

Thus

$$[[m \otimes w, m' \otimes w']a] + [[m' \otimes w', a]m \otimes w] + [[a, m \otimes w]m' \otimes w']]$$

$$= \varkappa\, q_2(m,m')[h^-(w,w'),a] - [m' \otimes aw', m \otimes w]$$

$$+ [m \otimes aw, m' \otimes w'],$$

whose components in $\mathfrak{g}_2$, $[UU]$, $[VV]$, $U \otimes V$ are, respectively,

$$\varkappa\, q_2(m,m')\{[h^-(w,w'),a] + h^-(aw',w) + h^-(aw,w')\}; \tag{14}$$

$$\lambda(Q(aw,w') - Q(aw',w))s_2(m,m'); \tag{15}$$

$$\lambda q_2(m,m')\ \{S(aw,w') + S(aw',w)\}; \tag{16}$$

$$-\lambda\, p_2(m,m') \otimes \{P(aw,w') + P(aw',w)\}. \tag{17}$$

All four of these are clearly zero in Case 7). Otherwise $h(w,w') = Q(w,w') + h^-(w,w')$ is skew-hermitian with respect to the involution $b \to \bar{b}$ in $F(z)$ or $\mathcal{Q}$, so from $\bar{a} = -a$ we find $[h^-(w,w'),a] = [h(w,w'),a] = -h(w,aw') - h(aw,w') = Q(aw',w) - Q(aw,w') - h^-(aw',w) - h^-(aw,w')$. Using the fact that $[h^-(w,w'),a] \in \mathcal{Q}_0$ (or is zero in Fz), and comparing components in F and in $\mathcal{Q}_0$ (or in Fz) shows (14) and (15) to be zero in case 5) (resp. 3)).

That (17) is zero now follows from the fact that for all $v \in V$,

$$(v,P(aw,a')) = Q((aw)v,w') = Q(aw',wv) \qquad \text{(cf. (15))}$$

$$= -Q(wv,aw') = -(v,P(w,aw')) = -(v,P(aw',w)).$$

Then

$$\begin{aligned}[v, S(aw',w)] &= P((aw')v,w) + P(wv,aw') \\ &= P(a(w'v),w) - P(a(wv,w') \\ &= -P(aw,w'v) - P((aw)v,w') \\ &= -[v, S(aw,w')],\end{aligned}$$

using that (17) is zero and the fact that $a(wv) = a(wzv) = (aw)(zv) = (aw)v$, because $zv \in C\ell^+(V)$. This shows that (16) is also zero, and verifies the Jacobi identity in this case.

The Jacobi identity for triples $m \otimes w$, $m' \otimes w'$, $x \in [UU]$ follows from the $[UU]$-invariance of q_2, p_2, s_2, and for triples $m \otimes w$, $m' \otimes w'$, $y \in [VV]$, from the $[VV]$-invariance of h^-, Q, P, S. Finally,

$$[[m \otimes w, m' \otimes w']u \otimes v] + [[m' \otimes w', u \otimes v]m \otimes w]$$

$$+ [[u \otimes v, m \otimes w]m' \otimes w']$$

has as components in $\mathfrak{g}_2$, $[UU]$, $[VV]$, $U \otimes V$, respectively,

$$\varkappa\, q_2(m'u,m)h^-(w'v,w) - \varkappa\, q_2(mu,m')h^-(wv,w'); \tag{18}$$

$$-\lambda(P(w,w'),v)[p_2(m,m'),u] + \lambda Q(w'v,w)s_2(m'u,m) - \lambda Q(wv,w')s_2(mu,m'); \tag{19}$$

$$-\lambda(p_2(m,m'),u)[P(w,w'),v] + \lambda q_2(m'u,m)S(w'v,w) \tag{20}$$
$$-\lambda q_2(mu,m')S(wv,w');$$

$$-\lambda q_2(m,m')u \otimes [v,S(w,w')] - \lambda Q(w,w')[u,s_2(m,m')] \otimes v \tag{21}$$
$$-\lambda p_2(m'u,m) \otimes P(w'v,w) + \lambda p_2(mu,m') \otimes P(wv,w').$$

From 2) of §1, $q_2(m'u,m) = -q_2(mu,m')$, and $h(wv,w') = -h(w,w'v)$ in both 3) and 5), so that $h^-(w'v,w) = h^-(w,w'v) = -h^-(wv,w')$. Thus (18) is zero. The term (19) is, by $Q(w'v,w) = Q(wv,w') = (v,P(w,w'))$ (from §2), $\lambda(v,P(w,w'))\,\{[u,p_2(m,m')] - s_2(mu,m') + s_2(m'u,m)\}$, which is zero by (5)'. A similar argument, using $(u,p_2(m,m')) = q_2(mu,m') = -q_2(m'u,m)$ and (2) shows that (20) is zero.

In (21), we use (1) and (2) to substitute for $Q(w,w')v$ and $[v,S(w,w')]$ in terms of $P(wv,w')$ and $P(w'v,w)$. The "coefficient" of $P(wv,w')$ in (21) then becomes

$$-\lambda q_2(m,m')u - \frac{1}{2}\lambda[u,s_2(m,m')] + \lambda\, p_2(mu,m'),$$

which by (1') and (2') is equal to zero. The same holds for the terms involving $P(w'v,w)$.

We have shown that our skew-symmetric composition in $\mathfrak{g}$ satisfies the Jacobi identity for all choices of $\varkappa$ and λ, except perhaps when all three factors are in $M_U \otimes M_V$.

§5.B. Dim V = 11. A Construction For E_8.

In general, we have

$$[[m \otimes w, m' \otimes w']m'' \otimes w''] = \varkappa\, q_2(m,m')m'' \otimes h^-(w,w')w''$$
$$- \lambda q_2(m,m')m'' \otimes w''S(w,w') - \lambda\, m''s_2(m,m') \otimes Q(w,w')w''$$
$$+ \lambda m''p_2(m,m') \otimes w''\, P(w,w').$$

Permuting the factors cyclically and adding, while using results for $\ell = 2$ from §4 to eliminate all terms involving p_2 and s_2, we find our sum is equal to

$$q_2(m,m')m'' \otimes \{\varkappa\, h^-(w,w')w'' - \lambda\, w''S(w,w')$$
$$- \lambda\, w''P(w,w') + 2\,\lambda Q(w',w'')w - 2\lambda\, wP(w',w'')$$
$$- 2\,\lambda\, Q(w'',w)w' - 2\,\lambda\, w'P(w'',w)\}$$

plus two terms where m,m',m'' are permuted cyclically, along with w,w',w''. Thus it suffices to show that the element of M_V in braces above is zero, i.e., that

$$\varkappa\, h^-(w,w')w'' - \lambda\, w''S(w,w') - \lambda\, w''P(w,w') \tag{22}$$
$$-2\,\lambda(Q(w'',w)w' + Q(w'',w')w) - 2\,\lambda(wP(w',w'') + w'P(w,w'')) = 0.$$

(Recall that Q,P,S are as in cases 3), 5), 7).) In case 7), $g_2 = 0$ and the term in h^- is missing. Here we have $Q = q_5$, $P = p_5$, $S = s_5$, and the vanishing of (22) is just our previously derived relation giving i) for $\ell = 5$, "+", in §4. That is, <u>in case</u> V <u>is as in</u> 7), g <u>is a Lie algebra for each choice of</u> $\varkappa$ <u>and</u> λ. The choice of $\varkappa$ is irrelevant to the multiplication in λ in any case, and if $\lambda = 0$ then $M_U \otimes M_V$ is an ideal in g. Accordingly we fix some $\lambda \neq 0$.

From $\dim U = 5$, $\dim [UU] = 10$ and $\dim M_U = 4$. Likewise, $\dim V = 11$, $\dim [VV] = 55$ and $\dim M_V = 32$. Thus g has dimension $10 + 55 + (5 \times 11) + (4 \times 32) = 248$. The non-isomorphism of the $[UU]$-modules F, U, $[UU]$, M_U shows that each ideal J of g has the form $J_0 + J_1 + U \otimes V_0 + M_U \otimes W_0$, where $J_0 = J \cap [VV]$, $J_1 = J \cap [UU]$, $U \otimes V_0 = J \cap (U \otimes V)$, $(M_U \otimes W_0) = J \cap (M_U \otimes M_V)$. Moreover, each of these four summands must be $[VV]$-stable. From the $[VV]$-irreducibility of $[VV]$, V, M_V, as well as from the simplicity of $[UU]$, it follows that if $J \neq \{0\}$ then J contains at least one of $[UU]$, $[VV]$, $U \otimes V$, $M_U \otimes M_V$, and that in this case $J = g$. Thus g is simple; because all our conditions are fulfilled in the split case, g is absolutely simple. The uniqueness of the dimension, 248, among dimensions of absolutely simple Lie algebras shows that g is an F-form of the split simple Lie algebra E_8. (When U and V are split, g is split; we shall display this and other information about root-space decompositions in Chapter 6.)

§5.C. Dim V = 5. A Construction For E_6.

Next let V be as in 3). Here we take $g_2 = Fz$, and we fix $\varkappa = 3\lambda \neq 0$ in j). Then (22), the last step in the Jacobi identity, reduces to

$$3h^-(w,w')w'' - w''S(w,w') - w''P(w,w') \tag{23}$$
$$-2(Q(w'',w)w' + Q(w'',w')w) - 2(wP(w',w'') + w'P(w,w'')) = 0,$$

which we must verify for $w, w', w'' \in M_V = M$, as in 3).

One may pass to an extension field to assume $z^2 = 1$ and that $M = M^+ \oplus M^-$, the sum of two irreducible right $C\ell(V)$-modules of dimension 4, with $wz = \varepsilon w$, $w \in M^\varepsilon$.

These two are absolutely irreducible isomorphic $C\ell^+(V)$-modules. Over the extension field, we may assume V has splitting basis $v_{-2}, v_{-1}, v_0, v_1, v_2$ with $(v_i, v_j) = \delta_{i,-j}$, and that $z = \sum_{j \subseteq \{1,2\}} v_{-J} v_0 v_J$. Then M^+ is generated by $w^+ \neq 0$, $w^+ u_i = 0$, $i = 1,2$, $w^+ z = w^+ = (\sum_J (-2)^{|J|}) w^+ v_0 = w^+ v_0$, and there is a $C\ell^+$-isomorphism φ of M^+ onto M^-. Thus M^- is generated by $\varphi(w^+)$, with $\varphi(w^+) v_i v_0 = \varphi(w^+ v_i v_0) = 0$, $i = 1,2$, from which $\varphi(w^+) v_i = 0$. Thus $\varphi(w^+) z = - \varphi(w^+)$ means that $\varphi(w^+) v_0 = - \varphi(w^+)$. Therefore $\varphi(w^+ v_{-J}) = \varphi(w^+) v_{-J}$ if $|J|$ is even, $\varphi(w^+ v_{-J}) = \varphi(w^+ v_0 v_{-J}) = \varphi(w^+) v_0 v_{-J}$ $= - \varphi(w^+) v_{-J}$ if $|J|$ is odd: $\varphi(w^+ v_{-J}) = (-1)^{|J|} \varphi(w^+) v_{-J}$, for $J \subseteq L = \{1,2\}$.

For $w, w' \in M$, $h(wz, w') = \varepsilon\, h(w,w') = -h(w, w'z) = -\varepsilon\, h(w,w')$ shows that each M^ε is totally isotropic with respect to h. Also, as before, $h(w^+ u_{-J}, \varphi(w^+) u_{-K}) = 0$ unless $J \cup K$ is a partition of L. Now consider the form $Q(w, \varphi(w')) = h^+(w, \varphi(w'))$ on M^+. For $J, K \subseteq L$,

$$
\begin{aligned}
& h^+(w^+ u_{-J}, \varphi(w^+ u_{-K})) = (-1)^{|K|}\, h^+(w^+ u_{-J} u_{-K}^\tau, \varphi(w^+)) \\
&= h^+(w^+ u_{-J} u_{-K}^\sigma, \varphi(w^+)) = (-1)^{|J||K|}\, h^+(w^+ u_{-K}^\sigma u_{-J}, \varphi(w^+)) \\
&= (-1)^{|J|(|K|+1)}\, h^+(w^+ u_{-K}^\sigma, \varphi(w^+) u_{-J}^\sigma) \\
&= (-1)^{|J||K| + \binom{|J|}{2} + \binom{|K|}{2}}\, h^+(w^+ u_{-K}, \varphi(w^+ u_{-J})),
\end{aligned}
$$

using the fact that only terms where J and K partition L are nonzero. For such J, K, we have $|J||K| + \binom{|J|}{2} + \binom{|K|}{2} = \frac{1}{2}(|J| + |K|)^2 - \frac{1}{2}(|J| + |K|) = \frac{4}{2} - \frac{2}{2} = 1$. That is,

$$h^+(w', \varphi(w)) = -h^+(w, \varphi(w')).$$

Thus $h^+(w, \varphi(w'))$ is alternate, nondegenerate, and $[VV]$-invariant. From the absolute irreducibility of M^+ it follows that there is a nonzero scalar β such that

$$h^+(w, \varphi(w')) = \beta\, q_2(w,w') \tag{24}$$

for all $w, w' \in M^+$.

For such w, w', $h(zw, \varphi(w')) = h(w, \varphi(w')) = h^+(w, \varphi(w')) + h^-(w, \varphi(w'))$ is equal to $zh(w, \varphi(w')) = zh^-(w, \varphi(w')) + h^+(w, \varphi(w'))z$, where the former term is in (our aplitting extension of) F, the latter in Fz. By comparing terms in Fz, we have

$$h^-(w, \varphi(w')) = h^+(w, \varphi(w'))z = \beta\, q_2(w,w')z. \tag{25}$$

The pairing P was defined in this case by $(v, P(w,w')) = h^+(wv, w')$, all $v \in V$. It follows that P is zero on $M^\varepsilon \otimes M^\varepsilon$ $(\varepsilon = \pm)$. For $w, w' \in M^+$, $(v, P(w, \varphi(w'))) = h^+(wv, \varphi(w')) = \beta\, q_2(wv, w') = \beta(v, p_2(w,w'))$, where p_2 is as before. That is, for $w, w' \in M^+$,

$$P(w, \varphi(w')) = \beta\, p_2(w,w'). \tag{26}$$

From the skewness of p_2, $P(w', \varphi(w)) = -P(w, \varphi(w'))$, while in general $P(w_1,w_2) = P(w_2,w_1)$ for all $w_1, w_2 \in M$. Likewise, we find

$$S(w, \varphi(w')) = \beta\, s_2(w,w') \tag{27}$$

for all $w,w' \in M^+$.

To verify (23), we may assume $w,w',w'' \in M^+ \cup M^-$. When $w,w' \in M^\varepsilon$, (23) reduces to showing

$$wP(w',w'') + w'P(w,w'') + Q(w'',w)w' + Q(w'',w')w = 0. \tag{28}$$

If $w'' \in M^\varepsilon$ as well, all terms are zero. Otherwise, proving (28) reduces to showing that for all $w, w' \in M^+$,

$$wP(w, \varphi(w')) + Q(\varphi(w'),w)w = 0$$

and $\varphi(w')P(\varphi(w'),w) + Q(w, \varphi(w'))\, \varphi(w') = 0$. That is, we must show

$$\beta\, w\, p_2(w,w') - \beta\, q_2(w,w')w = 0 \tag{29}$$

and

$$0 = \beta\, \varphi\, (w')p_2(w,w') + \beta\, q_2(w,w')\, \varphi(w')$$
$$= \beta\, \varphi\, (-w'p_2(w,w') + q_2(w,w')w'),$$

which by the skewness of p_2, q_2 is the same as (29). The relation (29) is immediate from "b) = -a)" of §4 for $\ell = 2$.

Accordingly we may assume $w \in M^\varepsilon$, $w' \in M^{-\varepsilon}$, and by symmetry that $w'' \in M^\varepsilon$. First assume $w,w'' \in M^+$; then (23) amounts to saying that for all $w' \in M^+$,

$$2h^-(w, \varphi(w'))w'' = w''\, S(w, \varphi(w')) + w''\, P(w, \varphi(w')) \tag{30}$$
$$+ 2Q(w'', \varphi(w'))w = 3\, w\, P(\varphi(w'), w''), \quad \text{or that}$$

$$3q_2(w,w')zw'' = w''\, s_2(w,w') + w''\, p_2(w,w') \tag{31}$$
$$+ 2q_2(w'',w')w + 2\, w\, p_2(w'',w').$$

Here $zw'' = w''$. Substituting for $w''p_2(w,w')$ and $wp_2(w'',w')$ from $\ell = 2$, "-" of §4, and for $w''s_2(w,w')$ from $\ell = 2$, "+" show that (31) holds.

When $w,w'' \in M^-$, $w' \in M^+$, (23) amounts to showing that for $w, w', w'' \in M^+$,

$$3h^-(\varphi(w),w')\, \varphi(w'') = \varphi(w'')S(\varphi(w),w')$$

$$+ \varphi(w'')P(\varphi(w),w') + 2Q(\varphi(w''),w')\, \varphi(w) + 2\, \varphi(w)P(w', \varphi(w'')),$$

or that

$$-3\ \varphi(h^-(\varphi(w),w')w'') = \varphi\{w''\ S(\varphi(w),w') - w''\ P(\varphi(w),w')\}$$
$$+\ 2Q(\varphi(w''),w')w - 2\ w\ P(w',\ \varphi(w'')),$$

i.e., that

$$-3q_2(w',w)w'' = w''\ s_2(w',w) - w''\ p_2(w',w)$$
$$-\ 2q_2(w',w'')w - 2\ w\ p_2(w',w''),$$

which is the same as (31).

Thus our space g is a Lie algebra. Its dimension is seen from $\dim g_2 = 1$, $\dim [UU] = 10 = \dim [VV]$, $\dim M_U \otimes M_V = 4 \times 8 = 32$ to be $1 + 10 + 10 + (5 \times 5) + (4 \times 8) = 78$. It will be verified in Chapter 6 that g is absolutely simple, and an F-form of the split simple Lie algebra E_6.

§5.D. Dim V = 7. A Construction For E_7.

Here let V be as in 5), let $g_2 = \mathcal{Q}_0$, and fix $\varkappa = 2\lambda \neq 0$ in j), so the final step in the Jacobi identity, (22), becomes

$$2h^-(w,w')w'' - w''\ S(w,w') - w''\ P(w,w') \tag{32}$$
$$-2(Q(w'',w)w' + Q(w'',w')w) - 2(wP(w',w'') + w'P(w,w'')) = 0.$$

To verify this, we may assume $\mathcal{Q}$ is split, with basis $e_1 = e_{11}$, $e_2 = e_{22}$, e_{12}, e_{21}, such that $e_{ij}e_{k\ell} = \delta_{jk}e_{i\ell}$. In particular $M = M_V = e_1M \oplus e_2M$, the e_iM being isomorphic absolutely irreducible [VV]-modules. Left multiplication by e_{21} effects a [VV]-isomorphism of e_1M onto e_2M, its inverse being left multiplication by e_{12}. Because z acts as the identity on M, and because z and [VV] generate $C\ell(V)$, these [VV]-isomorphisms are isomorphisms of $C\ell(V)$-modules.

We may assume $z = \sum_{J\subseteq L} v_{-J}v_0v_J$, where $L = \{1,2,3\}$, and where v_{-3}, v_{-2}, v_{-1}, u_0, v_1, v_2, v_3, with $(v_i,v_j) = \delta_{i,-j}$, is a splitting basis for V. Then a generator $w^{(i)}$ for e_iM, such that $w^{(i)}v_j = 0$, all $j > 0$, must satisfy $w^{(i)}v_0 = -w^{(i)}$. A basis for e_iM consists of all $w^{(i)}v_{-J}$, $J \subseteq L$. Having fixed $w^{(1)}$ as above, we may fix $w^{(2)} = e_{21}w^{(1)}$, so that then $w^{(1)} = e_{12}w^{(2)}$. The fact that the traceless elements of $\mathcal{Q}$ are the skew elements with respect to the involution $b \to \bar{b}$ means that $\bar{e}_{12} = -e_{12}$, $\bar{e}_{21} = -e_{21}$, $\bar{e}_{11} = e_{22}$. Therefore $h(e_{11}M,e_{11}M) \subseteq e_{11}h(M,M)e_{22} = F\,e_{12}$, so that $h^+(e_{11}M,\ e_{11}M) = 0$; likewise $h^+(e_{22}M,\ e_{22}M) = 0$, and therefore h^+ defines a dual pairing of e_1M and e_2M.

For $w, w' \in e_1M$, consider the bilinear mapping $h^+(w,\ e_{21}w')$ from $e_1M \times e_1M$ to F. This is non-degenerate and [VV]-invariant on the absolutely irreducible module e_1M; as in C. above, it follows that $h^+(w, e_{21}w') = \beta\, q_3(w,w')$ for some $\beta \neq 0$. Then it follows as in C. that $P(w, e_{21}w') = \beta\, p_3(w,w')$, $S(w, e_{21}w') = \beta\, s_3(w,w')$, and that both P and S are zero on $e_iM \otimes e_iM$, $i = 1,2$.

We have seen that $h(e_1M, e_1M) \subseteq Fe_{12}$, so that $h^-(e_1M, e_1M) \subseteq Fe_{12}$, and $h^-(e_1M, e_1M)e_1M = 0$. Thus all terms of (32) are zero if $w,w',w'' \in e_1M$, and likewise if $w,w',w'' \in e_2M$. For $w,w' \in e_1M$ we have

$$h(w,w') = h(w, e_{12}e_{21}w') = -h(w, e_{21}w')e_{12}$$
$$= -h^+(w, e_{21}w')e_{12} - h^-(w, e_{21}e')e_{12},$$

and we know $h(w,w') \in Fe_{12}$. Thus we may write $h(w,w') = \zeta(w,w')e_{12}$, where $\zeta(w,w') \in F$. Then

$$h(w, e_{21}w') = -h(w,w')e_{21} = -\zeta(w,w')e_{11}$$
$$= -\frac{\zeta(w,w')}{2}(e_{11} + e_{22}) - \frac{\zeta(w,w')}{2}(e_{11} - e_{22}).$$

Thus $h^+(w, e_{21}w') = -\frac{1}{2}\zeta(w,w')$, and

$$\zeta(w,w') = -2\,\beta\, q_3(w,w'),\ h^-(w, e_{21}w') = \beta\, q_3(w,w')(e_{11} - e_{22}).$$

It follows that $h(w,w') = -2\,\beta\, q_3(w,w')e_{12}$.

For $w, w' \in e_1M$, it suffices to verify (32) when $w'' \in e_2M$. Thus we may replace w'' by $e_{21}w''$, $w'' \in e_1M$. In this case, showing (32) amounts to showing

$$-4q_3(w,w')w'' + 2\, q_3(w,w'')w' + 2\, q_3(w',w'')w$$
$$-2\, w\, p_3(w',w'') - 2\, w'\, p_3(w,w'') = 0$$

for all w, w', w'' in a module "M" for [VV] as in §4. From the symmetry of q_3, this is the same as "b) = a) - 2g)" for $\ell = 3$ in §4. Thus (32) holds in this case. An analogous argument applies when $w'' \in e_1M$ and when w, w' are replaced by $e_{21}w$, $e_{21}w'$, $w, w' \in e_1M$. Then $h^-(e_{21}w, e_{21}w') = h(e_{21}w, e_{21}w') = -e_{21}h(w,w')e_{21} = 2\beta\, q_3(w,w')e_{21}$, and the expression in (32) is of the form $\beta\, e_{21}Y$, where $Y \in e_1M$ is an expression which is zero as before.

It therefore remains to verify (32) when w, w' are in opposite members of e_1M, e_2M. By the symmetry of (32) in w, w', we may assume w'' and w are in the same one of e_1M, e_2M. First consider (32) where $w, w'' \in e_1M$, and where w' has the form $e_{21}w'$, $w' \in e_1M$. Then

$$h^-(w, e_{21}w') = \beta\, q_3(w,w')(e_{11} - e_{22}),$$

and showing (32) is equivalent to showing that

$$2\, q_3(w,w')w'' - w''\, s_3(w,w') - w''\, p_3(w,w')$$
$$-2\, q_3(w'',w')w - 2\, w\, p_3(w'',w') = 0,$$

which is equivalent, by "i) = 4d) + h)" of $\ell = 3$ of §4, to

$$2\, q_3(w,w')w'' - 4\, q_3(w'',w)w' + 2\, q_3(w'',w')w$$
$$- 2\, w''\, p_3(w,w') - 2\, w\, p_3(w'',w') = 0,$$

and this is the identity of the previous case with the roles of w' and w'' interchanged.

For the remaining case we may assume $w' \in e_1M$ and replace $w \in e_2M$ by $e_{21}w$, $w \in e_1M$, with $w'' \in e_2M$ also being replaced by $e_{21}w''$, $w'' \in e_1M$. Then

$$h^-(e_{21}w,w')e_{21}w'' = \beta\, q_3(w',w)(e_{11} - e_{22})e_{21}w''$$
$$= -\beta\, q_3(w',w)e_{21}w'',$$

and (32) becomes

$$-2\, q_3(w',w)e_{21}w'' - e_{21}w''s_3(w',w) - e_{21}w''p_3(w',w)$$
$$+ 2\, q_3(w',w'')e_{21}w - 2\, e_{21}w\, p_3(w',w'') = 0,$$

using the symmetry of P, S, and the skewness of Q. Now from the symmetry of q_3 and the skewness of p_3, s_3, this last amounts to

$$e_{21}\, \{-2\, q_3(w,w')w'' + w''\, s_3(w,w') + w''\, p_3(w,w')$$
$$+ 2\, q_3(w'',w')w + 2\, w\, p_3(w'',w')\},$$

and the quantity in braces has been shown above to be zero. Thus $\mathfrak{g}$ is a Lie algebra.

The simplicity of $\mathfrak{g}$ follows as in B., using the action of $\mathfrak{g}_2 = \mathcal{Q}_0$ as well as those of $[UU]$, $[VV]$ to show that any non-zero ideal contains one of the five summands, hence all of them. The dimension of $\mathfrak{g}$ is $3 + 10 + 21 + (5 \times 7) + (4 \times 16) = 133$, and the proof of simplicity yields absolute simplicity. By the uniqueness of its dimension, $\mathfrak{g}$ must be an F-form of the split exceptional Lie algebra E_7.

§6.A. Composites With a 3-Dimensional Quadratic Space.

For this construction, let U be a 3-dimensional space as in 1) of §1. Let V be as in 4), 6), 8) or 9) of that section, with M_U, M_V as in those cases. The form q_1 on M_U is alternate, while the associated p_1, s_1 are symmetric. On M_V, both Q and P are skew, while S and h^- (if present) are symmetric. Thus if $\mathfrak{g}_2 = \{0\}$ in cases 8), 9), $\mathfrak{g}_2 = Fz$ in case 4), $\mathfrak{g}_2 = \mathcal{Q}_0$ in case 6), the definitions a) - j) of §5.A may be used (with only the subscript "2" replaced by "1") to define a skew-symmetric bilinear product on

$$\mathfrak{g} = \mathfrak{g}_2 \oplus [UU] \oplus [VV] \oplus (u \otimes V) \oplus (M_U \otimes M_V).$$

Verification of the Jacobi identity proceeds as before, repeating the steps of that section (with only trivial modifications, because now $\theta = -1$ in (5)). The only significant new considerations arise in treating the Jacobi identity for three factors from $M_U \otimes M_V$, which factors may be assumed of the form $m \otimes w$, for $m \in M_U$, $w \in M_V$. Here

$$[[m \otimes w, m' \otimes w']m'' \otimes w''] = \varkappa\, q_1(m,m')m'' \otimes h^-(w,w')w''$$
$$= \lambda\, q_1(m,m')m'' \otimes w''S(w,w') - \lambda\, m''s_1(m,m') \otimes Q(w,w')w''$$
$$+ \lambda\, m''p_1(m,m') \otimes w''P(w,w'),$$

and the sum of the twelve terms obtained by cyclically permuting the three factors is to be zero.

The considerations for $\ell = 1$ of §4 permit expression of the factors in M_U in all these terms using only $X = m''p_1(m,m') = q_1(m'',m)m' + q_1(m'',m')m$ (see $\ell = 1$, "+" of §4) and $Y = q_1(m,m')m''$. Thus "i) = -h) = -a)" says that $m''s_1(m,m') = -X$, and from "c) = h)", "f) = -g)", $m\, s_1(m',m'') = \frac{1}{2}X - \frac{3}{2}Y$, $m's_1(m,m'') = \frac{1}{2}X + \frac{3}{2}Y$. "b) = -h)" and "e) = 3g)" give $m\, p_1(m',m'') = -\frac{1}{2}X + \frac{3}{2}Y$, $m'p_1(m,m'') = -\frac{1}{2}X - \frac{3}{2}Y$, and similarly $q_1(m'',m)m' = \frac{1}{2}X - \frac{1}{2}Y$, $q_1(m'',m')m = \frac{1}{2}X + \frac{1}{2}Y$. Making these substitutions in the first tensor factors in the quantity that must vanish for the Jacobi identity to hold, we see upon collecting coefficients of X and Y that the following must be shown to be zero:

$$X \otimes \{-\tfrac{1}{2}\varkappa\, h^-(w',w'')w + \tfrac{1}{2}\varkappa\, h^-(w'',w)w'$$
$$+ \tfrac{1}{2}\lambda\, wS(w',w'') - \tfrac{1}{2}\lambda\, w'S(w'',w) + \lambda\, Q(w,w')w''$$
$$- \tfrac{1}{2}\lambda\, Q(w',w'')w - \tfrac{1}{2}\lambda\, Q(w'',w)w' + \lambda\, w''P(w,w')$$
$$- \tfrac{1}{2}\lambda\, w\, P(w,w'') - \tfrac{1}{2}\lambda\, w'P(w'',w)\}$$
$$+ Y \otimes \{\varkappa\, h^-(w,w')w'' - \tfrac{1}{2}\varkappa\, h^-(w',w'')w - \tfrac{1}{2}\varkappa\, h^-(w'',w)w'$$
$$- \lambda\, w''S(w,w') + \tfrac{1}{2}\lambda\, w\, S(w',w'') + \tfrac{1}{2}\lambda\, w'S(w'',w)$$
$$+ \tfrac{3}{2}\lambda\, Q(w',w'')w - \tfrac{3}{2}\lambda\, Q(w'',w)w'$$
$$+ \tfrac{3}{2}\lambda\, w\, P(w',w'') - \tfrac{3}{2}\lambda\, w'P(w'',w)\},$$

which we write as $X \otimes R + Y \otimes T$, for short. We show that for arbitrary λ an appropriate choice of $\varkappa$ for each of our four cases makes $R = 0 = T$.

§6.B. Dim V = 13, 6. Constructions For E_8, F_4.

When V is as in 8), we have $h^- = 0$, $Q = q_6$, $P = p_6$, $S = s_6$. Take $\varkappa = 0$. Deleting the factor λ now common to all terms and using the symmetry of s_6 and skewness of q_6, p_6, we find

$$\begin{aligned} T = &- w''s_6(w,w') + \frac{1}{2} w\, s_6(w',w'') + \frac{1}{2} w's_6(w,w'') \\ &- \frac{3}{2} q_6(w'',w')w - \frac{3}{2} q_6(w'',w)w' \\ &+ \frac{3}{2} w\, p_6(w',w'') + \frac{3}{2} w'\, p_6(w,w''), \end{aligned}$$

and T = 0 by the relation of $\ell = 6$, "+" of §4. Likewise, with the same conventions,

$$\begin{aligned} R = &\frac{1}{2} w\, s_6(w',w'') - \frac{1}{2} w's_6(w,w'') + q_6(w,w')w'' \\ &+ \frac{1}{2} q_6(w'',w')w - \frac{1}{2} q_6(w'',w)w' + w''\, p_6(w,w') \\ &- \frac{1}{2} w\, p_6(w',w'') + \frac{1}{2} w'\, p_6(w,w''), \end{aligned}$$

which is zero by the relation for $\ell = 6$, "-".

Thus g is in this case a Lie algebra of dimension $3 + 78 + (3 \times 13) + (2 \times 64) = 248$, whose simplicity follows as before (first treating an ideal as [VV]-module), the argument applying equally well in the split case, so yielding absolute simplicity. Therefore, by dimensions, g is an F-form of the split exceptional Lie algebra E_8.

Next let V be 6-dimensional, satisfying the conditions of 9). Again $h^- = 0$, and here Q, P, S are respectively q, p, s. The vanishing of T and R is, respectively, (10) and (11). Once again, g is therefore a Lie algebra, of dimension

$$3 + 15 + (3 \times 6) + (2 \times 8) = 52,$$

and g is again absolutely simple, an F-form of the split exceptional algebra F_4. (More details in this regard will be given in Chapter 6.)

§6.C. Dim V = 7. A Construction For E_6.

Here fix $\lambda \neq 0$ in F, and set $\varkappa = 3\lambda$. The case 4) is handled as was 3), under §5.C above. Namely, the form h is antihermitian with respect to the non-trivial F-involution of the field F(z). Passing to a splitting field "F" for V, we may assume $z = e_1 - e_2$, where e_1, e_2 are orthogonal idempotents forming an F-basis for F(z), with $e_1^\sigma = e_2$. As F-vector space, $M_V = e_1M_V \oplus e_2M_V$, each summand an 8-dimensional F-space and a right module for $C\ell(V)$. By dimensions, each e_iM (writing M for M_V) is an (absolutely) irreducible module for $C\ell^+(V)$. Thus e_1M has basis $\{w^+v_{-J} | J \subseteq L = \{1,2,3\}\}$, where $w^+v_i = 0$, $i > 0$. Here $\{v_i | -3 \leq i \leq 3\}$ is a splitting basis for V, and we fix $z = \sum_{J \subseteq L} v_{-J}v_0v_J$. From

$ze_1 = e_1$, it follows that $w^+v_0 = -w^+$.

An isomorphism φ of $C\ell^+(V)$-modules maps e_1M to e_2M, sending w^+ to an element annihilated by all v_0v_i, $i > 0$. Any such element must be a multiple of the generating element of e_2M annihilated by all v_i, $i > 0$, so that $\varphi(w^+)$ is such an element. Then

$$\varphi(w^+v_{-J}) = \varphi((-1)^{|J|}\, w^+v_0{}^{|J|}\, v_{-J})$$

$$(-1)^{|J|}\, \varphi(w^+)\, v_0{}^{|J|}\, v_{-J} = (-1)^{|J|}\, \varphi(w^+)v_{-J},$$

because $\varphi(w^+)v_0 = \varphi(w^+)$ follows from $\varphi(w^+)z = -\varphi(w^+)$.

For $w, w' \in e_1M$, $h(w, \varphi(w'))$

$$= e_1h(w,\varphi(w'))e_2^{\sigma} = b(w,w')e_1, \quad \text{where } b(w,w') \in F.$$

Meanwhile $h(w,w') \in e_1F(z)e_2 = 0$, and $h(\varphi(w), \varphi(w')) = 0$. From $h(w, \varphi(w')) = b(w,w')e_1$ we find

$$h^+(w, \varphi(w')) = \frac{1}{2}\, b(w,w'),$$

$$h^-(w, \varphi(w')) = \frac{1}{2}\, b(w,w')(e_1 - e_2) = \frac{1}{2}\, b(w,w')z.$$

For $x \in [VV]$, $h(wx, \varphi(w')) = -h(w, \varphi(w')x) = -h(w, \varphi(w'x))$, so that $b(wx,w') = -b(w,w'x)$. It follows as in earlier cases that there is $\beta \neq 0$ in F such that $b(w,w') = \beta\, q_3(w,w')$ for all $w,w' \in e_1M$, where q_3 is the nondegenerate symmetric form on e_1M as in §3. Thus we have

$$Q(w, \varphi(w')) = \frac{1}{2}\, \beta\, q_3(w,w'),$$

$$h^-(w, \varphi(w')) = \frac{1}{2}\, \beta\, q_3(w,w')z,$$

and it follows that P, S vanish on $e_1M \times e_1M$, with

$$P(w, \varphi(w')) = \frac{1}{2}\, \beta\, p_3(w,w')\ , \quad S(w, \varphi(w')) = \frac{1}{2}\, \beta\, s_3(w,w')\ .$$

Now consider the quantities "R" and "T", which must vanish for the Jacobi identity to hold. Identifying R and T from section A., we first suppose $w, w' \in e_1M$. Then if $w'' \in e_1M$ as well, the above shows that all terms in R, T are zero. We may therefore replace w'' by $\varphi(w'')$, $w'' \in e_1M$. The vanishing of R is then equivalent to the vanishing of

$$-3\, q_3(w',w'')w + 3\, q_3(w,w'')w' + w\, s_3(w',w'') - w's_3(w,w'')$$

$$-\, q_3(w',w'')w + q_3(w,w'')w' - w\, p_3(w',w'') + w'\, p_3(w,w''),$$

which by $\ell = 3$, "$-$" of §4 is equal to (using the symmetry of q_3)

$$2\, w''\, p_3(w,w') - w\, p_3(w', w'') + w'\, p_3(w,w'')$$

$$+ 3\, q_3(w'',w)w' - 3\, q_3(w'',w')w,$$

and this is zero by the relation e) = 3d) + 2h) for $\ell = 3$.

The vanishing of T is equivalent to the vanishing of

$$-3\, q_3(w',w'')w - 3\, q_3(w,w'')w' + w\, s_3(w',w'') + w'\, s_3(w, w'')$$

$$+ 3\, q_3(w',w'')w + 3\, q_3(w,w'')w' + 3\, w\, p_3(w',w'') + 3\, w'\, p_3(w,w''),$$

and this is zero from relations for $\ell = 3$, "+".

The other case where w, w' are in the same summand (e_2M) is treated by replacing w by $\varphi(w)$, w' by $\varphi(w')$, w, w' $\in e_1M$. We may assume $w'' \in e_1M$, since all terms are zero if $w'' \in e_2M$. Here

$$h^-(\varphi(w'),w'')\, \varphi(w) = h^-(w'',\varphi(w'))\, \varphi(w)$$

$$= \frac{1}{2}\, \beta\, q_3(w'',w')z\, \varphi(w) = -\frac{1}{2}\, \beta\, q_3(w'',w')\, \varphi(w),$$

and similarly when w, w' are interchanged. Also,

$$\varphi(w)S(\varphi(w),w'') = \varphi(wS(w'', \varphi(w')) = \frac{1}{2}\, \beta\, \varphi(w\, s_3(w'',w')) = -\frac{1}{2}\, \beta\, \varphi(w\, s_3(w',w'')),$$

$$\varphi(w)P(\varphi(w'),w'') = -\varphi(w)P(w'',\varphi(w')) = \varphi(w)zP(w'',\varphi(w')) = \dot{\varphi}(wzP(w'', \varphi(w'))$$

$$= \varphi(w\, P(w'', \varphi(w')) = \frac{1}{2}\, \beta\, \varphi(w\, p_3(w'',w')) = -\frac{1}{2}\, \beta\, \varphi(w\, p_3(w',w'')).$$

It follows that the new R and T are the images under φ of the negatives of the "old" ones, so are zero.

Thus let (without loss of generality) $w \in e_1M$, and replace w' by $\varphi(w')$, $w' \in e_1M$. We may assume $w'' \in e_1M$; the case $w'' \in e_2M$ is deduced from this by replacing w'' by $\varphi(w'')$, $w'' \in e_1M$, and using the relations above. Here the vanishing of R is equivalent to that of

$$-3\, q_3(w'',w')w + w\, s_3(w'',w') + 3\, q_3(w,w')w''$$

$$+ q_3(w'',w')w + 2\, w''\, p_3(w,w') + w\, p_3(w'',w'),$$

which is equal by "i) = 4d) + h)" for $\ell = 3$ to

$$-2\, q_3(w'',w')w + 2\, q_3(w,w')w'' + 2\, w''\, p_3(w,w') + w\, p_3(w'',w')$$

$$+ 4(q_3(w,w'')w' - q_3(w,w')w'') + w\, p_3(w'',w').$$

Substituting for $w''p_3(w,w'') + w\, p_3(w'',w')$ from "b) = a) - 2g)" shows that $R = 0$. Likewise, T vanishes if and only if

$$6\, q_3(w,w')w'' - 3\, q_3(w'',w')w - 2\, w''\, s_3(w,w') + w\, s_3(w'',w')$$

$$-3\, q_3(w'',w')w - 3\, w\, p_3(w'',w') \quad \text{is zero,}$$

which expression is equal, by "i) = 4d) + h)", to

$$6(q_3(w',w)w'' - q_3(w',w'')w) - 3\, w\, p_3(w'',w')$$

$$- 8(q_3(w'',w)w' - q_3(w'',w')w) - 2\, w''\, p_3(w,w')$$

$$+ 4(q_3(w,w'')w' - q_3(w,w')w'') + w\, p_3(w'',w')$$

$$= -2(w\, p_3(w'',w') + w''\, p_3(w,w')) - 4\, q_3(w,w'')w'$$

$$+ 2(q_3(w',w)w'' + q_3(w',w'')w),$$

and this is zero by the last relation used to show $R = 0$. The case where w'' is replaced by $\varphi(w'')$, $w'' \in e_1M$, is left to the reader.

It follows that $\mathcal{g}$ is a Lie algebra, of dimension

$$1 + 3 + 21 + (3 \times 7) + (2 \times 16) = 78.$$

We shall see that $\mathcal{g}$ is an F-form of the split exceptional simple Lie algebra E_6.

§6.D. Dim V = 9. A Construction For E_7.

For our final construction, we let V be as in 6), and we take $\varkappa = 2\lambda$ in j). Again the involution in $\mathcal{Q}$ is the standard one. As in §5.D., an extension field splits $\mathcal{Q}$, decomposing $M_V = M$ into absolutely irreducible modules $e_{11}M$, $e_{22}M$, with $y \to e_{21}y$ being an isomorphism $e_{11}M \to e_{22}M$ of $C\ell(V)$-modules, each of dimension 16. Assuming F to be the extended base field, we have $C\ell^+(V) \cong \mathrm{End}_F(e_{11}M)$.

Here $h(e_{11}M, e_{11}M) = Fe_{12}$ as before, and for $w, w' \in e_{11}M$, $h(w,w') = b(w,w')e_{12} = h^-(w,w')$, where $b(w,w') \in F$. As in §5.D, $h(e_{21}w, e_{21}w') = -e_{21}\, h(w,w')e_{21} = -b(w,w')e_{21} = h^-(e_{21}w, e_{21}w')$ always with $w, w' \in e_{11}M)$. Now the nondegeneracy of h^+ implies that the form $h^+(w, e_{21}w')$ is nondegenerate on $e_{11}M$, and $[VV]$-invariant, so that $Q(w, e_{21}w') = h^+(w, e_{21}w') = \beta\, q_4(w,w')$ for all $w, w' \in e_{11}M$, where q_4 is the symmetric form on $e_{11}M$ of the case "$\ell = 4$" of §4. The corresponding p_4 is symmetric, with s_4 being skew. The scalar β is fixed and non-zero. We have $h(w,e_{21}w') = -h(w,w')e_{21} = -b(w,w')e_{11}$

$$= -\frac{1}{2}\, b(w,w')1_{\mathcal{Q}} - \frac{1}{2}\, b(w,w')(e_{11} - e_{22})$$

$$= h^+(w, e_{21}w') + h^-(w, e_{21}w').$$

Comparing components gives $\beta\, q_4(w,w') = h^+(w, e_{21}w') = -\frac{1}{2}\, b(w,w')$, Hence

$h^{-}(w, e_{21}w') = \beta\ q_4(w,w')(e_{11} - e_{22})$ for $w, w' \in e_{11}M$, $h(w,w') = -2\ \beta\ q_4(w,w')e_{12}$, $h(e_{21}w, e_{21}w') = 2\ \beta\ q_4(w,w')e_{21}$. As in the earlier case, P and S are zero on $e_{11}M \times e_{11}M$ and on $e_{22}M \times e_{22}M$, while for $w, w' \in e_{11}M$,

$$P(w, e_{21}w') = \beta\ p_4(w,w')$$

$$S(w, e_{21}w') = \beta\ s_4(w,w').$$

When $w, w' \in e_{11}M$, both R and T are zero if $w'' \in e_{11}M$; thus we may replace w'' by $e_{21}w''$, $w'' \in e_{11}M$. With $\varkappa = 2\lambda \neq 0$, the vanishing of R is here equivalent to that of

$$-2\ q_4(w',w'')w + 2\ q_4(w,w'')w' + w\ s_4(w',w'') - w'\ s_4(w,w'')$$

$$-\ q_4(w',w'')w + q_4(w,w'')w' - w\ p_4(w',w'') + w'\ p_4(w,w''),$$

which is zero by the relation "f) = -3d) + e)" of $\ell = 4$ of §4 and the symmetry of q_4.

The vanishing of T is equivalent to that of

$$-9\ q_4(w,w')w'' - 2\ q_4(w',w'')w - 2\ q_4(w,w'')w'$$

$$+ w\ s_4(w',w'') + w'\ s_4(w,w'') + 3\ q_4(w',w'')w + 3\ q_4(w,w'')w'$$

$$+ 3\ w\ p_4(w',w'') + 3\ w'\ p_4(w,w''),$$

which is equal, by eliminating c) as for $\ell = 4$, "+" of §4, to

$$-3\ q_4(w,w')w'' - 3\ q_4(w'',w)w' - 3\ q_4(w'',w')w$$

$$+ 3\ w''\ p_4(w,w') + 3\ w\ p_4(w',w'') + 3\ w'(w,w''),$$

and this is zero by "b) = a) + g) - h)" of the same reference. The cases where both $w, w' \in e_{22}M$ are deduced in the same fashion (see §5.D).

For the remaining cases, we treat only that where $w \in e_{11}M$, $w'' \in e_{11}M$, $w' \in e_{22}M$, the others following from this by symmetry or being completely parallel to it. Thus we replace w' by $e_{21}w'$, $w' \in e_{11}M$, and R becomes a multiple of

$$-2\ q_4(w'',w')w - 4\ q_4(w'',w)w'$$

$$+ w\ s_4(w'',w')w + 2\ q_4(w,w')w'' + q_4(w'',w')w$$

$$+ 2\ w''\ p_4(w,w') + w\ p_4(w'',w').$$

Substitution of the expression given for "i)" in $\ell = 4$, "-" of §4 for $w\ s_4(w'',w')$ shows this last to be equal to

$$-q_4(w'',w)w' - q_4(w,w')w'' - q_4(w',w'')w$$

$$+ w'' p_4(w,w') + w' p_4(w'',w) + w p_4(w',w''),$$

and this is zero by "b) = a) + g) - h)", as before.

In this case, T is a multiple of

$$4 q_4(w,w')w'' - 2 q_4(w'',w')w + 4 q_4(w'',w)w'$$

$$- 2 w'' s_4(w,w') + w s_4(w'',w') - 3 q_4(w'',w')w$$

$$- 3 w p_4(w'',w'),$$

which is equal by substitution as above for terms involving s_4, to

$$q_4(w'',w)w' + q_4(w,w')w'' + q_4(w'',w')w$$

$$- w p_4(w',w'') - w' p_4(w,w'') - w'' p_4(w,w'),$$

a quantity which is zero as before.

Accordingly our g is a Lie algebra of dimension

$$3 + 3 + 36 + (3 \times 9) + (2 \times 32) = 133,$$

and is absolutely simple as before. Thus g is an F-form of the split exceptional simple Lie algebra E_7.

§7. Reductive Dual Pairs.

In all the constructions, [UU] and [VV] are simple subalgebras of g, centralizing one another. When dim. U = 5, dim. V = 11; when dim. U = 3, dim. V = 6; and when dim. U = 3, dim. V = 13, they are precisely the centralizers of one another. As such, they constitute a <u>reductive dual pair</u> in g in a sense that adapts a meaning that Roger Howe has used to great advantage, particularly within symplectic groups [Ho]. Precisely we shall take a reductive dual pair in g to mean a pair (g_1, g_2) of subalgebras such that each is reductive in g (<u>i.e.</u>, its adjoint representation on g is completely reducible) and such that each is the centralizer in g of the other.

Reductive dual pairs natural to our constructions in the other cases are as follows:

$$\left.\begin{array}{c} \text{dim. } U = 5 = \text{dim. } V \\ \text{and} \\ \text{dim } U = 3, \text{ dim. } V = 7 \end{array}\right\} \qquad g_1 = [UU] + Fz, \quad g_2 = [VV] + Fz$$

$$\left.\begin{array}{c}\text{dim. } U = 5,\ \text{dim. } V = 7\\ \text{and}\\ \text{dim. } U = 3,\ \text{dim. } V = 9\end{array}\right\}\qquad \left\{\begin{array}{c}\mathfrak{g}_1 = [UU],\quad \mathfrak{g}_2 = \mathcal{Q}_0 + [VV]\\ \text{and}\\ \mathfrak{g}_1 = [UU] + \mathcal{Q}_0,\quad \mathfrak{g}_2 = [VV]\end{array}\right\}$$

(Here $\mathcal{Q}_0$ may pass from one member of the reductive dual pair to the other in each case, giving a second pair.)

CHAPTER 6: REPRESENTATIONS OF EXCEPTIONAL ALGEBRAS CONSTRUCTED FROM QUADRATIC FORMS

§1. Decomposition into Root-Spaces

Throughout this chapter, U will be assumed to have dimension 3 or 5, as in Chapter 5, and moreover to be <u>split</u>. Thus U has splitting basis $\{u_{-1}, u_0, u_1\}$ or $\{u_{-2},u_{-1},u_0,u_1,u_2\}$ with $(u_i,u_j) = \delta_{i,-j}$. The irreducible $C\ell(V)$-module M_U has basis $\{m^+, m^+u_{-1}\}$ resp. $\{m^+,m^+u_{-2}, m^+u_{-1}, m^+u_{-2}u_{-1}\}$, where $m^+u_i = 0$ for $i > 0$, and (without loss of generality) $m^+u_0 = m^+$.

Let $\mathfrak{t}_0$ be the subspace of $[UU] \subseteq \mathfrak{g}$ with basis $[u_{-1},u_1]$ resp. $\{[u_{-1},u_1], [u_{-2},u_2]\}$. Then $\mathfrak{t}_0$ is a commutative subalgebra, centralized by the following subspaces of $\mathfrak{g}$: $\mathfrak{g}_2$, $[VV]$, $\mathfrak{t}_0$, $u_0 \otimes V$. For $i > 0$, $v \in V$, $[u_i \otimes v, [u_{-j},u_j]] = 4\,\delta_{ij}\, u_i \otimes v$, while $[u_{-i} \otimes v, [u_{-j},u_j]] = -4\,\delta_{ij}\, u_i \otimes v$. Likewise $[[u_iu_0], [u_{-j}u_j]] = 4\,\delta_{ij}\,[u_iu_0]$, and $[[u_{-i}u_0], [u_{-j}u_j]] = -4\,\delta_{ij}\,[u_{-i}u_0]$, in each case for all $j > 0$. When $\dim U = 5$, we have for $i,k > 0$,

$$[[u_iu_k], [u_{-j}u_j]] = (\delta_{ij} + \delta_{kj})[u_iu_k];$$

$$[[u_iu_{-k}], [u_{-j},u_j]] = (\delta_{ij} - \delta_{kj})[u_iu_{-k}];$$

$$[[u_{-i}u_{-k}], [u_{-j},u_j]] = -(\delta_{ij} + \delta_{kj})[u_{-i}u_{-k}], \quad \text{for } j = 1,2.$$

If $J \subseteq L = \{1\}$ resp. $\{1,2\}$, and if $w \in M_V$, we have

$$[m^+u_{-J} \otimes w, [u_{-j}u_j]] = \begin{cases} 2\, m^+u_{-J} \otimes w, & \text{if } j \notin J, \\ -2\, m^+u_{-J} \otimes w, & \text{if } j \in J. \end{cases}$$

When $L = \{1\}$ it follows that the non-zero weights of the adjoint representation of $\mathfrak{t}_0$ on $\mathfrak{g}$ are a system of roots of type BC_1, with root-spaces

$$\mathfrak{g}_{\alpha_1} = m^+ \otimes M_V,\ \mathfrak{g}_{-\alpha_1} = m^+u_{-1} \otimes M_V,\ \mathfrak{g}_{2\alpha_1} = F[u_1u_0] + u_1 \otimes V,\ \mathfrak{g}_{-2\alpha_1} = F[u_{-1}u_0] +$$

$u_{-1} \otimes V$. The centralizer $\mathfrak{g}_0$ of $\mathfrak{t}_0$ is $\mathfrak{g}_2 + [VV] + \mathfrak{t}_0 + u_0 \otimes V$. The fundamental root α_1 has $\alpha_1([u_{-1}u_1]) = 2$.

When $L = \{1,2\}$, the weights of $\mathfrak{t}_0$ are a system of roots of type BC_2, with root-spaces

$$\mathfrak{g}_{\alpha_2} = m^+u_{-1} \otimes M_V,\quad \mathfrak{g}_{-\alpha_2} = m^+u_{-2} \otimes M_V,\quad \mathfrak{g}_{\alpha_1+\alpha_2} = m^+ \otimes M_V,$$

$$\mathfrak{g}_{-(\alpha_1+\alpha_2)} = m^+u_{-L} \otimes M_V,\quad \mathfrak{g}_{\alpha_1} = F[u_1u_0] + u_1 \otimes V,$$

$\mathfrak{g}_{-\alpha_1} = F[u_{-1}u_0] + u_{-1} \otimes V$, $\mathfrak{g}_{2\alpha_2} = F[u_2u_{-1}]$,

$\mathfrak{g}_{-2\alpha_2} = F[u_1u_{-2}]$, $\mathfrak{g}_{\alpha_1+2\alpha_2} = F[u_2u_0] + u_2 \otimes V$,

$\mathfrak{g}_{-(\alpha_1+2\alpha_2)} = F[u_{-2}u_0] + u_{-2} \otimes V$, $\mathfrak{g}_{2\alpha_1+2\alpha_2} = F[u_2u_1]$,

$\mathfrak{g}_{-(2\alpha_1+2\alpha_2)} = F[u_{-2}u_{-1}]$. The fundamental roots α_1, α_2 have $\alpha_1([u_{-1}u_1]) = 4$, $\alpha_1([u_{-2}u_2]) = 0$ and $\alpha_2([u_{-1}u_1]) = -2$, $\alpha_2([u_{-2}u_2]) = 2$. The centralizer $\mathfrak{g}_0$ of $\mathfrak{t}_0$ is again $\mathfrak{g}_2 + [VV] + \mathfrak{t}_0 + u_0 \otimes V$.

Now consider the case where V is split as well, a splitting basis being $v_{-\ell}$, $v_{-(\ell-1)},\ldots,v_0,\ldots,v_\ell$, with $L = \{1,2,\ldots,\ell\}$, v_0 being absent if $\dim V = 6$.

F_4, E_8: First with $\dim U = 3$, $\dim. V = 6$ resp. 13, we have a commutative subalgebra $\mathfrak{h}$ of $\mathfrak{g}$ with basis $[u_{-1},u_1]$, $\{[v_{-j}v_j] \mid 1 \le j \le \ell\}$, $u_0 \otimes v_0$. (The last of these is of course absent when $\dim. V = 6$.) If $w^+ \in M_V$ has $w^+v_i = 0$ for all $i > 0$, and $w^+v_0 = w^+$ when $\dim V = 13$, then if w^+v_{-J} $(J \subseteq L)$ is defined as was "w^+u_{-J}" in §5.3, we have

$$[(m^+ \otimes w^+v_{-J}), [v_{-j}v_j]] = 2\,\varepsilon_{j,J}\,(m^+ \otimes w^+v_{-J}),$$

$$[m^+u_{-1} \otimes w^+v_{-J}, [v_{-j}v_j]] = 2\,\varepsilon_{j,J}(m^+u_{-1} \otimes w^+v_{-J}),$$

where $\varepsilon_{j,J} = 1$ if $j \notin J$, $\varepsilon_{j,J} = -1$ if $j \in J$. When $\dim. V = 13$,

$$[m^+u_{-K} \otimes w^+v_{-J}, u_0 \otimes v_0] = (-1)^{|J|+|K|}\, m^+u_{-K} \otimes w^+v_{-J},$$

where $K \subseteq \{1\}$. Thus each $m^+u_{-K} \otimes w^+v_{-J}$ is a weight vector for the (right) adjoint action of $\mathfrak{h}$ on $\mathfrak{g}$, and the weights are distinct and non-zero. The elements $[u_{-1}u_0]$ and $[u_1u_0]$ centralize all $[v_{-j}v_j]$ and are weight-vectors if $\dim. V = 6$. When $\dim. V = 13$, we have $[[u_iu_0], u_0 \otimes v_0] = 4u_i \otimes v_0$, $[u_i \otimes v_0, u_0 \otimes v_0] = [u_iu_0]$, so that each $[u_iu_0] \pm 2u_i \otimes v_0$, $i = \pm 1$, is a weight vector. Similarly, each $[v_jv_0] \pm 2u_0 \otimes v_j$ is a weight vector in $\mathfrak{g}_0$, in dimension 13, while each $u_0 \otimes v_j$ is such a weight vector in dimension 6. Further weight vectors in $\mathfrak{g}_0$ are the $[v_iv_j]$, $ij \ne 0$, $i + j \ne 0$, and there are also weight vectors $u_i \otimes v_j$, $i = \pm 1$, $j \ne 0$. The totality of weight vectors so obtained is equal to the codimension of $\mathfrak{h}$ in $\mathfrak{g}$, and all weights are distinct and non-zero. It follows that $\mathfrak{h}$ is a split toral Cartan subalgebra of $\mathfrak{g}$, of dimension 4 resp. 8, and we are entitled to refer to the weights as <u>roots</u>.

When $\dim V = 6$, let roots γ_1, γ_2, γ_3, γ_4 relative to $\mathfrak{h}$ be defined by having root-vectors $[v_{-2},v_1]$, $[v_{-3},v_2]$, $u_0 \otimes v_3$, $m^+ \otimes w^+v_{-L} = m^+ \otimes w^+v_{-3}v_{-2}v_{-1}$, respectively. If λ_1, λ_2, λ_3, λ_4 is a basis for $\mathfrak{h}^*$ dual to $\frac{1}{4}[u_1,u_1]$, $\frac{1}{4}[v_{-1}v_1]$, $\frac{1}{4}[v_{-2},v_2]$, $\frac{1}{4}[v_{-3}v_3]$, then $\gamma_1 = \lambda_2 - \lambda_3$, $\gamma_2 = \lambda_3 - \lambda_4$, $\gamma_3 = \lambda_4$, $\gamma_4 = \frac{1}{2}(\lambda_1 - \lambda_2 - \lambda_3 - \lambda_4)$, and comparison with Planche VIII, pp. 272-3 of [Bo] shows that γ_1, γ_2, γ_3, γ_4 are the simple roots of a system of roots of type F_4, which one verifies from the data above to be the system of roots of $\mathfrak{g}$ relative to $\mathfrak{h}$. The Dynkin diagram is

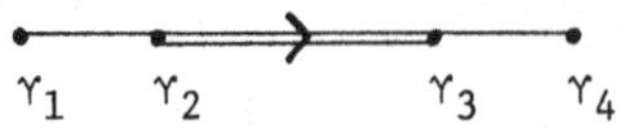

,

with $\gamma_i\big|_{\mathfrak{t}_0} = 0$ for $i < 4$, $\gamma_4\big|_{\mathfrak{t}_0} = \alpha_1$.

The centralizer $\mathfrak{g}_0$ of $\mathfrak{t}_0$ is the sum of $\mathfrak{h}$ and those root-spaces for roots that are combinations of $\gamma_1, \gamma_2, \gamma_3$. The derived algebra $[\mathfrak{g}_0\, \mathfrak{g}_0]$ has $[\mathfrak{g}_0\, \mathfrak{g}_0] \cap \mathfrak{h}$ spanned by all $[v_{-j}v_j]$ $(= [u_0 \otimes v_{-j}, u_0 \otimes v_j])$, $1 \le j \le 3$, the center of $\mathfrak{g}_0$ being $\mathfrak{t}_0$. One has $\gamma_4\big|_{[\mathfrak{g}_0\mathfrak{g}_0] \cap \mathfrak{h}}$

$$= -\frac{1}{2}(\lambda_2 + \lambda_3 + \lambda_4)\Big|_{[\mathfrak{g}_0\mathfrak{g}_0] \cap \mathfrak{h}} = -\frac{1}{2}(\gamma_1 + 2\gamma_2 + 3\gamma_3)\Big|_{[\mathfrak{g}_0\mathfrak{g}_0] \cap \mathfrak{h}}.$$

Thus $[\mathfrak{g}_0\mathfrak{g}_0]$ is split simple of type B_3, with splitting Cartan subalgebra $[\mathfrak{g}_0\mathfrak{g}_0] \cap \mathfrak{h} = \mathfrak{h}_0$ and simple roots $\beta_1, \beta_2, \beta_3$ the restrictions to $\mathfrak{h}_0$ of $\gamma_1, \gamma_2, \gamma_3$, respectively.

When dim. $V = 13$, let roots $\gamma_1,\ldots,\gamma_8$ relative to $\mathfrak{h}$ have root-vectors $[v_{-5},v_6]$, $[v_{-4}v_5]$, $[v_{-3}v_4]$, $[v_{-2},v_3]$, $[v_{-1}v_2]$, $[v_1v_0] - 2u_0 \otimes v_1$, $m^+ \otimes w^+v_{-L}$, $[v_1v_0] + 2u_0 \otimes v_1$, respectively. With $\lambda_1,\ldots,\lambda_8$ dual to $\frac{1}{4}[u_{-1}u_1]$, $\frac{1}{4}[v_{-6}v_6],\ldots,$ $\frac{1}{4}[v_{-1}v_1]$, $\frac{1}{2}u_0 \otimes v_0$ as before, one finds $\gamma_1 = \lambda_2 - \lambda_3$, $\gamma_2 = \lambda_3 - \lambda_4$, $\gamma_3 = \lambda_4 - \lambda_5$, $\gamma_4 = \lambda_5 - \lambda_6$, $\gamma_5 = \lambda_6 - \lambda_7$, $\gamma_6 = \lambda_7 - \lambda_8$, $\gamma_8 = \lambda_7 + \lambda_8$, $\gamma_7 =$ $\frac{1}{2}(\lambda_1 - \lambda_2 - \lambda_3 - \lambda_4 - \lambda_5 - \lambda_6 - \lambda_7 + \lambda_8)$. The data above display a system of roots of type E_8 relative to $\mathfrak{h}$, coinciding in our labeling with that of p. 268 (Planche VII) of [Bo]. With a simple relabeling, our roots $\gamma_1,\ldots,\gamma_8$ are the "base" chosen in that reference, and have Dynkin diagram

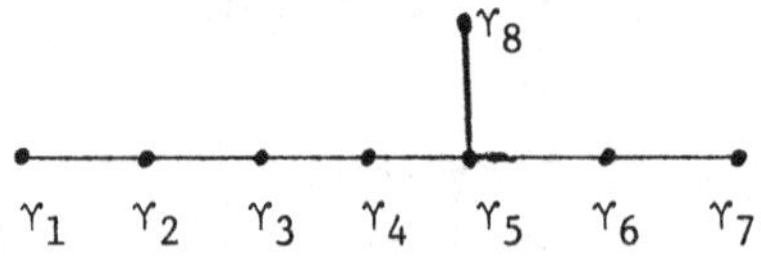

.

We have $\gamma_i\big|_{\mathfrak{t}_0} = 0$ for all $i \ne 7$, $\gamma_7\big|_{\mathfrak{t}_0} = \alpha_1$. Here $\mathfrak{g}_0$ is the sum of $\mathfrak{h}$ and the root-spaces for roots that are combinations of $\gamma_1,\ldots,\gamma_6, \gamma_8$. The center of $\mathfrak{g}_0$ is $\mathfrak{t}_0$, $\mathfrak{h}_0 = [\mathfrak{g}_0\mathfrak{g}_0] \cap \mathfrak{h}$ is a splitting Cartan subalgebra for $[\mathfrak{g}_0\, \mathfrak{g}_0]$, a simple Lie algebra of type D_7 with simple roots $\beta_1,\ldots,\beta_7$ relative to $\mathfrak{h}_0$, where $\beta_i = \gamma_i\big|_{\mathfrak{h}_0}$, $1 \le i \le 6$, $\beta_7 = \gamma_8\big|_{\mathfrak{h}_0}$. From the fact that $\mathfrak{h}_0$ contains $[[v_iv_0] + 2u_0 \otimes v_i, [v_{-i}v_0] - 2u_0 \otimes v_{-i}]$ for $1 \le |i| \le 6$, we see that all $[v_{-i}v_i]$, $1 \le i \le 6$, and $u_0 \otimes v_0$ form a basis for $\mathfrak{h}_0$, and thus that $\gamma_7\big|_{\mathfrak{h}_0} =$ $\frac{1}{2}(\lambda_8 - \lambda_7 - \ldots - \lambda_2)\big|_{\mathfrak{h}_0} =$

$-\frac{1}{2}(\beta_1 + 2\beta_2 + 3\beta_3 + 4\beta_4 + 5\beta_5 + \frac{7}{2}\beta_6 + \frac{5}{2}\beta_7)$.

E_6: Still with dim. $U = 3$, let dim. $V = 7$. Let V be split, so that $F(z)$ is split, $z^2 = 1$. Since we are regarding V as obtained by passing to a splitting field from the situation 4) of Chapter 5, the module M_V is a free $F(z)$-module of rank 8, with basis $\{w_i v_{-J} \mid i = 1,2,\ J \subseteq L = \{1,2,3\}\}$, such that $w_i v_j = 0$ for all $j > 0$, $w_i v_0 = (-1)^{i+1} w_i = -w_i z = -zw_i$. (Compare §5.6C, from which the above differs only by labels.) Here we take h to be the commutative subalgebra of g with basis $[u_{-1}, u_1]$, $[v_{-j}\ v_j]$ $(1 \le j \le 3)$, $u_0 \otimes v_0$, and z. Root-vectors relative to h are as when dim. $V = 13$, the $m^+ u_{-K} \otimes w^+ v_{-J}$ being replaced by the $m^+ u_{-K} \otimes w_i v_{-J}$, $i = 1,2$. If we take $\lambda_1, \ldots, \lambda_6$ to be the basis for h^* dual to the basis $\frac{1}{4}[u_{-1}u_1]$, $\frac{1}{4}[v_{-3}v_3]$, $\frac{1}{4}[v_{-2}v_2]$, $\frac{1}{4}[v_{-1}v_1]$, $\frac{1}{2}(u_0 \otimes v_0)$, z (in that order) for h, the root-vectors not in $M_U \otimes M_V$ are 40 in number, the corresponding roots being $\pm \lambda_i \pm \lambda_j$, $i \ne j$, $1 \le i, j \le 5$, and the 32 root-vectors from $M_U \otimes M_V$ correspond to roots of the form $\frac{1}{2}\sum_{i=1}^{6} \varepsilon_i \lambda_i$, $\varepsilon_i = \pm 1$, $\prod_{i=1}^{6} \varepsilon_i = 1$. (Here our conventions depart from those of [Bo] - cf. Planche V, p. 260.) In particular, we have the following root-vectors:

$m^+ \otimes w_2 v_{-L}$, for $\gamma_1 = \frac{1}{2}(\lambda_1 - \lambda_2 - \lambda_3 - \lambda_4 - \lambda_5 + \lambda_6)$:

$[v_1 v_0] + 2u_0 \otimes v_1$, for $\gamma_2 = \lambda_4 + \lambda_5$;

$[v_{-1} v_2]$, for $\gamma_3 = \lambda_3 - \lambda_4$;

$[v_1 v_0] - 2u_0 \otimes v_1$, for $\gamma_4 = \lambda_4 - \lambda_5$;

$m^+ \otimes w_1 v_{-L}$, for $\gamma_5 = \frac{1}{2}(\lambda_1 - \lambda_2 - \lambda_3 - \lambda_4 + \lambda_5 - \lambda_6)$;

$[v_{-2} v_3]$, for $\gamma_6 = \lambda_2 - \lambda_3$.

The corresponding roots are simple roots for a lexicographic ordering according to the coefficients of $\lambda_1, \ldots, \lambda_6$ (in that order). That is, h is a splitting Cartan subalgebra for g, which is seen, by such considerations as in Chapter 4 of [J3], to be simple of type E_6. The restrictions to t_0 of γ_1 and γ_5 are each equal to α_1, while those of $\gamma_2, \gamma_3, \gamma_4, \gamma_6$ are zero. The centralizer g_0 of t_0 is the sum of h and of the root-spaces for roots not involving λ_1, thus for combinations of $\gamma_2, \gamma_3, \gamma_4, \gamma_6$, and its derived algebra $[g_0 g_0]$ is simple of type D_4. A fundamental system of roots for $[g_0 g_0]$ relative to the splitting Cartan subalgebra $[g_0 g_0] \cap h = h_0$ is the set $\beta_1, \beta_2, \beta_3, \beta_4$ of respective restrictions of $\gamma_2, \gamma_3, \gamma_4, \gamma_6$, and a basis for h_0 is the set of the $[v_{-j} v_j]$, $1 \le j \le 3$, together with $u_0 \otimes v_0$, as before. Here the center of g_0 is $Fz + t_0$. The restrictions to h_0 of the remaining simple roots γ_1, γ_5 are respectively equal to

$-\frac{1}{2}(2\beta_1 + 2\beta_2 + \beta_3 + \beta_4), \quad -\frac{1}{2}(\beta_1 + 2\beta_2 + 2\beta_3 + \beta_4).$

E_7: Here we consider dim. $U = 3$, dim. $V = 9$, and split the situation where V and M_V are as in 6) of §5.1. Thus the quaternion algebra $\mathcal{Q}$ is assumed split, and the structure of M_V is as in the split case of §5.6.D. That is, M_V is a direct sum of two irreducible $C\ell(V)$-submodules W_1, W_2, which are isomorphic and generated by elements w_1, w_2, with $w_i v_j = 0$, all $j > 0$, $w_i v_0 = w_i$, so with respective bases $\{w_i v_{-J} \mid J \subseteq L = \{1,2,3,4\}\}$.

Let h be the commutative subalgebra with basis $e_{22} - e_{11}$ (from $g_2 = \mathcal{Q}_0$) $[u_{-1}u_1]$, $[v_{-j}v_j]$ $(1 \le j \le 4)$, and $u_0 \otimes v_0$. As in the last two cases we have a family of root-vectors relative to h in $[UU]$, $[VV]$, $U \otimes V$, $M_U \otimes M_V$, to which we adjoin $e_{21}, e_{12} \in \mathcal{Q}_0$. The last two centralize all of our basis for h except $e_{22} - e_{11}$, where $[e_{21}, e_{22} - e_{11}] = -2e_{21}$, $[e_{12}, e_{22} - e_{11}] = 2e_{12}$. Again we let $\lambda_1, \ldots, \lambda_7$ be a basis for h^*, dual to the basis $\frac{1}{4}[u_{-1}u_1]$, $\frac{1}{4}[v_{-4}v_4]$, $\frac{1}{4}[v_{-3}v_3]$, $\frac{1}{4}[v_{-2}v_2]$, $\frac{1}{4}[v_{-1}v_1]$, $\frac{1}{2}u_0 \otimes v_0$, $\frac{1}{2}(e_{22} - e_{11})$ for h (in that order). There are 60 roots with root-vectors in $[UU] + [VV] + U \otimes V$, corresponding to roots $\pm\lambda_i \pm \lambda_j$, $i \ne j$, $1 \le i, j \le 6$. The two roots with root-vectors in $\mathcal{Q}_0$ are $\pm\lambda_7$, and the root-vectors from $M_U \otimes M_V$, 64 in number, correspond to roots $\frac{1}{2}\sum_{i=1}^{7} \varepsilon_i \lambda_i$, $\varepsilon_i = \pm 1$, $\prod_{i=1}^{6} \varepsilon_i = 1$. (Compare [Bo], Planche VI, p. 264, where the conventions differ slightly.)

From lexicographic orderings, as in the last case, we find the simple roots, which we label as follows, with corresponding root-vectors:

$\gamma_1 = \lambda_7$: e_{12};

$\lambda_2 = \frac{1}{2}(\lambda_1 - \lambda_2 - \lambda_3 - \lambda_4 - \lambda_5 - \lambda_6 - \lambda_7)$: $m^+ \otimes w_1 v_{-L}$;

$\gamma_3 = \lambda_5 - \lambda_6$: $[v_1 v_0] - 2u_0 \otimes v_0$;

$\gamma_4 = \lambda_4 - \lambda_5$: $[v_{-1}v_2]$;

$\gamma_5 = \lambda_3 - \lambda_4$: $[v_{-2}v_3]$;

$\gamma_6 = \lambda_2 - \lambda_3$: $[v_{-3}v_4]$;

$\gamma_7 = \lambda_5 + \lambda_6$: $[v_1 v_0] + 2u_0 \otimes v_1$.

The Dynkin diagram is of type E_7:

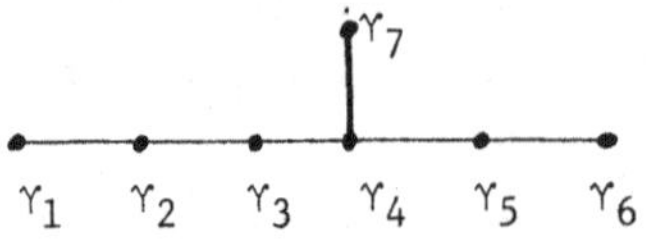

Here $\gamma_i\big|_{t_0} = 0$ for $i \ne 2$, $\gamma_2\big|_{t_0} = \alpha_1$. The centralizer g_0 of t_0 is the sum of h and of root-spaces for roots that are combinations of $\gamma_1, \gamma_3, \ldots, \gamma_7$, i.e., for $\pm\gamma_1$ and for combinations of $\gamma_3, \ldots, \gamma_7$. A basis for the splitting Cartan subalgebra

$h_0 = [g_0 g_0] \cap h$ of $[g_0 g_0]$ consists of $e_{22} - e_{11}$, the $[v_{-j}v_j]$, $1 \le j \le 4$, and $u_0 \otimes v_0$, as before, and the semisimple Lie algebra $[g_0 g_0]$ is of type $A_1 \oplus D_5$, a simple system of roots relative to h_0 consisting of the restriction β_0 of γ_1, and respectively, $\beta_1,\dots,\beta_5$ of $\gamma_6, \gamma_5, \gamma_4, \gamma_7, \gamma_3$:

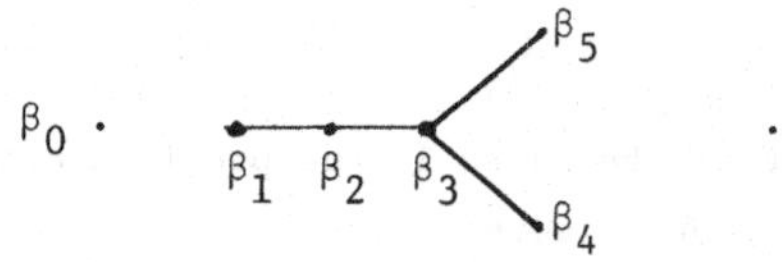

The restriction to h_0 of γ_2 is equal to

$$-\frac{1}{2}\beta_0 - \frac{1}{2}(\beta_1 + 2\beta_2 + 3\beta_3 + \frac{3}{2}\beta_4 + \frac{5}{2}\beta_5),$$

and the center of g_0 is t_0.

$E_{7,2}$: The case here is that with dim. U = 5, dim. V = 7, as in §5.5.D. We retain all the notations of the last section for the split case, with $L = \{1,2,3\}$, and take h to be the commutative subalgebra of g with basis $e_{22} - e_{11}$, $[u_{-1}u_1]$, $[u_{-2}u_2]$, $[v_{-j}v_j]$ $(1 \le j \le 3)$, $u_0 \otimes v_0$. The root-vectors of g relative to h are the $[u_iu_j]$ $ij \ne 0$, $i + j \ne 0$, the $[v_iv_j]$, with similar restrictions, the $[u_iu_0] \pm 2u_i \otimes v_0$ $(i \ne 0)$, the $[v_iv_0] \pm 2u_0 \otimes v_i$ $(i \ne 0)$, the $\overset{+}{m}u_{-K} \otimes w_iv_{-J}$, $i = 1,2$, $K \subseteq \{1,2\}$, $J \subseteq L$, and the two elements e_{12}, e_{21} of Q_0. If $\lambda_1,\dots,\lambda_7$ is the basis for h^* dual (in order) to $\frac{1}{4}[u_{-2}u_2]$, $\frac{1}{4}[u_{-1}u_1]$, $\frac{1}{4}[v_{-3}v_3]$, $\frac{1}{2}[v_{-2}v_2]$, $\frac{1}{4}[v_{-1}v_1]$, $\frac{1}{2}u_0 \otimes v_0$, $\frac{1}{2}(e_{22} - e_{11})$, then the roots of g relative to h are exactly as in the preceding case, a set of root-vectors for the simple roots so determined being:

$$\gamma_1 = \lambda_7 : e_{12}$$

$$\gamma_2 = \frac{1}{2}(\lambda_1 - \lambda_2 - \lambda_3 - \lambda_4 - \lambda_5 + \lambda_6 - \lambda_7): \ \overset{+}{m}u_{-1} \otimes w_1v_{-2};$$

$$\gamma_3 = \lambda_5 - \lambda_6: \ [v_1v_0] - 2\,u_0 \otimes v_1;$$

$$\gamma_4 = \lambda_4 - \lambda_5: \ [v_{-1},v_2];$$

$$\gamma_5 = \lambda_3 - \lambda_4: \ [v_{-2},v_3];$$

$$\gamma_6 = \lambda_2 - \lambda_3: \ u_1 \otimes v_{-3};$$

$$\gamma_7 = \lambda_5 + \lambda_6: \ [v_1v_0] + 2\,u_0 \otimes v_1.$$

Thus $\gamma_i\big|_{t_0} = 0$ if $i \ne 2, 6$, $\gamma_6\big|_{t_0} = \alpha_1$, $\gamma_2\big|_{t_0} = \alpha_2$. The Dynkin diagram is of type E_7, labeled as in the preceding case. The centralizer g_0 of t_0 is split with $[g_0g_0]$ of type $A_1 + D_4$ relative to the splitting Cartan subalgebra $h_0 = [g_0g_0] \cap h$, with basis the $[v_{-j}v_j]$, $u_0 \otimes v_0$ and $e_{22} - e_{11}$, and with t_0 as the center of g_0. Restrictions to h_0 of $\gamma_1, \gamma_3, \gamma_4, \gamma_6, \gamma_7$ form simple roots $\beta_0, \beta_1, \beta_2, \beta_3, \beta_4$ respectively for $[g_0g_0]$, with diagram

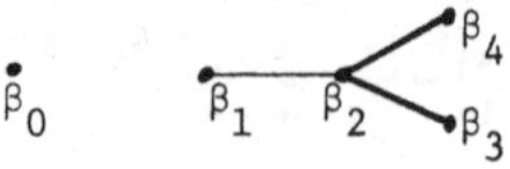

.

Finally, $\gamma_2\big|_{h_0} = -\frac{1}{2}\beta_0 - \frac{1}{2}(2\beta_1 + 2\beta_2 + \beta_3 + \beta_4)$, and $\gamma_6\big|_{h_0} = -\frac{1}{2}(\beta_1 + 2\beta_2 + 3\beta_3 + \beta_4)$.

$E_{6,2}$: Here dim. U = 5 = dim. V, V being split, F(z) being as in the previous case "E_6" of this section, M_V a free F(z)-module of rank 4, with basis $\{w_i v_{-J} | i = 1,2,\ J \subseteq L = \{1,2\}\}$, $w_i v_j = 0$ for $j > 0$, $w_i v_0 = (-1)^i w_i = w_i z = z w_i$. Again h is to be the commutative subalgebra of g with basis $[u_{-i}u_i]$, $[v_{-j}v_j]$ $(i,j = 1,2)$, $u_0 \otimes v_0$ and z. Root-vectors relative to h are as in the earlier E_6, and if $\lambda_1,\ldots,\lambda_6$ is the basis for h^* dual to $\frac{1}{4}[u_{-2}u_2]$, $\frac{1}{4}[u_{-1}u_1]$, $\frac{1}{4}[v_{-2}v_2]$, $\frac{1}{4}[v_{-1}v_1]$, $\frac{1}{2}[u_0 \otimes v_0)$, z (in order), we again have 40 root-vectors not in $M_U \otimes M_V$ and 32 in $M_U \otimes M_V$, corresponding to roots as before. A set of root-vectors corresponding to the simple roots relative to the lexicographic ordering determined by $\lambda_1,\ldots,\lambda_6$ is:

$$\gamma_1 = \frac{1}{2}(\lambda_1 - \lambda_2 - \lambda_3 - \lambda_4 - \lambda_5 + \lambda_6):\ m^+u_{-1} \otimes w_2 v_{-L};$$

$$\gamma_2 = \lambda_4 + \lambda_5 :\ [v_1v_0] + 2\, u_0 \otimes v_1;$$

$$\gamma_3 = \lambda_3 - \lambda_4 :\ [v_{-1},v_2];$$

$$\gamma_4 = \lambda_4 - \lambda_5 :\ [v_1v_0] - 2\, u_0 \otimes v_1;$$

$$\gamma_5 = \frac{1}{2}(\lambda_1 - \lambda_2 - \lambda_3 - \lambda_4 + \lambda_5 - \lambda_6) :\ m^+u_{-1} \otimes w_1 v_{-L};$$

$$\gamma_6 = \lambda_2 - \lambda_3 :\ u_1 \otimes v_{-2}.$$

The Dynkin diagram is of type E_6 as before.

The restrictions of $\gamma_2, \gamma_3, \gamma_4$ to t_0 are zero, while $\gamma_1\big|_{t_0} = \gamma_5\big|_{t_0} = \alpha_2$, $\gamma_6\big|_{t_0} = \alpha_1$. The centralizer g_0 of t_0 is spanned by h and the root-vectors for roots that are combinations of $\gamma_2, \gamma_3, \gamma_4$, and has center $t_0 + Fz$. Relative to $h_0 = [g_0 g_0] \cap h$, whose basis is $[v_{-1}v_1]$, $[v_{-2}v_2]$, $u_0 \otimes v_0$, $[g_0 g_0]$ is split of type A_3, with fundamental roots $\beta_1 = \gamma_2\big|_{h_0}$, $\beta_2 = \gamma_3\big|_{h_0}$, $\beta_3 = \gamma_4\big|_{h_0}$, and diagram

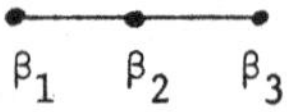

.

The restrictions to h_0 of $\gamma_1, \gamma_5, \gamma_6$ are as follows:

$$\gamma_1\big|_{h_0} = -\frac{3}{4}\beta_1 - \frac{1}{2}\beta_2 - \frac{1}{4}\beta_3;$$

$$\gamma_5\Big|_{h_0} = -\frac{1}{4}\beta_1 - \frac{1}{2}\beta_2 - \frac{3}{4}\beta_3;$$

$$\gamma_6\Big|_{h_0} = -\frac{1}{2}\beta_1 - \beta_2 - \frac{1}{2}\beta_3.$$

$E_{8,2}$: Here dim. U = 5, dim. V = 11, as in §5.5.B, and we split as in the previous case "E_8" of this section, taking as h the commutative subalgebra of g with basis $[u_{-i}, u_i]$, $i = 1,2$; $[v_{-j}v_j]$, $1 \le j \le 5$, and $u_0 \otimes v_0$. Relative to h, $M_U \otimes M_V$ contains 128 root-vectors, and $[UU] + [VV] + U \otimes V$ contains 112 more. If $\lambda_1, \ldots, \lambda_8$ is the ordered basis for h^* dual to $\frac{1}{4}[u_{-2}u_2]$, $\frac{1}{4}[u_{-1}u_1]$, $\frac{1}{4}[v_{-5}v_5], \ldots, \frac{1}{4}[v_{-1}v_1]$, $\frac{1}{2}(u_0 \otimes v_0)$ as before, the former roots are those of the form $\frac{1}{2}\sum_{i=1}^{8} \varepsilon_i \lambda_i$, $\varepsilon_i = \pm 1$, $\prod_{i=1}^{8} \varepsilon_i = 1$, while the latter are the $\pm\lambda_i \pm \lambda_j$, $i \ne j$, $1 \le i, j \le 8$. The simple roots γ_i are taken as in the previous case of E_8, and have root-vectors as follows:

$$\gamma_1 = \lambda_2 - \lambda_3 : u_1 \otimes v_{-5};$$

$$\gamma_2 = \lambda_3 - \lambda_4 : [v_{-4}v_5];$$

$$\gamma_3 = \lambda_4 - \lambda_5 : [v_{-3}v_4];$$

$$\gamma_4 = \lambda_5 - \lambda_6 : [v_{-2}v_3];$$

$$\gamma_5 = \lambda_6 - \lambda_7 : [v_{-1}v_2];$$

$$\gamma_6 = \lambda_7 - \lambda_8 : [v_1v_0] - 2\, u_0 \otimes v_1;$$

$$\gamma_7 = \frac{1}{2}(\lambda_1 - \lambda_2 - \lambda_3 - \lambda_4 - \lambda_5 - \lambda_6 - \lambda_7 + \lambda_8) : m^+u_{-1} \otimes w^+v_{-L};$$

$$\gamma_8 = \lambda_7 + \lambda_8 : [v_1v_0] + 2\, u_0 \otimes v_1.$$

One has $\gamma_i\Big|_{t_0} = 0$ for $i \ne 1, 7$; $\gamma_1\Big|_{t_0} = \alpha_1$, $\gamma_7\Big|_{t_0} = \alpha_2$. Here $g_0 = t_0 + [g_0g_0]$, a basis for the splitting Cartan subalgebra $h_0 = [g_0g_0] \cap h$ of $[g_0g_0]$ being the $[v_{-j}v_j]$, $1 \le j \le 5$, along with $u_0 \otimes v_0$, and fundamental roots $\beta_1, \ldots, \beta_6$ being the respective restrictions to h_0 of $\gamma_2, \ldots, \gamma_6, \gamma_8$. Thus $[g_0g_0]$ is split simple of type D_6:

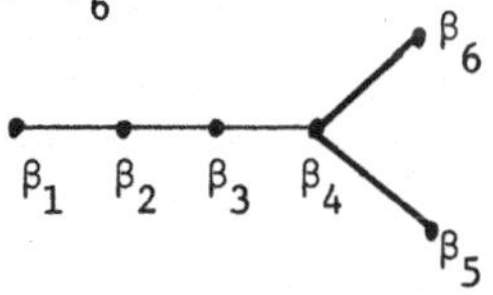

Finally, one has $\gamma_1\Big|_{h_0} = -\frac{1}{2}(2\beta_1 + 2\beta_2 + 2\beta_3 + 2\beta_4 + \beta_5 + \beta_6)$,

$$\gamma_7\Big|_{h_0} = -\frac{1}{2}(\beta_1 + 2\beta_2 + 3\beta_3 + 4\beta_4 + 3\beta_5 + 2\beta_6).$$

§2. Fundamental Weights and Their Restrictions.

In general, for a split semi-simple Lie algebra $\mathfrak{g}$ with splitting Cartan subalgebra $\mathfrak{h}$ and fundamental system of roots $\gamma_1,\ldots,\gamma_r$ with respect to $\mathfrak{h}$, the fundamental weights (with respect to all these data) are the elements of a basis for $\mathfrak{h}^*$ dual to the basis $h_1,\ldots,h_r$ for $\mathfrak{h}$ such that: i) $h_i \in [\mathfrak{g}_{\gamma_i}\ \mathfrak{g}_{-\gamma_i}]$; ii) $\gamma_i(h_i) = 2$. For $\mathfrak{g}$, $\mathfrak{h}$, $\gamma_1,\ldots,\gamma_r$ as in the last section $(r = 4,6,7,8)$ we denote the corresponding fundamental weights by $\pi_1,\ldots,\pi_r$. For the appropriate $[\mathfrak{g}_0\mathfrak{g}_0]$, $\mathfrak{h}_0$, (β_0), $\beta_1,\ldots,\beta_t$, we denote the corresponding fundamental weights by (ω_0), $\omega_1,\ldots,\omega_t$.

The present section is nothing more than a transcription from Planches V-VIII of [Bo] of the expressions for the π_i in terms of the γ_j, then an interpretation of their restrictions to $\mathfrak{h}_0$, to $\mathfrak{t}_0$, and (if different from $\mathfrak{t}_0$) to the center $\mathfrak{t}_0 + Fz$ of $\mathfrak{g}_0$. Labeling is as in §1.

F_4: From Planche VIII,

$$\pi_1 = 2\gamma_1 + 3\gamma_2 + 4\gamma_3 + 2\gamma_4;$$
$$\pi_2 = 3\gamma_1 + 6\gamma_2 + 8\gamma_3 + 4\gamma_4;$$
$$\pi_3 = 2\gamma_1 + 4\gamma_2 + 6\gamma_3 + 3\gamma_4;$$
$$\pi_4 = \gamma_1 + 2\gamma_2 + 3\gamma_3 + 2\gamma_4 .$$

Thus $\pi_1\big|_{\mathfrak{t}_0} = 2\alpha_1 = \pi_4\big|_{\mathfrak{t}_0}$; $\pi_3\big|_{\mathfrak{t}_0} = 3\alpha_1$; $\pi_2\big|_{\mathfrak{t}_0} = 4\alpha_1$, and $\mathfrak{t}_0$ is the center of $\mathfrak{g}_0$. The restrictions to $\mathfrak{h}_0$ are as follows:

$$\pi_1\big|_{\mathfrak{h}_0} = \omega_1;$$
$$\pi_2\big|_{\mathfrak{h}_0} = \omega_2;$$
$$\pi_3\big|_{\mathfrak{h}_0} = \omega_3;$$
$$\pi_4\big|_{\mathfrak{h}_0} = 0 .$$

E_8: From Planche VII,

$$\pi_1 = 2\gamma_1 + 3\gamma_2 + 4\gamma_3 + 5\gamma_4 + 6\gamma_5 + 4\gamma_6 + 2\gamma_7 + 3\gamma_8 ,$$
$$\pi_2 = 3\gamma_1 + 6\gamma_2 + 8\gamma_3 + 10\gamma_4 + 12\gamma_5 + 8\gamma_6 + 4\gamma_7 + 6\gamma_8 ,$$
$$\pi_3 = 4\gamma_1 + 8\gamma_2 + 12\gamma_3 + 15\gamma_4 + 18\gamma_5 + 12\gamma_6 + 6\gamma_7 + 9\gamma_8 ,$$
$$\pi_4 = 5\gamma_1 + 10\gamma_2 + 15\gamma_3 + 20\gamma_4 + 24\gamma_5 + 16\gamma_6 + 8\gamma_7 + 12\gamma_8 ,$$
$$\pi_5 = 6\gamma_1 + 12\gamma_2 + 18\gamma_3 + 24\gamma_4 + 30\gamma_5 + 20\gamma_6 + 10\gamma_7 + 15\gamma_8 ,$$

$$\pi_6 = 4\gamma_1 + 8\gamma_2 + 12\gamma_3 + 16\gamma_4 + 20\gamma_5 + 14\gamma_6 + 7\gamma_7 + 10\gamma_8 ,$$

$$\pi_7 = 2\gamma_1 + 4\gamma_2 + 6\gamma_3 + 8\gamma_4 + 10\gamma_5 + 7\gamma_6 + 4\gamma_7 + 5\gamma_8 ,$$

$$\pi_8 = 3\gamma_1 + 6\gamma_2 + 9\gamma_3 + 12\gamma_4 + 15\gamma_5 + 10\gamma_6 + 5\gamma_7 + 8\gamma_8 .$$

From §1, $\pi_1\big|_{t_0} = 2\alpha_1$; $\pi_2\big|_{t_0} = 4\alpha_1 = \pi_7\big|_{t_0}$; $\pi_8\big|_{t_0} = 5\alpha_1$; $\pi_3\big|_{t_0} = 6\alpha_1$; $\pi_6\big|_{t_0} = 7\alpha_1$; $\pi_4\big|_{t_0} = 8\alpha_1$; $\pi_5\big|_{t_0} = 10\alpha_1$, and t_0 is the center of g_0.

The restrictions of the π_i to h_0 are as follows:

$$\pi_1\big|_{h_0} = \omega_1 .$$

$$\pi_2\big|_{h_0} = \omega_2 .$$

$$\pi_3\big|_{h_0} = \omega_3 .$$

$$\pi_4\big|_{h_0} = \omega_4; \quad \pi_5\big|_{h_0} = \omega_5; \quad \pi_6\big|_{h_0} = \omega_6 ;$$

$$\pi_7\big|_{h_0} = 0; \quad \pi_8\big|_{h_0} = \omega_7.$$

E_6: From Planche V,

$$\pi_1 = \tfrac{1}{3}(4\gamma_1 + 5\gamma_2 + 6\gamma_3 + 4\gamma_4 + 2\gamma_5 + 3\gamma_6),$$

$$\pi_2 = \tfrac{1}{3}(5\gamma_1 + 10\gamma_2 + 12\gamma_3 + 8\gamma_4 + 4\gamma_5 + 6\gamma_6),$$

$$\pi_3 = 2\gamma_1 + 4\gamma_2 + 6\gamma_3 + 4\gamma_4 + 2\gamma_5 + 3\gamma_6,$$

$$\pi_4 = \tfrac{1}{3}(4\gamma_1 + 8\gamma_2 + 12\gamma_3 + 10\gamma_4 + 5\gamma_5 + 6\gamma_6),$$

$$\pi_5 = \tfrac{1}{3}(2\gamma_1 + 4\gamma_2 + 6\gamma_3 + 5\gamma_4 + 4\gamma_6 + 3\gamma_6),$$

$$\pi_6 = \gamma_1 + 2\gamma_2 + 3\gamma_3 + 2\gamma_4 + \gamma_5 + 2\gamma_6.$$

Thus $\pi_1\big|_{t_0} = 2\alpha_1 = \pi_5\big|_{t_0} = \pi_6\big|_{t_0}$; $\pi_2\big|_{t_0} = 3\alpha_1 = \pi_4\big|_{t_0}$; $\pi_3\big|_{t_0} = 4\alpha_1$. For the restrictions to h_0,

$$\pi_1\big|_{h_0} = \pi_5\big|_{h_0} = 0; \quad \pi_2\big|_{h_0} = \omega_1; \quad \pi_3\big|_{h_0} = \omega_2; \quad \pi_4\big|_{h_0} = \omega_3; \quad \pi_6\big|_{h_0} = \omega_4.$$

The values at z are given by $\pi_i(z) = 0$, $i = 3,6$; $\pi_1(z) = \frac{2}{3}$, $\pi_5(z) = -\frac{2}{3}$, $\pi_2(z) = \frac{1}{3}$, $\pi_4(z) = -\frac{1}{3}$.

E_7: From Planche VI,

$$\pi_1 = \tfrac{1}{2}(3\gamma_1 + 4\gamma_2 + 5\gamma_3 + 6\gamma_4 + 4\gamma_5 + 2\gamma_6 + 3\gamma_7),$$

$\pi_2 = 2\gamma_1 + 4\gamma_2 + 5\gamma_3 + 6\gamma_4 + 4\gamma_5 + 2\gamma_6 + 3\gamma_7$,

$\pi_3 = \frac{1}{2}(5\gamma_1 + 10\gamma_2 + 15\gamma_3 + 18\gamma_4 + 12\gamma_5 + 6\gamma_6 + 9\gamma_7)$,

$\pi_4 = 2\gamma_1 + 6\gamma_2 + 9\gamma_3 + 12\gamma_4 + 8\gamma_5 + 4\gamma_6 + 6\gamma_7$,

$\pi_5 = 2\gamma_1 + 4\gamma_2 + 6\gamma_3 + 8\gamma_4 + 6\gamma_5 + 3\gamma_6 + 4\gamma_7$,

$\pi_6 = \gamma_1 + 2\gamma_2 + 3\gamma_3 + 4\gamma_4 + 3\gamma_5 + 2\gamma_6 + 2\gamma_7$,

$\pi_7 = \frac{1}{2}(3\gamma_1 + 6\gamma_2 + 9\gamma_3 + 12\gamma_4 + 8\gamma_5 + 4\gamma_6 + 7\gamma_7)$.

The center of $\mathfrak{g}_0$ is $\mathfrak{t}_0$, to which the π_i have restrictions as follows: $\pi_1\big|_{\mathfrak{t}_0} = \pi_6\big|_{\mathfrak{t}_0} = 2\alpha_1$; $\pi_7\big|_{\mathfrak{t}_0} = 3\alpha$; $\pi_2\big|_{\mathfrak{t}_0} = \pi_5\big|_{\mathfrak{t}_0} = 4\alpha_1$; $\pi_3\big|_{\mathfrak{t}_0} = 5\alpha_1$; $\pi_4\big|_{\mathfrak{t}_9} = 6\alpha_1$.

The restrictions of the π_i to $\mathfrak{h}_0$ are:

$\pi_1\big|_{\mathfrak{h}_0} = \frac{1}{2}\beta_0 = \omega_0$; $\pi_2\big|_{\mathfrak{h}_0} = 0$; $\pi_3\big|_{\mathfrak{h}_0} = \omega_5$; $\pi_4\big|_{\mathfrak{h}_0} = \omega_3$; $\pi_5\big|_{\mathfrak{h}_0} = \omega_2$; $\pi_6\big|_{\mathfrak{h}_0} = \omega_1$;
$\pi_7\big|_{\mathfrak{h}_0} = \omega_4$.

$E_{7,2}$: Again the center of $\mathfrak{g}_0$ is $\mathfrak{t}_0$, and we have

$\pi_1\big|_{\mathfrak{t}_0} = \alpha_1 + 2\alpha_2$, $\pi_2\big|_{\mathfrak{t}_0} = 2\alpha_1 + 4\alpha_2$, $\pi_3\big|_{\mathfrak{t}_0} = 3\alpha_1 + 5\alpha_2$, $\pi_4\big|_{\mathfrak{t}_0} = 4\alpha_1 + 6\alpha_2$,

$\pi_5\big|_{\mathfrak{t}_0} = 3\alpha_1 + 4\alpha_2$, $\pi_6\big|_{\mathfrak{t}_0} = 2\alpha_1 + 2\alpha_2$, $\pi_7\big|_{\mathfrak{t}_0} = 2\alpha_1 + 3\alpha_2$.

Restricting to $\mathfrak{h}_0$,

$\pi_1\big|_{\mathfrak{h}_0} = \omega_0$; $\pi_2\big|_{\mathfrak{h}_0} = 0 = \pi_6\big|_{\mathfrak{h}_0}$; $\pi_3\big|_{\mathfrak{h}_0} = \omega_1$, $\pi_4\big|_{\mathfrak{h}_0} = \omega_2$; $\pi_5\big|_{\mathfrak{h}_0} = \omega_3$; $\pi_7\big|_{\mathfrak{h}_0} = \omega_4$.

$E_{6,2}$: From the list for E_6 above and from $E_{6,2}$ in §1,

$\pi_1\big|_{\mathfrak{t}_0} = \pi_5\big|_{\mathfrak{t}_0} = \alpha_1 + 2\alpha_2$; $\pi_2\big|_{\mathfrak{t}_0} = \pi_4\big|_{\mathfrak{t}_0} = 2\alpha_1 + 3\alpha_2$; $\pi_6\big|_{\mathfrak{t}_0} = 2\alpha_1 + 2\alpha_2$;

$\pi_3\big|_{\mathfrak{t}_0} = 3\alpha_1 + 4\alpha_2$.

For the values at z, $\pi_i(z) = 0$ for $i = 3,6$; $\pi_1(z) = \frac{2}{3} = -\pi_5(z)$, $\pi_2(z) = \frac{1}{3} = -\pi_4(z)$. The restrictions to $\mathfrak{h}_0$ are as follows:

$\pi_1\big|_{\mathfrak{h}_0} = 0 = \pi_5\big|_{\mathfrak{h}_0} = \pi_6\big|_{\mathfrak{h}_0}$; $\pi_2\big|_{\mathfrak{h}_0} = \omega_1$;

$\pi_3\big|_{\mathfrak{h}_0} = \omega_2$; $\pi_4\big|_{\mathfrak{h}_0} = \omega_3$.

$E_{8,2}$: From the list for E_8 above and from $E_{8,2}$ in §1,

$$\pi_1\Big|_{t_0} = 2\alpha_1 + 2\alpha_2; \quad \pi_2\Big|_{t_0} = 3\alpha_1 + 4\alpha_2; \quad \pi_3\Big|_{t_0} = 4\alpha_1 + 6\alpha_2 ;$$

$$\pi_4\Big|_{t_0} = 5\alpha_1 + 8\alpha_2; \quad \pi_5\Big|_{t_0} = 6\alpha_1 + 10\alpha_2; \quad \pi_6\Big|_{t_0} = 4\alpha_1 + 7\alpha_2 ;$$

$$\pi_7\Big|_{t_0} = 2\alpha_1 + 4\alpha_2; \quad \pi_8\Big|_{t_0} = 3\alpha_1 + 5\alpha_2.$$

Moreover, t_0 is the center of g_0.

The restrictions of the π to h_0 are:

$$\pi_1\Big|_{h_0} = 0 = \pi_7\Big|_{h_0} ; \pi_2\Big|_{h_0} = \omega_1; \pi_3\Big|_{h_0} = \omega_2; \pi_4\Big|_{h_0} = \omega_3; \pi_5\Big|_{h_0} = \omega_4;$$

$$\pi_6\Big|_{h_0} = \omega_5; \pi_8\Big|_{h_0} = \omega_6.$$

§3. Restrictions on Relative Rank. Existence.

Here we assume U is split. If g_0 is the centralizer of t_0, then $[VV] + u_0 \otimes V \subseteq [g_0g_0]$ identifies with $[V_0V_0]$, where $V_0 = V \oplus Fe_0$, an orthogonal sum with $(e_0,e_0) = -1$. The identification associates $x \in [VV]$ with its image in $C\ell(V_0)$ under the canonical embedding of $C\ell(V)$ in $C\ell(V_0)$, and $u_0 \otimes v$ $(v \in V)$ with $\frac{1}{2}[e_0,v]$. The subalgebra $[g_0g_0]$ is either isomorphic to $[V_0V_0]$, or to $[V_0V_0] + Q_0$, the latter only in cases of type E_7. We study the restrictions imposed by insisting that t_0 be a <u>maximal</u> F-split torus in g.

These restrictions, that g_0 contain no element $h \notin t_0$ such that ad h acts in F-diagonalizable fashion on g, may also be formulated as follows: i) the <u>center</u> of g_0 shall contain no $h \notin t_0$ with the last property; <u>and</u> ii) $[g_0g_0]$ shall contain no nonzero element that acts nilpotently (but not as zero) in some finite-dimensional module. The first of these conditions applies only to cases of type E_6, where it is equivalent to the condition that ad z not be F-diagonalizable, or that F(z) be a quadratic extension field of F. The second says that neither $[V_0V_0]$ nor Q_0 shall contain nilpotent elements other than zero. Thus, in cases of type E_7, Q must be a division algebra, and in all cases, V_0 must be an anisotropic quadratic space. The effect of the last condition is to say that $(v,v) \neq 1$ for all $v \in V$. Because (by the theorem of Jacobson-Morozov, for instance) an ad-nilpotent element of a semisimple algebra acts nilpotently on every finite-dimensional module, these conditions on Q and V are sufficient to yield i). We thus have

<u>Proposition 6.1. The toral subalgebra t_0, of dimension one or two, is a maximal split toral subalgebra of g as constructed in Chapter 5 if and only if the given quadratic form (v,v) on V fails to represent 1, and either: a) if the dimension of V is 7 resp. 5 then the center F(z) of $C\ell(V)$ is a quadratic field extension of F (algebras of type E_6); or b) if the dimension of V is 9 resp. 7 then Q is a quaternionic division algebra over F (algebras of type E_7).</u>

Thus for $\dim t_0 = 2$, V is either as in 3) of §5.1, with F(z) a field, as in 5) with Q a division algebra, or as in 7). If $\dim t_0 = 1$, either V is as in 4) with F(z) a field, as in 6) with Q a division algebra, as in 8), or as in 9). In all cases $(v,v) \neq 1$ for all $v \in V$. (That these constraints are necessary, i.e. that every exceptional central simple Lie algebra of F-type BC_2, and every one of F-type BC_1 with highest root-space of dimension greater than one, results from our constructions of Chapter 5, with V so constrained, is proved in V.7. c) and V.7. d) ii) of [Sel].

For example, with $F = \mathbb{R}$ and V a 5-dimensional $\mathbb{R}$-space with a negative-definite quadratic form we have that the even Clifford algebra splits and the center of the full Clifford algebra is isomorphic to $\mathbb{C}$. Thus, when $\dim U = 5$, we are in the situation of Prop. 6.1, and the central simple Lie algebra of type E_6 resulting from our construction is of $\mathbb{R}$-rank 2 and has $\mathbb{R}$-root system of type BC_2. In the notations of Tits (cf. [T2], p. 59), the type of g is ${}^2E^{16'}_{6,2}$.

With $F = \mathbb{R}$ and V a 6-dimensional negative definite quadratic $\mathbb{R}$-space, we have $C\ell(V) \simeq (-1,-1) \otimes (1,1) \otimes (-1,-1)$, by Lemma 5, p. 233 of [J5], where the notation for quaternion algebras (α,β) is as in [O'M], §5.7. Thus $(-1,-1) = \mathbb{H}$, $(1,1) = M_2(\mathbb{R})$, $C\ell(V) \simeq M_2(\mathbb{H} \otimes \mathbb{H}) \simeq M_8(\mathbb{R})$. For $\dim U = 3$, our pair U, V yields by the construction a central simple $\mathbb{R}$-Lie algebra g of type F_4, relative rank 1, and with non-reduced root system, corresponding to Tits' diagram $F^{21}_{4,1}$ (p. 60). As shown in §V.8 of [Sel], these are the only cases with $F = \mathbb{R}$ where t_0 is a maximal split torus.

Next let $F = \mathbb{Q}$, and let $e_1,\ldots,e_7$ be a basis for a vector space over $\mathbb{Q}$, with a symmetric bilinear form (v_1,v_2) relative to which the e_i are orthogonal and $(e_i,e_i) = -1$ for all $i < 7$, while $(e_7,e_7) = -2$. By the factorization cited above, $C\ell(V) \simeq (-1,-1) \otimes (1,1) \otimes (-1,-1) \otimes \mathbb{Q}(\sqrt{2}) \simeq M_8(\mathbb{Q}(\sqrt{2}))$. The form is negative definite, so cannot represent 1. Thus if we combine this V and our 3-dimensional space U (always assumed to be split), the resulting absolutely simple form of E_6 over $\mathbb{Q}$ has 1-dimensional maximal split torus t_0, and is of type BC_1 relative to t_0.

With $F = \mathbb{Q}$, let $e_1,\ldots,e_7$ be a basis for V as above, but now with $(e_6,e_6) = -3 = (e_7,e_7)$, $(e_i,e_i) = -1$ for $i < 6$. Here $C\ell(V) \simeq (-1,-1) \otimes (1,1) \otimes (-1,-3) \otimes (\mathbb{Q} \times \mathbb{Q})$, and $(-1,-1) \otimes (-1,-3) \simeq (-1,3) \otimes (1,-1)$, by [O'M], §57.10. Thus $C\ell(V) \simeq M_4((-1,3)) \otimes (\mathbb{Q} \times \mathbb{Q}) \simeq M_4((-1,3)) \times M_4((-1,3))$, and it is an easy matter of considering congruences mod 3 to show that the quadratic form $\langle -1 \rangle \perp \langle 3 \rangle$ does not represent 1 over $\mathbb{Q}$. Thus $(-1,3) = Q$ is a quaternionic division algebra over $\mathbb{Q}$ ([O'M], §57.9), $C\ell^+(V) \simeq M_4(Q)$, the form on V does not represent 1, and the center of $C\ell(V)$ is split. Thus we may combine this V with a 5-dimensional split space U over $\mathbb{Q}$ to obtain an absolutely simple Lie algebra g of dimension 133 over $\mathbb{Q}$, having maximal $\mathbb{Q}$-split toral subalgebra t_0 of dimension 2, relative to which g is of type BC_2.

The remaining three cases: E_7, E_8, $E_{8,2}$ require more exotic ground fields for their existence. This is indicated by the "-" in the columns corresponding to $\mathbb{R}$

and to number fields in the lines $E^{48}_{7,1}$, $E^{91}_{8,1}$, $E^{66}_{8,2}$ in the tables for "Types E_7 and E_8" on pp. 59, 60 of [T2]. These lines correspond to absolutely simple algebraic groups defined over F whose Lie algebras have our prescribed split toral and root structure, and the tables state that there is no such group for $F = \mathbb{R}$ or F a number field. The statement is tantamount to saying that there is no quadratic space V with all the appropriate properties over such a field. (For $F = \mathbb{R}$, see [Sel], §V.8.)

For the following examples, I am indebted to Professor Albrecht Pfister.

Let $F = \mathbb{Q}(t)$, where t is an indeterminate, and consider the symmetric bilinear form in 9 variables on $V = F^9$ with diagonal matrix $\mathrm{diag}\{-1,-1,-1,-1,-1,-1,-2,t,-2t\}$. If we assume this form represents 1, then there are polynomials $p_0(t),\ldots,p_9(t)$ with integral coefficients, not all zero, such that

$$-\sum_{i=0}^{6} p_i(t)^2 - 2p_7(t)^2 + t(p_8(t)^2 - 2p_9(t)^2) = 0.$$

By degrees, it follows that this can only be the case if the terms of highest degree m, say $a_{mi}t^m$, in any of the $p_i(t)$, $i \le 7$, satisfy $\Sigma_{i=0}^{6} a_{mi}^2 + 2a_{m7}^2 = 0$, or $a_{m8}^2 - 2a_{m9}^2 = 0$, where the notation should be clear, and neither of these equations has nontrivial integral solutions. Thus the form <u>does</u> <u>not</u> <u>represent</u> 1.

The discriminant of the form is $(-1)^{\binom{9}{2}} 4t^2 = 4t^2$, so the center of $C\ell(V)$ is split. From [J5], p. 237, the even Clifford algebra $C\ell^+(V)$ is $C\ell(V_0)$, where V_0 has form $\mathrm{diag}\{-1,-1,-1,-1,-1,-2,t,-2t\}$, and $C\ell(V_0)$ is isomorphic to $(-1,-1) \otimes (1,1) \otimes (-1,-2) \otimes (-2t,4t) \cong (1,1) \otimes (-1,2) \otimes (1,-1) \otimes (t,-2t)$. Each of the first three factors is split, because the binary form with each of the above as diagonal matrix represents 1 ([J5], p. 146), so $C\ell^+(V) \cong M_8((t,-2t))$. To say that $(t,-2t)$ is not a quaternionic division algebra is to say that the form $\mathrm{diag}\{t,-2t\}$ over F represents 1, or that there are polynomials $p(t)$, $q(t)$, $r(t) \in Z[t]$, not all zero, such that

$$r(t)^2 = t(p(t)^2 - 2q(t)^2).$$

If the term of highest degree in $q(t)$ or $p(t)$ has degree at least equal to that of $r(t)$, this would make 2 be a square in $\mathbb{Z}[t]$, a contradiction; otherwise the left-hand side has higher degree than the right, again a contradiction. Thus V and $C\ell(V)$ have our requisite structure to yield an F-form of E_7 of relative rank 1 and type BC_1.

Next let V have dimension 11 over $F = \mathbb{Q}(t)$, carrying a symmetric bilinear form with diagonal matrix $\mathrm{diag}\{-1,-1,-1,-1,-1,-3,-3,t,t,-3t,-3t\}$. The discriminant is $(-1)^{\binom{11}{2}} \cdot (-81)t^4) = 81t^4$, so the center of the Clifford algebra is split. As above,

$$C\ell^+(V) \cong (-1,-1) \otimes (1,1) \otimes (-3,-3) \otimes (-t,-t) \otimes (-3t,-3t)$$

$$\cong (1,1) \otimes (-1,-1) \otimes (-1,-3) \otimes (-1,-t) \otimes (-1,-3t)$$

(here using (3) of 57.10 of [O'M] on each of the last three factors). Combining the second and third factors, as well as the fourth and fifth, according to (4) of the same proposition in [O'M], and dropping a factor "t^2" by (2) of that proposition, we find

$$C\ell^+(V) \cong (1,1) \otimes (-1,3) \otimes (1,-1) \otimes (-1,3) \otimes (1,-1),$$

from which $C\ell^+(V) \cong M_{32}(F)$.

To see that the form fails to represent 1, the assumption that it does represent 1 is seen to yield non-trivial integral polynomials $p_0(t),\ldots,p_{11}(t)$ with

$$\sum_{i=0}^{5} p_i(t)^2 + 3(p_6(t)^2 + p_7(t)^2)$$

$$= t(p_8(t)^2 + p_9(t)^2 - 3p_{10}(t)^2 - 3p_{11}(t)^2).$$

If there is a degree among those of $p_0(t),\ldots,p_7(t)$ greater than that of any of $p_8(t),\ldots,p_{11}(t)$, we get a contradiction. Otherwise the highest coefficients of $p_8(t),\ldots,p_{11}(t)$ give a relation $a_8^2 + a_9^2 = 3(a_{10}^2 + a_{11}^2)$ in integers. By inspecting congruences mod 3 in a minimal solution to this last equation, we see there are no nontrivial solutions, thus that the form does not represent 1. Accordingly, V and $C\ell(V)$ satisfy all our conditions for constructing an F-form of E_8 of relative rank 2 and type BC_2.

Finally, let t and u be independent indeterminates over $\mathbb{Q}$ and consider the diagonal form in 13 variables over $F = \mathbb{Q}(t,u)$ with matrix diag{-1,-1,-1,-1,-1,-1, -2,t,-2t,u,-2u,ut,-2ut}. By first comparing degrees in u, then those in t, we see as in the 9-dimensional case that the form does not represent 1. The discriminant is $(-1)^{\binom{13}{2}}\cdot 16u^4t^4 = 16u^4t^4$, so the center of the full Clifford algebra is split, and

$$C\ell^+(V) \cong (-1,-1) \otimes (1,1) \otimes (-1,-2) \otimes (t,-2t) \otimes (-u,2u) \otimes (ut, -2ut)$$

$$\cong (1,1) \otimes (-1,-1) \otimes (-1,-2) \otimes (2,t) \otimes (2,-u) \otimes (2,ut)$$

$$\cong (1,1) \otimes (-1,2) \otimes (1,-1) \otimes (1,-1) \otimes (2,-ut) \otimes (2,ut)$$

$$\cong (1,1) \otimes (-1,2) \otimes (1,-1) \otimes (1,-1) \otimes (2,-1) \otimes (1,-1) \cong M_{64}(F).$$

Thus V and $C\ell(V)$ satisfy all conditions for our construction with $\dim U = 3$ to yield an F-form of E_8 of type BC_1.

§4. Fundamental Modules: The Case of Relative Rank One.

We continue to assume that t_0 is a maximal split torus in g, one of the exceptional central simple algebras constructed as above. In this section we concentrate on algebras of relative rank one, where $\dim U = 3$ and $H_1 = [u_{-1}u_1]$ is a fundamental coroot corresponding to the simple root α_1. The information derived in §2 tells us, in each case, which weights λ must have their λ-admissible g_0-modules constructed in order to obtain all μ-admissible g_0-modules, for all possible highest t_0-weights μ, by Cartan multiplication. We proceed case-by-case.

F_4: Here $g_0 = FH_1 + [VV] + u_0 \otimes V$, where V is a 6-dimensional quadratic space with split Clifford algebra and form not representing 1. From §2, we must consider $\lambda = 2\alpha_1, 3\alpha_1, 4\alpha_1$.

<u>Proposition</u> F_4. <u>There are two</u> $2\alpha_1$-<u>admissible</u> g_0-<u>modules</u>:

a) <u>The base field</u> F, <u>with</u> $[g_0g_0]$ <u>acting trivially</u>, H_1 <u>as multiplication by</u> 4, <u>and</u>

b) <u>The vector space</u> V_0, <u>with</u> $[g_0g_0]$ <u>acting by</u> $[V_0V_0]$, H_1 <u>as multiplication by</u> 4.

<u>There is one</u> $3\alpha_1$-<u>admissible</u> g_0-<u>module, namely the space</u> M_V, <u>of dimension</u> 8, <u>the action of</u> H_1 <u>being multiplication by</u> 6, <u>and with</u>

$$w(x + u_0 \otimes v) = wx + wv$$

<u>for all</u> $w \in M_V$, $x \in [VV]$, $v \in V$.

<u>All</u> $4\alpha_1$-<u>admissible</u> g_0-<u>modules are submodules of tensor products of two</u> $2\alpha_1$-<u>admissible modules</u>.

<u>Proof</u>. It is immediate that all the modules in question are irreducible g_0-modules with the appropriate action of H_1. (Indeed, they are absolutely irreducible for $\lambda = 2\alpha_1, 3\alpha_1$. To show their admissibility and completeness, let K be a finite extension field of F splitting V, W_i the unique irreducible g_K module of highest weight π_i, $1 \le i \le 4$. Here n_K is the sum of the root-spaces of g_K relative to h (as in §2) for which the root γ has a positive coefficient of γ_4. In particular, n annihilates a highest weight-vector w_i of W_i, and W_i is a finite-dimensional g-module. As in Chapter 2, all irreducible g-submodules of W_i are isomorphic, and W_i is their sum. The elements of W_i annihilated by n are all eigenvectors for H_1, belonging to the same eigenvalue, and w_i is among them. Thus the common eigenvalue is $\pi_i(H_1)$, and each irreducible g-submodule of W_i has the property that its subspace annihilated by n is an irreducible admissible g_0-module with $m_i = \pi_i(H_1)$, where $m_i = \lambda_i(H_1)$.

In the case of π_4, we have seen in §2 that the restriction of π_4 to a splitting Cartan subalgebra of $[g_0g_0]_K$ is zero. Thus the $(g_0)_K$-submodule of W_4

generated by w_4 is one-dimensional, so is surely irreducible. Every element of $\mathfrak{g}_0$ maps w_4 to a K-multiple of w_4, $[\mathfrak{g}_0\mathfrak{g}_0]$ acts trivially on w_4, and $w_4H_1 = 4w_4$. That is, the elements W_4^n of W_4 annihilated by n contain a $\mathfrak{g}_0$-submodule as in a). Because all $\mathfrak{g}_0$-submodules of the set W_4^n are isomorphic, it follows that the module of a) satisfies our condition of admissibility. The irreducible $\mathfrak{g}$-module is absolutely irreducible, of dimension 26.

Next consider π_1. From §2, the $[\mathfrak{g}_0\mathfrak{g}_0]$-submodule of W_1 generated by w_1 is irreducible of highest weight ω_1. The unique $[\mathfrak{g}_0\mathfrak{g}_0]_K$-irreducible module of this kind is $(V_0)_K$. Thus the $\mathfrak{g}_0$-module V_0 of b) has the property that $(V_0)_K$ is $(\mathfrak{g}_0)_K$-isomorphic to the subspace of W_1 where H_1 has the eigenvalue 4. This space in turn is a sum of irreducible admissible $\mathfrak{g}_0$-modules. As in Chapter 2, it follows that these $\mathfrak{g}_0$-modules are isomorphic to V_0. Thus the module of b) is irreducible. The corresponding $\mathfrak{g}$-module is the adjoint module.

To show that the module M_V is $3\alpha_1$-admissible, it suffices as in the last paragraph to display a $[\mathfrak{g}_0\mathfrak{g}_0]_K$-isomorphism between $(M_V)_K$ and the $[\mathfrak{g}_0\mathfrak{g}_0]_K$-irreducible module of highest weight ω_3, as in §2. When $[\mathfrak{g}_0\mathfrak{g}_0]_K$ is identified with $[V_0V_0]_K$, the module of highest weight ω_3 is identified with an irreducible module for the even Clifford algebra $C\ell^+(V_{0K}) = C\ell^+(V_0)_K$. From $(e_0v_1)(e_0v_2) = -(e_0e_0)v_1v_2 = v_1v_2$, for $v_1, v_2 \in V$, it follows that there is an F-algebra homomorphism $C\ell(V) \to C\ell^+(V_0)$ mapping $v \in V$ to $e_0v \in C\ell^+(V_0)$. By the simplicity of $C\ell(V)$ and the equality of dimensions, this map is an isomorphism.

We may assume that K contains an element ζ with $\zeta^2 = -1$. Then if $v_{-3}, v_{-2}, \ldots, v_3$ is a splitting basis for V_K, a splitting basis for $(V_0)_K$ is $v_{-3}, v_{-2}, v_{-1}, v_0 = \zeta e_0, v_1, v_2, v_3$. We may take a generator w^+ for $(M_V)_K$ with $w^+v_i = 0$ for all $i > 0$ and we may extend the action of $C\ell^+(V_0)_K$ to an action of $C\ell(V_{0K})$ on $(M_{V_0})_K$, an 8-dimensional K-space, in such a way that a generator y^+ has $y^+v_i = 0$ for all $i > 0$, $y^+v_0 = y^+$. Then the K-linear mapping sending w^+v_{-J} to y^+v_{-J} for $|J|$ even and w^+v_{-J} to ζy^+v_{-J} for $|J|$ odd is an isomorphism of $C\ell(V)_K$-modules, the structure of $C\ell(V)_K$-module on $(M_{V_0})_K$ being given by the isomorphism $C\ell(V) \to C\ell^+(V_0)$ as above. In particular, we obtain isomorphic structures of $[V_0V_0]_K$-module on $(M_V)_K$ and $(M_{V_0})_K$. With this structure, each irreducible $\mathfrak{g}_0$-submodule of $(M_V)_K$ is $3\alpha_1$-admissible, by Chapter 2. The structure is such that the action on $w \in (M_V)_K$ of e_0v, $v \in V$, is that of $v \in V$, so that of $[e_0v, e_0v'] \in [V_0V_0]$ $(v,v' \in V)$ is that of $[vv'] \in [VV]$. Thus M_V, being stable under such elements, is a $\mathfrak{g}_0$-submodule, with action as specified in the statement of the Proposition. Because M_V is absolutely irreducible, the corresponding $\mathfrak{g}$-module is, too, and has dimension 273.

From Chapter 2, the $\mathfrak{g}_0$-irreducible constituents of the tensor product of two modules as in a) or b) are $4\alpha_1$-admissible. Thus all listed modules for $\mathfrak{g}_0$ are admissible. In fact, we could have associated as above with W_2 the irreducible $[\mathfrak{g}_0\mathfrak{g}_0]_K$-module $\Lambda^2(V_0)_K \approx [\mathfrak{g}_0\mathfrak{g}_0]_K$, of which $\Lambda^2(V_0) \approx [\mathfrak{g}_0\mathfrak{g}_0]$ is an F-form. Clearly

$\Lambda^2(V_0)$ is a $\mathfrak{g}_0$-irreducible submodule of $V_0 \otimes V_0$, and is therefore redundant in terms of the processes of the Proposition.

To show that the modules of the Proposition are exhaustive, first let M be an irreducible $\mathfrak{g}$-module whose highest $\mathfrak{t}_0$-weight λ has $\lambda(H_1) = 4$. With K as above, M_K is a completely reducible $\mathfrak{g}_K$-module, the highest weights of whose irreducible constituents all have the value 4 at H_1. If $\pi = \sum_{i=1}^{4} n_i \pi_i$ is one of these highest weights, $\pi(H_i) \neq 4$ unless $\pi = \pi_1, \pi_4$. Thus all highest weights of M_K are either π_1 or π_4, and it follows from the above that every irreducible $\mathfrak{g}$-submodule of M_K has as highest $\mathfrak{t}_0$-weight space one of a) or b). This applies in particular to M. A similar argument applies for $\lambda(H_1) = 6$, where we must have $\pi = \pi_3$, and for $\lambda(H_1) = 0$, where $\pi = 0$, and M is one-dimensional. For $\lambda(H_1) = 8$, there are four possibilities: $\pi = 2\pi_i$, $\pi = \pi_1 + \pi_4$, $\pi = 2\pi_4$ and $\pi = \pi_2$. In each case, the $(\mathfrak{g}_0)_K$-module is contained in a tensor product: For $2\pi_4$, one may take the tensor square of the trivial $2\alpha_1$-admissible $[\mathfrak{g}_0\mathfrak{g}_0]$-module of a), or its extension to K; for $\pi_1 + \pi_4$, the tensor product of a) and b) is absolutely irreducible, and the absolutely irreducible $\mathfrak{g}$-module with this $\mathfrak{g}_0$-module as highest $\mathfrak{t}_0$-weight has highest $\mathfrak{h}$-weight $\pi_1 + \pi_4$ over K. As $\mathfrak{g}_0$-module,

$$V_0 \otimes V_0 = (V_0 \otimes V_0)^+ \oplus (V_0 \otimes V_0)^-$$

$$= F \oplus (V_0 \otimes V_0)_0^+ \oplus (V_0 \otimes V_0)^-,$$

where the three summands are absolutely irreducible, $4\alpha_1$-admissible and non-isomorphic, the dimensions of the second and third being 27 and 21, respectively. We have seen that there are at most four non-isomorphic $4\alpha_1$-admissible $\mathfrak{g}_0$-modules, and thus they are realized, with dimensions 1, 7, 21, 27. Here $(V_0 \otimes V_0)^-$ corresponds to π_2 and $(V_0 \otimes V_0)_0^+$ to $2\pi_1$. This completes the proof.

E_8: The setting is as for F_4, but with V of dimension 13, both the center of $C\ell(V)$ and the simple algebra $C\ell^+(V)$ being split. From §2, the fundamental $\mathfrak{t}_0$-weights are among: $2\alpha_1$ (from π_1); $4\alpha_1$ (from π_2, π_7 as well as from $2\pi_1$); $5\alpha_1$ (from π_8); $6\alpha_1$ (from $\pi_3, \pi_1 + \pi_2, \pi_1 + \pi_7$ and $3\pi_1$); $7\alpha_1$ (from $\pi_6, \pi_1 + \pi_8$); $8\alpha_1$ (from $\pi_4, \pi_1 + \pi_3, 2\pi_1 + \pi_2, 2\pi_1 + \pi_7, 4\pi_1, 2\pi_2, \pi_2 + \pi_7$ and $2\pi_7$); $10\alpha_1$ (from $\pi_5, \pi_1 + \pi_4, 2\pi_1 + \pi_3, 3\pi_1 + \pi_2, 3\pi_1 + \pi_7, 5\pi_1, \pi_1 + 2\pi_2, \pi_1 + \pi_2 + \pi_7, \pi_2 + \pi_3, \pi_2 + \pi_7, 2\pi_8$). The weight $m_1 = 0$ is only associated with $\pi_0 = 0$.

<u>Proposition</u> E_8. <u>There is one</u> $2\alpha_1$-<u>admissible</u> $\mathfrak{g}_0$-<u>module, namely the vector space</u> V_0 <u>with</u> $[\mathfrak{g}_0\mathfrak{g}_0]$ <u>acting as</u> $[V_0V_0]$, H_1 <u>as multiplication by</u> 4.

<u>There are two</u> $4\alpha_1$-<u>admissible</u> $\mathfrak{g}_0$-<u>modules, namely the adjoint module</u> $[\mathfrak{g}_0\mathfrak{g}_0] = [V_0V_0] = \Lambda^2(V_0)$, <u>and the one-dimensional trivial</u> $[\mathfrak{g}_0\mathfrak{g}_0]$-<u>module</u> F, <u>in each case with</u> H_1 <u>acting by multiplication by</u> 8.

<u>There is one</u> $5\alpha_1$-<u>admissible</u> $\mathfrak{g}_0$-<u>module, namely the 64-dimensional space</u> M_V, <u>with</u> H_1 <u>acting as multiplication by</u> 10, <u>and</u> $[\mathfrak{g}_0\mathfrak{g}_0] = [VV] + u_0 \otimes V$ <u>acting as in</u>

the case $m_1 = 6$ *for* F_4 *in Proposition* F_4.

All other admissible g_0*-modules are submodules of tensor products of two or more in the above list. For* $\lambda = 0$, *one has only the trivial* g_0*-module* F.

Proof. The argument is essentially the same as that of Proposition F_4. For $2\alpha_1$, one sees that the $[g_0g_0]$-module V_0 is an F-form of the $[g_0g_0]_K$-submodule of W_1 where H_1 acts as the scalar 4, using §2. Because π_1 is the only highest weight for g_K with restriction $2\alpha_1$ to t_0, V_0 is the only $2\alpha_1$-admissible g_0-module. The absolute irreducibility of V_0 means that the corresponding representation of g is absolutely irreducible, a form of W_1, the adjoint representation of g_K. Thus our representation of g with highest weight module V_0 is the adjoint representation.

For $4\alpha_1$, consideration of the irreducible g_K-module W_7 yields, as in a) of Proposition F_4, the one-dimensional g_0-module of our proposition. Consideration of W_2 and identification of the $[g_0g_0]_K$-module of highest weight ω_2 with the subspace of W_2 belonging to the eigenvalue 8 for H_1 (cf. §2), and this $[g_0g_0]_K$-module with the adjoint module $[g_0g_0]_K \approx [V_0V_0]_K \approx \Lambda^2(V_0)_K$ shows that $\Lambda^2(V_0)$ is an F-form of the $[g_0g_0]_K$-module in question, and thus is $4\alpha_1$-admissible. This latter module is of course a submodule of the tensor product with itself of the module V_0 for $2\alpha_1$. The weight $2\pi_1$, the remaining one having restriction $4\alpha_1$ to t_0, is a composite $\pi_1 + \pi_1$, and as such gives rise only to irreducible g_0-modules arising by "Cartan multiplication" of that for π_1, namely V_0 (cf. Chapter 2). All the g-modules here are absolutely irreducible.

For $5\alpha_1$, only W_8 enters the picture. In the notations of §2, the subspace of W_8 belonging to the eigenvalue 10 for H_1 is the $[g_0g_0]_K$-module of highest weight ω_7, or one of the half-spin modules for $[V_0V_0]_K$. The two half-spin modules may be realized as follows, over the original ground field F:

Let J be a minimal right ideal in $C\ell^+(V_0)$, regarded as subalgebra of $C\ell(V_0)$. The isomorphism of $C\ell^+(V_0)$ with $C\ell(V)$ shows that $C\ell^+(V_0)$ is the sum of two ideals, each a split algebra of dimension 2^{12}, and therefore that J has dimension 2^6, the image of $C\ell^+(V_0)$ in $\mathrm{End}_F(J)$ being the full algebra $\mathrm{End}_F(J)$. Now $Je_0 \subseteq C\ell^-(V_0)$ is another $C\ell^+(V_0)$-module, of the same dimension 2^6, the central element $z_0 = e_0z$ of $C\ell^+(V_0)$ corresponds to $z \in C\ell(V)$ under our isomorphism, $z_0^2 = 1$, and we may assume $wz_0 = -w$ for all $w \in J$. From $e_0z = -ze_0$ we have $e_0z_0 = -e_0ze_0 = -z_0e_0$, so that $we_0z_0 = -wz_0e_0 = we_0$ for all $w \in J$.

If $w \in J$, $v \in V_0$, consider the map $J \otimes V_0 \to C\ell(V_0)$ sending $w \otimes v$ to wv. For $v = e_0$, the image is in Je_0; for $v \in V$, $we_0v \in J$, $we_0v \otimes e_0 \to we_0ve_0 \in Je_0$. But $we_0ve_0 = -we_0^2v = wv$. Thus the image JV_0 is equal to Je_0, and the map above is clearly a morphism of right $[V_0V_0]$-modules. In particular, the irreducible $[V_0V_0]$-module Je_0 is a submodule of $J \otimes V_0$.

With the action $w(x + u_0 \otimes v) = wx + wv$ of $[VV] + u_0 \otimes V$ on M_V, we see upon extension of F to a splitting field K that w^+ is a highest weight vector relative to $\beta_1,\ldots,\beta_7$, the corresponding weight being ω_7. Now it follows as above that M_V, with the given action, is the unique $5\,\alpha_1$-admissible $\mathfrak{g}_0$-module. We may assume the right ideal J above is isomorphic to M_V as $[V_0V_0]$-module, so that $wz_0 = w$ for all $w \in J$. Then $V_0 \otimes J$ contains an irreducible (in fact absolutely irreducible) submodule isomorphic to Je_0, hence not isomorphic to M_V, but rather having highest weight ω_6 upon extension of the base field to K. (The module may be identified with the vector space M_V, the action of $x + u_0 \otimes v$ on $w \in M_V$ now being $wx - wv$.) That is, there is an absolutely irreducible $7\,\alpha_1$-admissible $\mathfrak{g}_0$-module corresponding to W_6, and it is a submodule of the tensor product of such a module for $2\,\alpha_1$ and the one for $5\,\alpha_1$. The (absolutely) irreducible $\mathfrak{g}$-module of highest $\mathfrak{t}_0$-weight $5\,\alpha_1$ has dimension 147, 250.

In the remaining cases of fundamental representations for $\mathfrak{g}_K$, those of W_3, W_4, W_5 the corresponding highest weights ω_3, ω_4, ω_5 for $[\mathfrak{g}_0\mathfrak{g}_0]_K$ are those in the standard representations of $[\mathfrak{g}_0\mathfrak{g}_0]_K = [V_0V_0]_K$ on $\Lambda^3(V_0)_K$, $\Lambda^4(V_0)_K$, $\Lambda^5(V_0)_K$ respectively. The unique F-forms of these are the representations on $\Lambda^3(V_0)$, $\Lambda^4(V_0)$, $\Lambda^5(V_0)$ of $[\mathfrak{g}_0\mathfrak{g}_0]$. They are absolutely irreducible and when extended to $\mathfrak{t}_0$ so that H_1 acts as 12, 16, 20, respectively, are submodules of $\otimes^3 V_0$, $\otimes^4 V_0$, $\otimes^5 V_0$, with the $\mathfrak{g}_0$-module structure of V_0 as for $2\,\alpha_1$. As such, they are λ-admissible for $\lambda = 6\,\alpha_1$, $8\,\alpha_1$, $10\,\alpha_1$. All the conclusions of Proposition E_8 now follow from Chapter 2, except for the comments about the case $\lambda = 0$. Clearly the trivial $\mathfrak{g}$-module F has $\lambda = 0$ and the trivial $\mathfrak{g}_0$-module F as highest weight space. Conversely, the only dominant integral function π on $\mathfrak{h}$ which has $\pi(H_1) = 0$ is $\pi = 0$, and it follows that the trivial representation of $\mathfrak{g}$ is the only irreducible one with $\lambda = 0$.

E_7: Here V has dimension 9, the center of $C\ell(V)$ is split, and $C\ell^+(V) \cong \mathrm{End}\,(M_V)$, where $\mathcal{Q}$ is a quaternionic division algebra, M a left $\mathcal{Q}$-module of dimension 4. From §2, besides $\lambda = 0$ we have the fundamental $\mathfrak{t}_0$-weights λ and associated weights relative to $\mathfrak{h}$ as follows: $\lambda = 2\,\alpha_1$ (π_1, π_6); $\lambda = 3\,\alpha_1(\pi_7)$; $\lambda = 4\,\alpha_1(\pi_2,\ \pi_5,\ 2\,\pi_1,\ \pi_1 + \pi_6,\ 2\,\pi_6)$; $\lambda = 5\,\alpha_1$ $(\pi_3,\ \pi_1 + \pi_7,\ \pi_6 + \pi_7)$; $\lambda = 6\,\alpha_1$ $(\pi_4,\ 2\,\pi_7,\ \pi_1 + \pi_3,\ \pi_1 + \pi_5,\ 3\,\pi_1,\ 2\,\pi_1 + \pi_6,\ \pi_1 + 2\,\pi_6,\ 3\,\pi_6,\ \pi_5 + \pi_6,\ \pi_3 + \pi_6)$.

Proposition E_7: There are two $2\,\alpha_1$-admissible $\mathfrak{g}_0$-modules, namely
i) the vector space V_0 with H_1 acting as multiplication by 4, $\mathfrak{g}_2 = \mathcal{Q}_0$ annihilating V_0 and $[\mathfrak{g}_0\mathfrak{g}_0]$ acting as $[V_0, V_0]$, and
ii) the vector space $\mathcal{Q}$, with H_1 acting as multiplication by 4, $\mathcal{Q}_0$ by right multiplication as elements of $\mathcal{Q}$, and with $[V_0V_0]$ annihilating $\mathcal{Q}$.

There is one $3\,\alpha_1$-admissible $\mathfrak{g}_0$-module, namely a $C\ell(V)$-module isomorphic as $C\ell^+(V)$-module to M_V, with H_1 acting as multiplication by 6, and with

$$w(a + x + u_0 \otimes v) = wx + wv$$

for all $w \in M_V$, $a \in \mathcal{Q}_0$, $x \in [VV]$, $v \in V$.

All other admissible $\mathfrak{g}_0$-modules are submodules of tensor products of two or more

<u>of these three. For $\lambda = 0$, one has only the trivial $\mathfrak{g}_0$-module</u> F.

<u>Proof</u>. As in previous cases, V_0 is an absolutely irreducible $\mathfrak{g}_0$-module, and an F-form of the $[\mathfrak{g}_0\mathfrak{g}_0]_K$-submodule of W_6 where H_1 acts as the scalar 4. W_6 is the adjoint module for $\mathfrak{g}_K$, and V_0 is the highest weight space relative to $\mathfrak{t}_0$ of the adjoint module for $\mathfrak{g}$. Thus V_0 is $2\,\alpha_1$-admissible.

Clearly $\mathcal{Q}$, with $\mathfrak{g}_0$-module structure as above, is irreducible, and decomposes over K into the direct sum of two isomorphic 2-dimensional modules annihilated by $[V_0V_0]_K$, each affording the unique 2-dimensional irreducible representation of $(\mathcal{Q}_0)_K$. The restriction of π_1 to $\mathfrak{h}_0$ is the fundamental weight ω_0, and the representation of $[\mathfrak{g}_0\mathfrak{g}_0]_K$ on the subspace of W_1 where H_1 acts as 4 has $[V_0V_0]_K$ as its kernel and restricts to the 2-dimensional representation of $(\mathcal{Q}_0)_K$. Accordingly all of its $[\mathfrak{g}_0\mathfrak{g}_0]$-submodules annihilate $[V_0V_0]$, and their irreducible $\mathcal{Q}_0$-summands decompose over K into copies of this module. By the "completeness test" of §2.3 (for instance) it follows that they are all isomorphic to $\mathcal{Q}$. Thus the modules of i) and ii) are admissible for $\lambda = 2\,\alpha_1$, and we see as in earlier cases that they are the only ones with this property. The corresponding irreducible $\mathfrak{g}$-module decomposes over K into two isomorphic modules of dimension 56.

For $3\alpha_1$, M_V with the prescribed structure has all the asserted properties, except perhaps for admissibility. Here consider the $\mathfrak{g}_K$-module W_7 (the only irreducible $\mathfrak{g}_K$-module whose highest weight π has $\pi(H_1) = 6$). The values of π_7 at the various $[v_{-j}v_j]$, $1 \leq j \leq 4$, are all 2, and at $u_0 \otimes v_0$, the value is 1; meanwhile $\mathfrak{h}_0 \cap (\mathcal{Q}_0)_K$ is annihilated by π_7. Accordingly the subspace of W_7 where H_1 acts as multiplication by 6 is a half-spin module for our subalgebra $[V_0V_0]_K$, the module such that in our canonical isomorphism of $C\ell^+(V_0)$ with $C\ell(V)$ (extended over K), the element e_0v_0 of $C\ell^+(V_0)$ corresponding to v_0 (or to $u_0 \otimes v_0$) in the map $[VV]_K \oplus u_0 \otimes V_K \to C\ell^+(V_0)_K$, acts on a highest weight vector w^+ to fix w^+. By assumption, there are two inequivalent irreducible right $C\ell^+(V_0)$-modules, and likewise two for $C\ell^+(V_0)_K$. If we fix a central element z_0, $z_0^\sigma = -z_0$, $z_0^2 = 1$, then z_0 is unique up to sign, and the two modules are distinguished by whether z_0 acts as 1 or -1. In $C\ell^+(V_0)_K$ we may realize z_0 as $\pm e_0 z$, where $z = \sum_{J\subseteq\{1,2,3,4\}} v_{-J}v_0v_J$ is central in $C\ell(V)_K$, and v_0 acts on a highest weight vector for $(M_V)_K$ by the same scalar as does z. Thus there is exactly one of the two irreducible $C\ell^+(V_0)$-modules whose irreducible constituents over K are our specified half-spin module for $[V_0V_0]_K$, and this one must be chosen to resolve the ambiguity presented by the fact that there are two $C\ell(V)$-structures on M_V extending its $C\ell^+(V)$-structure. Once a choice of sign $z_0 = \pm e_0 z$, relative to a splitting is made, our choice of $[V_0V_0]$-module structure on M_V is fixed. This one is the "M_V" to be used in connection with $3\alpha_1$. Thus it is admissible and unique for $\lambda = 3\alpha_1$. The corresponding $\mathfrak{g}$-module decomposes over K into the sum of two isomorphic irreducible modules of dimension 912.

As in the case of E_8, $M_V \otimes V_0$ contains as submodule the unique irreducible $[g_0g_0]$-module which is an F-form of the set of elements of W_3 where H_1 has the eigenvalue 10. (This may be identified with the vector space M_V, with $w(a + x + u_0 \otimes v) = wx - wv$.) The analogous subspaces for W_5, W_4 have as F-forms, respectively, the absolutely irreducible $[g_0g_0]$-modules $\Lambda^2(V_0)$, $\Lambda^3(V_0)$, contained in $V_0 \otimes V_0$, $V_0 \otimes V_0 \otimes V_0$, as before with E_8. Finally, the space in W_2 where H_1 acts as the scalar 8 is a one-dimensional trivial $[g_0g_0]_K$-module. With V_0 as for $\lambda = 2\alpha_1$, the bilinear form $v_1 \otimes v_2 \to (v_1,v_2)$ on V_0 affords a homomorphism $V_0 \otimes V_0 \to F$ of g_0-modules mapping $V_0 \otimes V_0$ onto the unique F-form of this $(g_0)_K$-module. Thus all $4\alpha_1$-admissible g_0-modules are submodules of tensor products of two for $2\alpha_1$, and Proposition E_7 follows by our "completeness test".

E_6: As before, it is only the weights $\pi_1,\ldots,\pi_6$ and their corresponding modules $W_1,\ldots,W_6$ for g_K whose F-constituents must be realized. The corresponding values for λ are: $2\alpha_1$, for π_1, π_5, π_6; $3\alpha_1$, for π_2, π_4; $4\alpha_1$, for π_3.

<u>Proposition</u> E_6: <u>There are two</u> $2\alpha_1$<u>-admissible</u> g_0<u>-modules</u>: i) <u>The space</u> V_0, <u>of dimension</u> 8, <u>with</u> z <u>annihilating</u> V_0, H_1 <u>multiplying elements of</u> V_0 <u>by</u> 4, <u>and</u> $[V_0V_0]$ <u>acting by commutation within</u> $C\ell(V_0)$. ii) <u>The space</u> F(z) <u>of dimension</u> 2, <u>annihilated by</u> $[V_0V_0]$ <u>with</u> H_1 <u>acting as the scalar</u> 4, <u>and with</u> $z \in F(z)$ (a field), <u>acting as multiplication by</u> $\frac{2}{3}$ z.

<u>There is one</u> $3\alpha_1$<u>-admissible</u> g_0<u>-module, namely the unique irreducible right</u> $C\ell(V)$<u>-module</u> M_V, <u>a vector space of dimension</u> 8 <u>over</u> F(z), <u>with</u> $w H_1 = 6w$, $w(z + x + u_0 \otimes v) = -\frac{1}{3} zw + wx + wv$ for $w \in M_V$, $x \in [VV]$, $v \in V$.

<u>All other admissible</u> g_0<u>-modules are submodules of tensor products of two or more of these three</u>. <u>For</u> $\lambda = 0$, <u>only the trivial</u> g_0<u>-module</u> F <u>is admissible</u>.

<u>Proof</u>. Each of the modules presented is an irreducible g_0-module with H_1 acting as the appropriate scalar. If $\gamma \in K$ is such that $\gamma^2 = z^2 \in F$, the module of ii) splits into two one-dimensional subspaces over K, where $z_0 = \gamma^{-1}z$ has $z_0^2 = 1$, so acts as $\frac{2}{3}$ in one space and as $-\frac{2}{3}$ in the other.

With the appropriate choice of γ, we may assume $\gamma^{-1}z = \sum_{J\subseteq\{1,2,3\}} v_{-J}v_0v_J$ in $C\ell(V)_K$, and that $\pi_1(\gamma^{-1}z) = \frac{2}{3}$, $\pi_5(\gamma^{-1}z) = -\frac{2}{3}$, $\pi_2(\gamma^{-1}z) = \frac{1}{3}$, $\pi_4(\gamma^{-1}z) = -\frac{1}{3}$, while $\pi_3(z) = 0 = \pi_6(z)$, from §2.

Now the subspace of W_1 where H_1 acts as multiplication by 4 has dimension 1 over K, and there z acts as multiplication by $\frac{2}{3}\gamma$, while $[g_0g_0]_K$ annihilates the space. Accordingly the action of z satisfies the polynomial equation $X^2 - (\frac{2}{3}\gamma)^2 = 0$ there, with coefficients in F. This is an irreducible F-polynomial, and it follows that each irreducible g_0-constituent of the space is 2-dimensional over F, annihilated by $[g_0g_0]$, with z acting to satisfy the irreducible polynomial above. The same applies to W_5, with $\frac{2}{3}\gamma$ replaced by $-\frac{2}{3}\gamma$, so all g_0-irreducible constituents of the space of either W_1 or W_5 where H_1 acts as 4 are isomorphic to our module of ii). The corresponding irreducible $\mathfrak{g}$-module has dimension 54, and decomposes into $W_1 + W_5$ upon field extension.

That the absolutely irreducible $\mathfrak{g}_0$-module V_0 of i) is the unique type of irreducible $\mathfrak{g}_0$-module occurring in the subspace of W_6 where H_1 acts as 4 is seen just as for E_7 and E_8. It follows that the modules of i) and ii) are admissible, and that they exhaust all $2\alpha_1$-admissible $\mathfrak{g}_0$-modules. The latter irreducible $\mathfrak{g}$-module is the adjoint module.

In the cases of W_2, W_4, the highest weights as $[\mathfrak{g}_0\mathfrak{g}_0]_K$-modules for the subspaces where H_1 acts as 6 are ω_1, ω_3 respectively, and correspond to the two nonisomorphic half-spin modules for $C\ell^+(V_0)_K \cong C\ell(V)_K$. From the values $\pi_2(z) = \frac{1}{3}\gamma$, $\pi_4(z) = -\frac{1}{3}\gamma$, the representing transformation of $z \in C\ell(V)$ satisfies the F-polynomial equation $X^2 - \frac{1}{9}\gamma^2 = 0$, irreducible over F, and centralizes the action of $[\mathfrak{g}_0\mathfrak{g}_0]$. Thus each $\mathfrak{g}_0$-submodule of the space where $\pi(H_1) = 6$ in either W_2 or W_4 is a sum of 2-dimensional irreducible z-spaces, on each of which z has <u>both</u> eigenvalues $\pm\frac{1}{3}\gamma$ (when the ground field is extended to K). Accordingly extension of the ground field to K for the submodule in question yields two nonisomorphic $(\mathfrak{g}_0)_K$-summands of this space. These are exactly the two summands into which the given $\mathfrak{g}_0$-module splits upon extension of the base field to K. Thus the module in question is $3\alpha_1$-admissible, and is unique for this $\mathfrak{t}_0$-weight. Its dimension is $702 = 2 \times 351$, with 351 being the common dimension of W_2 and W_4.

To see the assertion about splitting of the module, note that if K splits V, $(M_V)_K$ decomposes into $e_1M_V \oplus e_2M_V$ (now regarding K as the ground field F), as in §5.6.C. Then each e_iM_V is generated as $C\ell(V)$-module by an element w_i with $w_iv_j = 0$ for all $j > 0$, $w_iv_0 = (-1)^i w_i$. Clearly each e_iM_V is an (absolutely) irreducible $(\mathfrak{g}_0)_K$-submodule in our action, in which z acts as a scalar, and with w_i as highest weight vector for $[\mathfrak{g}_0\mathfrak{g}_0]_K$, the two highest weights being ω_1 and ω_3, in the notation of "E_6" of §2. Both ω_1 and ω_3 have the same values at $[v_{-j}v_j]$, $1 \le j \le 3$; we have $w_iv_0 = (-1)^i w_i$, if the sign of v_0 is so chosen that $\gamma^{-1}z = \sum_{\{J\subseteq 1,2,3\}} v_{-J}v_0v_J$, and $w_i(\gamma^{-1}z) = (-1)^{i+1}w_i$. Now $\omega_1(u_0 \otimes v_0) = \pi_2(u_0 \otimes v_0) = 1$, $\gamma^{-1}z = e_1 - e_2$, and so our module action of $(\mathfrak{g}_0)_K$ on e_1M_V gives $w_2 \cdot (\gamma^{-1}z) = -\frac{1}{3}(e_1 - e_2)w_2 = \frac{1}{3}w_2$. That is, the highest weight ω_1 combines with the eigenvalue $\frac{1}{3}$ for $\gamma^{-1}z$, and the representation of $(\mathfrak{g}_0)_K$ on e_2M_V is equivalent to that on the subspace of W_2 where H_1 acts by 6. The corresponding considerations apply to W_4 and ω_3, with e_2M_V replaced by e_1M_V.

For $\lambda = 4\alpha_1$, the subspace of W_3 where H_1 acts by 8 has highest $[\mathfrak{g}_0\mathfrak{g}_0]_K$-weight ω_2, and is annihilated by z. Thus it is isomorphic to the $[V_0V_0]_K$-module $\Lambda^2(V_0)_K$, with trivial z-action, having $\Lambda^2(V_0)$ as (absolutely) irreducible F-form. It follows as before that all $4\alpha_1$-admissible $\mathfrak{g}_0$-modules are submodules of tensor products of those for $2\alpha_1$, and the proof of Proposition E_6 is complete.

§5. Fundamental Modules: Relative Rank Two.

$E_{6,2}$: Here $H_1 = \frac{1}{2}[u_{-1}u_1]$, $H_2([u_{-2}u_2] - [u_{-1}u_1])$, $\mathfrak{g}_0 = FH_1 + FH_2 + Fz + [VV] + u_0 \otimes V$, where V is a 5-dimensional space as in 3) of §5.1, with the further stipulation that $(v,v) \neq 1$ for all $v \in V$. Thus $V_0 = Fe_0 \oplus V$, as before, is an anisotropic 6-dimensional space, and $[\mathfrak{g}_0\mathfrak{g}_0] \approx [V_0V_0]$. The element z is central and in $C\ell^-(V)$, with $z^2 \in F$, $z^2 \notin F^2$. From §2, the values of $\pi_1,\ldots,\pi_6$ at H_1, H_2 are (m_1, m_2), where: $(m_1,m_2) = (0,2)$ for π_1 and π_5; $(m_1,m_2) = (1,2)$ for π_2 and π_4; $(m_1,m_2) = (2,0)$ for π_6; $(m_1,m_2) = (2,2)$ for π_3. Over the splitting field K, the representation of $[\mathfrak{g}_0\mathfrak{g}_0]_K$ on the subspace of W_i of t_0-weight $\pi_i|_{t_0}$ has highest weight 0 for W_1,W_5,W_6; ω_i, $1 \le i \le 3$, for W_{i+1}. We describe highest t_0-weights λ in terms of $m_1 = \lambda(H_1)$, $m_2 = \lambda(H_2)$.

<u>Proposition</u> $E_{6,2}$: <u>There is one admissible $\mathfrak{g}_0$-module when $(m_1,m_2) = (0,2)$, namely the field $F(z)$ with trivial action of $[V_0V_0]$, with H_1 annihilating $F(z)$, H_2 acting as multiplication by 2, and $a \in Fz$ acting as multiplication by $\frac{2}{3}a$.</u>

<u>There is one admissible $\mathfrak{g}_0$-module when $(m_1,m_2) = (1,2)$, namely the irreducible $C\ell(V)$-module M_V, the actions of H_1 and H_2 being multiplications by 1 and 2 respectively, and with</u>

$$w(a + x + u_0 \otimes v) = \frac{1}{3}\, aw + wx + wv$$

<u>for</u> $w \in M_V$, $a \in Fz$, $x \in [VV]$, $v \in V$.

<u>There is one admissible $\mathfrak{g}_0$-module when $(m_1,m_2) = (2,0)$, namely the field F, annihilated by $[V_0V_0]$, Fz, H_2, and with H_1 acting as 2.</u>

<u>There is an admissible $\mathfrak{g}_0$-module when $(m_1,m_2) = (2,2)$, namely the space V_0, with H_1, H_2 acting by 2, Fz annihilating V_0, and with $[V_0V_0] \approx [\mathfrak{g}_0\mathfrak{g}_0]$ acting by its natural action on V_0.</u>

<u>The only module for $(m_1,m_2) = (0,0)$ is the trivial one.</u>

<u>All other admissible $\mathfrak{g}_0$-modules are submodules of tensor products of the above.</u>

<u>Proof</u>. The proof is as for E_6 in the proposition of that title in §4 above. The only difference to be noted is the change of sign in the modules associated with W_2 and W_4.

Here the associated irreducible $\mathfrak{g}$-module has dimension 54 and 702 in the first cases, splitting into two nonisomorphic absolutely irreducible modules of dimensions 27 resp. 351 over K. In the cases $(m_1,m_2) = (2,0)$ resp. $(2,2)$ it is absolutely irreducible of dimensions 78 (adjoint module) resp. 2925.

$E_{7,2}$: Here V is a 7-dimensional space, with $(v,v) \neq 1$, the center of $C\ell(V)$ being split, $C\ell^+(V) \approx \mathrm{End}_Q(M_V)$, M_V a 2-dimensional left module for the quaternionic division algebra Q, and $\mathfrak{g}_0 = FH_1 + FH_2 + Q_0 + [VV] + u_0 \otimes V$, $[\mathfrak{g}_0\mathfrak{g}_0] = Q_0 + [VV] + u_0 \otimes V \approx Q_0 + [V_0V_0]$. The pairs (m_1,m_2) coming from $\pi_1,\ldots,\pi_7$ are: $(0,2)$, from

π_1; (0,4), from π_2; (1,4), from π_3; (2,4), from π_4; (2,2), from π_5; (2,0), from π_6; (1,2), from π_7. The highest weights for the representations of $[g_0 g_0]_K$ on the respective subspaces of $W_1,\ldots,W_7$ of these t_0-weights are: ω_0, 0, ω_1, ω_2, ω_3, 0, ω_4, in the notations of §2. The admissible g_0-modules are given in the following proposition.

Proposition $E_{7,2}$: For $(m_1,m_2) = (0,2)$ there is one admissible g_0-module, namely $\mathcal{Q}$, annihilated by $[V_0V_0]$ and with $b \in \mathcal{Q}_0$ sending $a \in \mathcal{Q}$ to $ab \in \mathcal{Q}$.

For $(m_1,m_2) = (2,0)$ there is a unique admissible g_0-module, namely the one-dimensional space F.

For $(m_1,m_2) = (2,2)$, there is, in addition to the tensor product $\mathcal{Q}$ of the two above, one more admissible g_0-module, namely V_0 with the usual action of $[V_0V_0]$ and annihilation by $\mathcal{Q}_0$.

For $(m_1,m_2) = (1,2)$ there is a unique admissible g_0-module, namely M_V, with the actions of H_1 and H_2 being 1 and 2 respectively, and with

$$w(a + z + u_0 \otimes v) = - aw + wx + wv$$

for $w \in M_V$, $a \in \mathcal{Q}_0$, $x \in [VV]$, $v \in V$.

For $(m_1,m_2) = (1,4)$, there are two admissible g_0-modules, one of them being the tensor product of those for $(m_1,m_2) = (0,2)$ and $(m_1,m_2) = (1,2)$. The other is M_V, with appropriate actions of H_1 and H_2, and with

$$w(a + x + u_0 \otimes v) = - aw + wx - wv.$$

The only admissible module for $(m_1,m_2) = (0,0)$ is the trivial module. All other admissible g_0-modules are submodules of appropriate tensor products of those above.

Proof. The case $(m_1,m_2) = (0,2)$, corresponding to the fundamental module W_1 for g_K, is exactly as in the earlier Proposition E_7. Likewise for $(m_1,m_2) = (2,0)$ and W_6, and for $(m_1,m_2) = (2,2)$ and W_5. In the split case, if W_1 has highest weight vector w_0, then $w_0 \otimes w_0$ is highest weight vector for a submodule of $W_1 \otimes W_1$ of highest weight $2\pi_1$, while $w_0 e_{-\gamma_1} \otimes w_0 - w_0 e_{-\gamma_1}$, where $0 \neq e_{-\gamma_i} \in g_{-\gamma_i}$, is highest weight vector for a submodule of highest weight π_2. Over F, the module $\mathcal{Q}$ for g_0 corresponding to $(m_1,m_2) = (0,2)$ has $\mathcal{Q} \otimes \mathcal{Q} \cong \mathrm{End}_F(\mathcal{Q})$ as associative F-algebra, an isomorphism φ sending $b \otimes c$ to $(d \to \bar{b}\, dc)$. If $a \in \mathcal{Q}_0$, the image of $ab \otimes c + b \otimes ac$ sends d to $\bar{b}[da]c$. That is, the right $\mathcal{Q}_0$-module structure on $\mathrm{End}_F(\mathcal{Q})$ obtained by transporting the tensor product module structure on $\mathcal{Q} \otimes \mathcal{Q}$ (using φ) has $a \in \mathcal{Q}_0$ sending $T \in \mathrm{End}_F\mathcal{Q}$) to $-$ (ad a)T (the composite, acting on the right, of F-endomorphisms of $\mathcal{Q}$). Accordingly, every left ideal in $\mathrm{End}_F(\mathcal{Q})$ is a $\mathcal{Q}_0$-submodule, and minimal left ideals are the subspaces of $\mathrm{End}_F(\mathcal{Q})$ mapping $\mathcal{Q}$ onto given 1-dimensional subspaces of $\mathcal{Q}$. In each such left ideal there is a 1-dimensional subspace annihilating $\mathcal{Q}_0$, and (ad a)T = 0 for all $\in \mathcal{Q}_0$ if T is in such a subspace. Thus $\mathcal{Q} \otimes \mathcal{Q}$ contains a trivial $\mathcal{Q}_0$-module of multiplicity at least 4. A $\mathcal{Q}_0$-complement in each minimal left ideal of $\mathrm{End}_F(\mathcal{Q})$ cannot annihilate $\mathcal{Q}_0$ and has dimension 3, so must be irreducible under our action of $\mathcal{Q}_0$. That is, it is isomorphic to $\mathcal{Q}_0$, we have $\mathcal{Q} \otimes \mathcal{Q} \approx 4F + 4\,\mathcal{Q}_0$, and the tensor square of the admissible (0,2)-admissible g_0-module contains two nonisomorphic (0,4)-admissible g_0-modules, both absolutely irreducible. That corresponding to F is seen from the splitting in §2 to be associated with W_2, giving a g-module of dimension 1539. That corresponding to $\mathcal{Q}_0$ is given by restricting the h-weight $2\pi_1$. In particular, π_2 is redundant from the point of view of our constructions.

For $(m_1,m_2) = (1,2)$, it is seen as in other cases that there is a unique admissible g_0-module, namely the subspace of an irreducible g-submodule of the g_K-module W_7 where H_1,H_2 have the given eigenvalues (1,2). In the labeling of §2, the $[g_0g_0]_K$-action on the subspace of W_7 corresponding to this pair of eigenvalues is irreducible with highest weight ω_4. If we fix a splitting basis $v_{-3},\ldots,v_0,\ldots,v_3$ for V_K and extend the action of $C\ell^+(V)$ on M_V to one of $C\ell(V)$ such that $z = -\sum_{J\subseteq\{1,2,3\}} v_{-J}v_0v_J$ fixes M_V, then $(M_V)_K$ is a direct sum of two copies of the irreducible $[g_0g_0]_K$-module of highest weight ω_4. Now the assertion for this case follows by the "completeness test" as in other cases. The corresponding irreducible g-module has dimension $2 \cdot \dim W_7 = 1824$.

(The dimensions for the fundamental g-modules first discussed are: (0,2): 112; (2,0): 133 (adjoint module); (2,2): 8645. The first splits over K into two copies of W_1, while the second and third are absolutely irreducible.)

For $(m_1,m_2) = (1,4)$, the only dominant integral functions on h with this restriction to t_0 are $\pi_1 + \pi_7$ and π_3. The tensor product of the g_0-modules corresponding to π_1 and to π_7, as in the statement of the proposition, is irreducible and decomposes over K into four absolutely irreducible isomorphic modules, each of them isomorphic to the subspace of the g_K-module of highest weight $\pi_1 + \pi_7$ where t_0 has the weight (1,4).

That M_V, with the new action, has the property that $(M_V)_K$, as $(g_0)_K$-module, is the direct sum of two copies of the subspace of W_3 corresponding to our weight is seen as for $(m_1,m_2) = (1,2)$, here with ω_1 replacing ω_4. The assertions for $(m_1,m_2) = (1,4)$ follow. The case $(m_1,m_2) = (0,0)$ is trivial as usual. The new irreducible g-module of highest t_0-weight (1,4) has dimension $2\cdot\dim W_3 = 55{,}328$.

The only "fundamental" g-modules still to consider are those occurring in the g_K-module W_4, where the space corresponding to $(m_1,m_2) = (2,4)$ is an irreducible $[g_0g_0]_K$-module of highest weight ω_2, and accordingly identifies with $\Lambda^2(V_0)_K$, the adjoint module of $[V_0V_0]_K$. Now $\Lambda^2(V_0)$ is in the tensor square of V_0 as $[g_0g_0]$-module, but the t_0-weight of the tensor square of the module V_0 corresponding to W_5 above is (4,4), not (2,4) as required. Instead, we consider the tensor square of the module M_V corresponding to W_7. In this M_V, the t_0-weight is (1,2), so that of the tensor square has the correct value (2,4).

As $[g_0g_0]_K$-module, $(M_V)_K \approx M_4 \oplus M_4$, where M_4 is the absolutely irreducible half-spin module of highest weight ω_4, here for $[V_0V_0]_K$, and $M_4 \otimes_K M_4 \approx M_{2\omega_4} + M_{\omega_2} + M_0$, the subscripts denoting highest weights. Here $M_{\omega_2} \approx \Lambda^2(V_0)_K$, so the absolutely irreducible g_0-module $\Lambda^2(V_0)$, with appropriate t_0-weight, has the property that $\Lambda^2(V_0)_K$ is an irreducible constituent of $(M_V \otimes M_V)_K$. It follows from the "completeness test" of Chapter 2, that $\Lambda^2(V_0)$ is a g_0-submodule of $M_V \otimes M_V$, and the proof of Proposition $E_{7,2}$ is complete.

$E_{8,2}$: Here both the center of $C\ell(V)$, where V has dimension 11, $(v,v) \neq 1$ for all $v \in V$, and the even Clifford algebra $C\ell^+(V)$ are split. Corresponding to $W_1,\ldots,W_8$ for g_K, we have the following pairs (m_1,m_2) of values for $\pi_i(H_1)$, $\pi_i(H_2)$, respectively: (2,0), (2,2), (2,4), (2,6), (2,8), (1,6), (0,4), (1,4). The representations of $[g_0g_0]_K$ on the corresponding t_0-weight subspaces have highest weights, respectively: 0, ω_1, ω_2, ω_3, ω_4, ω_5, 0, ω_6. Again we assume the "v_0" of our splitting to be so chosen that our fixed z, with $z^2 = 1$, has $z = -\sum_{J\subseteq\{1,2,3\}} v_{-J}v_0v_J$, and that our (absolutely) irreducible $C\ell^+(V)$-module M_V is fixed by z. Here $g_0 = t_0 + [V_0V_0]$.

Proposition $E_{8,2}$: For $(m_1,m_2) = (2,0)$, there is one admissible g_0-module, namely F, annihilated by $[V_0V_0]$ and with the prescribed action of t_0. In fact, for $0 \le k \le 3$, there is one admissible g_0-module for $(m_1,m_2) = (2,2k)$, namely $\Lambda^k(V_0)$ with the standard action of $[V_0V_0]$ and with the prescribed action of t_0.

For $(m_1,m_2) = (0,4)$, there is one admissible g_0-module, namely F, now with this action of t_0.

For $(m_1,m_2) = (1,4)$, there is one admissible g_0-module, namely M_V, with this action of t_0 and with

$$w(x + u_0 \otimes v) = wx + wv \quad \text{for } w \in M_V,\ x \in [VV],\ v \in V.$$

For $(m_1,m_2) = (1,6)$, there is one admissible g_0-module, again M_V with this action of t_0 and with

$$w(x + u_0 \otimes v) = wx - wv.$$

The only admissible g_0-module for $(m_1,m_2) = (0,0)$ is the trivial one.

All other admissible g_0-modules are submodules of suitable tensor products of those above.

Proof. Except for the last assertion, all the statements of the proposition are proved as in earlier cases, the situation here being simpler in that all the displayed admissible g_0-modules are absolutely irreducible. For the last assertion, in view of the fact that the listed cases yield (absolutely) irreducible g-modules which are forms of the $\mathfrak{g}_K$-modules W_i, $i \neq 5$, and of the trivial $\mathfrak{g}_K$-module, it is enough to show that $\Lambda^4(V_0)$, an F-form of the $[\mathfrak{g}_0\mathfrak{g}_0]_K$-submodule of W_5 where t_0 acts by $(m_1,m_2) = (2,8)$, and where the highest $[\mathfrak{g}_0\mathfrak{g}_0]_K$-weight is ω_4, is a $[V_0V_0]$-submodule of $M_V \otimes M_V$, where M_V is as for $(m_1,m_2) = (1,4)$.

Here $(M_V)_K \simeq M_{\omega_6}$, the irreducible half-spin module for $[V_0V_0]_K$ of highest weight ω_6, and if w_0 is a highest weight-vector for M_{ω_6}, then $w_0e_{-\gamma_6} \otimes w_0 - w_0 \otimes w_0e_{-\gamma_6}$ is a highest weight-vector for a submodule of $M_{\omega_6} \otimes M_{\omega_6}$ isomorphic to M_{ω_4}. Now it follows as in the proof of Proposition $E_{7,2}$ that $\Lambda^4(V_0)$ is a $[V_0V_0]$-submodule of $M_V \otimes M_V$. (One may also decompose this tensor product by the techniques of §5.3.) This completes the proof.

The dimensions of the fundamental irreducible $\mathfrak{g}$-modules for these highest t_0-weights are easy. All fundamental representations of $\mathfrak{g}_K$ have F-forms, and the dimensions are as in the split case (for the values, see, e.g. [T4], p. 50 - this work may be used as a reference for the other dimensions given above).

§6. Bizarre Identities.

Having interpreted the $\mathfrak{g}_0$-module that is the highest t_0-weight space of an irreducible $\mathfrak{g}$-module, as in §§4, 5 above, we can interpret the meaning of the annihilation of that module by certain elements of the universal enveloping algebra $U(\mathfrak{g}_0)$, as discussed in Chapter 2.

For example, consider the case F_4, with $\pi(H_1) = 4$. Here the elements of $\mathfrak{g}_{\alpha_1}$ are the elements of $m^+ \otimes M_V$, $\mathfrak{g}_{-\alpha_1} = m^+u_{-1} \otimes M_V$ (cf. §1), and the projection on $U(\mathfrak{g}_0)$ of each

$$f_1 \cdots f_5 e_1 \cdots e_5, \tag{1}$$

$f_j \in \mathfrak{g}_{-\alpha_1}$, $e_j \in \mathfrak{g}_{\alpha_1}$, must annihilate the module F for $\mathfrak{g}_0$ of a) of Proposition F_4, as well as the module V_0 of b) of that Proposition. Now the projection of (1) on $U(\mathfrak{g}_0)$ is the same as that of

$$\sum_{i=1}^{5} f_1 \cdots [f_i e_1] \cdots f_5 e_2 \cdots e_5,$$

and this element of $U(\mathfrak{g})$ is equal to

$$\sum_{i=1}^{5} [f_i e_1]\, f_1 \cdots \hat{f}_i \cdots f_5 e_2 \cdots e_5 \tag{2}$$

$$+ \sum_{j<i} f_1 \cdots [f_j[f_i e_1]] \cdots \hat{f}_i \cdots f_5 e_2 \cdots e_5,$$

with the factor $[f_j[f_i e_1]]$ coming in the j-th position.

From calculations in M_U, and from the definition of the bracket of two elements of $M_U \otimes M_V$, we have for $x, y \in M_V$, and a normalization such that $q_1(\overset{+}{m}, \overset{+}{m}u_{-1}) = 1$, $\overset{+}{m}u_0 = \overset{+}{m}$,

$$[\overset{+}{m}u_{-1} \otimes x, \overset{+}{m} \otimes y] = - s(x,y) - u_0 \otimes p(x,y) + \frac{1}{2} q(x,y)[u_{-1}u_1] \tag{3}$$

$$= q(x,y)\, H_1 - s(x,y) - u_0 \otimes p(x,y).$$

For $w \in M_V$,

$$[\overset{+}{m}u_{-1} \otimes w\, [\overset{+}{m}u_{-1} \otimes x, \overset{+}{m} \otimes y]]$$

$$= -q(x,y)\, \overset{+}{m}u_{-1} \otimes w - \overset{+}{m}u_{-1} \otimes ws(x,y)$$

$$+ \overset{+}{m}u_{-1} \otimes w\, p(x,y)$$

$$= \overset{+}{m}u_{-1} \otimes [w,x,y],$$

where $[w,x,y] = wp(x,y) - q(x,y)w + ws(x,y)$.

The effect of (3) on $\xi \in F$, the module of a) of Proposition F_4, is to send ξ to $4q(x,y)\xi$, while its effect on $\xi e_0 + v$, $\xi \in F$, $v \in V$, in the module V_0 of b) of Proposition F_4, sends $\xi e_0 + v$ to

$$(4q(x,y)\ \ \xi + 2(v,p(x,y)))\ e_0 \tag{4}$$

$$+ 4q(x,y)v - [v,s(x,y)] + 2\ \xi\ p(x,y).$$

From (2), if $\varphi_t(x_1,\ldots,x_t;\ y_1,\ldots,y_t)$ denotes the mapping of our (right) $\mathfrak{g}_0$-module given by the action of the projection on $\mathcal{U}(\mathfrak{g}_0)$ of the element (1), where $f_i = \overset{+}{m}u_{-1} \otimes x_i$, $e_i = \overset{+}{m} \otimes y_i$, we have that if

$$\varphi_t(x_1,\ldots,x_t;y_1,\ldots,y_t) = \sum_{i=1}^{t} \varphi_1(x_i;y_1)\ \varphi_{t-1}(x_1,\ldots,\hat{x}_i,\ldots,x_t;y_2,\ldots,y_t) \tag{5}$$

$$+ \sum_{j<i} \varphi_{t-1}(x_1,\ldots,[x_j,x_i,y_1],\ldots,\hat{x}_i,\ldots,x_t;y_2,\ldots,y_t),$$

then $\varphi_5 = 0$.

In the case (a), the meaning of (5) is as follows: Let $\varphi(x,y) \in F$ be defined as $4q(x,y)$, and define $\varphi_t(x_1,\ldots,x_t;y_1,\ldots,y_t) \in F$ inductively by (5), with $\varphi_1(x_1;y_1) = \varphi(x_1,y_1)$. Then $\varphi_5 = 0$. The result is a certain 10-linear identity relating the compositions q,p,s on M_V.

The case b) of Proposition F_4 has, as in (3), for $\varphi(x,y)$ the endomorphism

$$4q(x,y)I - s(x,y) - \frac{1}{2}\ [e_0,p(x,y)]$$

of $V_0 = Fe_0 + V$. Here all $\varphi_t(x_1,\ldots,x_t;y_1,\ldots,y_t)$, defined inductively as before, are in $\mathrm{End}_F(V_0)$ (acting on the right of V_0), and again we have the 10-linear identity $\varphi_5 = 0$.

The case $m_1 = 6$ for this Proposition yields for $\varphi(x,y)$ the F-endomorphism

$$6q(x,y)1 - s(x,y) - p(x,y) \text{ of } M_V, \tag{6}$$

an irreducible right module for $C\ell(V)$, which algebra contains all the terms of (6); the action of $\varphi(x,y)$ is just the right-operation by (6). Then φ_t is defined by the same inductive process as before, having values in $C\ell(V) - \mathrm{End}_F(M_V)$, and the resulting identity is the 14-linear one $\varphi_7 = 0$.

One has analogous identities for the other fundamental modules. Some samples when $\dim U = 5$ will be considered below. When $\mathfrak{t}_0$ is a maximal split toral subalgebra of $\mathfrak{g}$, the identities of the form $\varphi_{m+1} \equiv 0$ associated with an irreducible finite-dimensional $\mathfrak{g}_0$-module X, where $\varphi_1(x;y)$ is the transformation of X giving the action of $[m^+u_{-1} \otimes x,\ m^+ \otimes y]$ (when $\dim U = 3$; there will be such conditions for <u>both</u> α_1 and α_2 when $\dim U = 5$), are the essential ones in assuring that X is the highest $\mathfrak{t}_0$-weight module of an irreducible finite-dimensional $\mathfrak{g}$-module of highest $\mathfrak{t}_0$-weight λ. (Here $\lambda(H_1) = m$.)

When V is split, the highest $\mathfrak{t}_0$-weight-space in any of our fundamental modules is annihilated by all

$$f_1 \ldots f_{m_i+1},$$

where $\lambda(H_i) = m_i$ and $f_j \in \mathfrak{g}_{-\alpha_j}$ as above. The corresponding identities are derived as before.

We offer some further illustrations with $\dim U = 5$. First let V be a 5-dimensional real vector space with a nondegenerate negative definite symmetric bilinear form (v_1,v_2). Then $C\ell(V) \approx \mathbb{H} \otimes M_2(\mathbb{R}) \otimes \mathbb{C} \approx M_4(\mathbb{C})$ (see, <u>e.g.</u>, [J5], Lemma 5, p. 233), where $\mathbb{H}$ is the usual Hamiltonian quaternions. Thus, over $\mathbb{R}$, V satisfies the conditions of Proposition $E_{6,2}$, with $K = \mathbb{C}$, and V_0 is a negative-definite 6-dimensional space. We may take $z = i \in \mathbb{C}$, $i^2 = -1$.

In the case $(m_1,m_2) = (0,2)$, where $\mathfrak{g}_{\alpha_1} = \mathbb{R}\,[u_1u_0] + u_1 \otimes V$, $\mathfrak{g}_{-\alpha_1} = \mathbb{R}\,[u_{-1}u_0] + u_{-1} \otimes V$, we see that the irreducible admissible module $\mathbb{C}$ is annihilated by $[\mathfrak{g}_{-\alpha_1}\ \mathfrak{g}_{\alpha_1}]$, and that $[m^+u_{-2} \otimes w,\ m^+u_{-1} \otimes w'] \in [\mathfrak{g}_{-\alpha_2}\ \mathfrak{g}_{\alpha_2}]$ acts as does $Q(w,w')\ s_2(m^+u_{-2},m^+u_{-1}) + 3h^-(w,w')q_2(m^+u_{-2},m^+u_{-1}) = -Q(w,w')H_2 - 3h^-(w,w')$, which acts on $\mathbb{C}$ as multiplication by $-2(Q(w,w') + h^-(w,w')) = -2h(w,w')$, where $h{:}M_V \times M_V \to \mathbb{C}$ is the skew-hermitian form of §5.6, 3). Here we have

$$[m^+u_{-2} \otimes w[m^+u_{-2} \otimes x,\ m^+u_{-1} \otimes y]]$$

$$= m^+u_{-2} \otimes (2Q(x,y)w + 3h^-(x,y)w + wP(x,y)$$

$$- wS(x,y)),$$

so that $[w,x,y] =$

$$2Q(x,y)w + 3h^-(x,y)w + wP(x,y) - wS(x,y).$$

Thus with φ_t defined recursively as a 2t-linear function on M_V with values in $\mathbb{C}$, starting with $\varphi_1(x_1:y_1) = -2h(x_1,y_1)$, by the recursion (5), we have $\varphi_3 \equiv 0$, a 6-linear identity (over $\mathbb{R}$) on $M_V = \mathbb{C}^4$, with compositions as in §5.1.

Likewise one obtains identities corresponding to the other modules of Proposition $E_{6,2}$. Those resulting from the annihilation of the module by elements of the form $f_1 \cdots f_{m_1+1} e_1 \cdots e_{m_1+1}$, $e_i \in \mathfrak{g}_{\alpha_1}$, $f_i \in \mathfrak{g}_{-\alpha_1}$, fall into a class already considered in [Se3]. The others correspond to identities of the form $\varphi_1 = 0$ or $\varphi_3 = 0$, where φ_t is real 2t-linear from M_V to the endomorphisms of F, M_V or V_0, [w,x,y] is as above, and $\varphi_1(x,y)$ is the action on the module of $[m^+ u_{-2} \otimes x, m^+ u_{-1} \otimes y] \in \mathfrak{g}_0$.

Proposition $E_{8,2}$ provides identities of the form $\varphi_1 = 0$, $\varphi_3 = 0$, $\varphi_5 = 0$, $\varphi_7 = 0$. For instance, the case $(m_1,m_2) = (0,4)$ yields a 10-linear identity involving the triple product [w,x,y] on M_V and the bilinear map $Q: M_V \times M_V \to F$. The case $(m_1,m_2) = (1,4)$ yields a 10-linear identity for a bilinear map $M_V \times M_V \to \mathrm{End}\, M_V$, and the case $(m_1,m_2) = (1,6)$ yields a 14-linear identity for a second bilinear map $M_V \times M_V \to \mathrm{End}\, M_V$. In view of their complexity, it seems highly unlikely that these identities could have been brought to light by more direct means. It remains to be seen whether they will have any significance beyond their representation-theoretical interpretation.

CHAPTER 7. NON-REDUCED EXCEPTIONAL ALGEBRAS WITH A ONE-DIMENSIONAL ROOT SPACE

The final collection of isotropic central simple Lie algebras consists of algebras where the system of roots relative to a maximal F-split torus is of the form $\{\pm\alpha, \pm 2\alpha\}$, the root-spaces belonging to $\pm 2\alpha$ having dimension one. In [Sel], a coordinatization of these was given in terms of Freudenthal triple systems, following work of Freudenthal [Fr], Springer [Sp], Faulkner [Fa], Faulkner and Ferrar [FF] and Meyberg [Me]. Other versions have been offered by Kantor [Ka] and by Allison [A4] (see also Brown [Br]). The setting in these cases is that of a simple Lie algebra L, with a split 3-dimensional subalgebra $\mathfrak{s}$, such that the irreducible constituents of the adjoint representation of $\mathfrak{s}$ on L have dimensions 1, 3 or 5. The restriction above implies that the 5-dimensional representation occurs with multiplicity one. It also requires that a splitting Cartan subalgebra of $\mathfrak{s}$ be a maximal split toral subalgebra of L.

Implications of the last condition for the coordinatizing system seem only to have been given serious attention in [Sel] and in the works of Allison [A1], [A4], [A5]. The author finds the approach of Allison more satisfying than the one based on Freudenthal triple systems. The version offered here gains something in generality, because it may yield anisotropic Lie algebras when $\mathfrak{s}$ is not taken to be split.

§1. Conventions on 3-Dimensional Algebras.

Let $\mathfrak{s}$ be a 3-dimensional central simple Lie algebra over F. There is a 3-dimensional vector space V over F with a non-degenerate symmetric bilinear form (u,v), such that $\mathfrak{s}$ may be viewed as the Lie algebra of F-endomorphisms X of V (acting on the right) with

$$(uX,v) + (u, vX) = 0$$

for all u,v. (For instance, take V to be the underlying space of $\mathfrak{s}$, the form to be the Killing form and the action to be the right adjoint action of $\mathfrak{s}$ on V.) The algebra $\mathfrak{s}$ is split or anisotropic according as the Witt index of the form is one or zero. In the former case, we may multiply the form by a scalar as necessary to make its (normalized) discriminant a square; that is, we may assume V has a basis v_{-1}, v_0, v_1 with $(v_i,v_j) = \delta_{i,-j}$.

The (right) adjoint representation of $\mathfrak{s}$ on $\mathrm{End}_F(V)$ breaks up into three irreducible submodules:

i) the one-dimensional module $F\cdot I_V = m_1$;

ii) $\mathfrak{s}$ itself, the adjoint module m_3;

iii) the 5-dimensional module m_5 of those $T \in \mathrm{End}_F(V)$ such that $\mathrm{Tr}(T) = 0$ and $(uT,v) = (u,vT)$ for all u,v.

Each of these is absolutely irreducible. In the split case, their tensor products are classically known to decompose as follows:

$$m_1 \otimes m_i \cong m_i\,, \quad i = 1,3,5;$$

$$m_3 \otimes m_3 \cong \overset{-}{m_3} \oplus (\overbrace{m_1 \oplus m_5}^{+});$$

$$m_3 \otimes m_5 \cong m_3 \oplus m_5 \oplus m_7:$$

$$m_5 \otimes m_5 \cong \overbrace{m_3 \oplus m_7}^{-} \oplus (\overbrace{m_1 \oplus m_5 \oplus m_9}^{+}).$$

Here m_i denotes the unique irreducible $\mathfrak{s}$-module of dimension i, again absolutely irreducible. Where the symbols $\pm$ occur above summands in the tensor squares $m_i \otimes m_i$, they indicate that those summands make up the symmetric tensors $(m_i \otimes m_i)^+$ or the skew-symmetric tensors $(m_i \otimes m_i)^-$. Thus, for example, $(m_5 \otimes m_5)^- \cong m_3 \otimes m_7$. Because all summands in the split cases have different dimensions, it follows (for example by the action of the Galois group of a finite splitting extension) that the same decompositions hold when $\mathfrak{s}$ is anisotropic.

Likewise, we have

$$m_3 \otimes m_7 \cong m_5 \oplus m_7 \oplus m_9;$$

$$m_3 \otimes m_9 \cong m_7 \oplus m_9 \oplus m_{11};$$

$$m_5 \otimes m_7 \cong m_3 \oplus m_5 \oplus m_7 \oplus m_9 \oplus m_{11};$$

$$m_5 \otimes m_9 \cong m_5 \oplus m_7 \oplus m_9 \oplus m_{11} \oplus m_{13};$$

all summands being absolutely irreducible.

Let a,b,c be odd positive integers, $\varepsilon = \pm$. We use (a,b,c), (a,a,ε,b), and (a,a,b,ε,c) as respective abbreviations for the spaces $\mathrm{Hom}_{\mathfrak{s}}(m_a \otimes m_b, m_c)$, $\mathrm{Hom}_{\mathfrak{s}}((m_a \otimes m_a)^{\varepsilon}, m_b)$, $\mathrm{Hom}\,((m_a \otimes m_a)^{\varepsilon} \otimes m_b, m_c)$. Then the observations above yield the following table of dimensions:

(3,3,3,−,3):1	(3,3,3,−,5):1
(3,3,3,+,3):2	(3,3,3,+,5):1
(3,3,5,−,3):1	(3,3,5,−,5):1
(3,3,5,+,3):1	(3,3,5,+,5):2
(5,5,3,−,3):1	(5,5,3,−,5):2
(5,5,3,+,3):2	(5,5,3,+,5):1
(5,5,5,−,3):2	(5,5,5,−,5):2

(5,5,5,+,3):1 (5,5,5,+,5):3.

Write X,Y,Z for elements of $\mathfrak{s} = m_3$, regarded as endomorphisms of V, and likewise R,S,T for elements of m_5. Thus, for instance, $[S[XY]] \in m_5$, and there is an $\mathfrak{s}$-homomorphism of $m_3 \otimes m_3 \otimes m_5$ to m_5 sending $X \otimes Y \otimes S$ to [S[XY]] for all S,X,Y. The kernel of this $\mathfrak{s}$-morphism contains $(m_3 \otimes m_3)^+ \otimes m_5$, so the map may be identified with an element of the space (3,3,5,-,5). All this is abbreviated in the notation "[S[XY]] $\in$ (3,3,5,-,5)". The abuse of notation will be abandoned after having served its immediate purpose, namely to give a concise way of writing bases and relations in the various Hom-spaces, as listed below:

(3,3,3,-,1): Basis Tr([XY]Z);
relations Tr((XZ)Y) - Tr((YZ)X) = -Tr([XY]Z).

(3,3,3,+,1) = {0}.
(3,3,5,-,1) = {0}.
(3,3,5,+,1): Basis Tr((XY + YX)S);
relations Tr((XS)Y) + Tr((YS)X) = Tr((XY + YX)S).

(5,5,3,-,1): Basis Tr([ST]X);
relations Tr((SX)T) - Tr((TX)S) = -Tr([ST]X).

(5,5,3,+,1) = {0}.
(5,5,5,-,1) = {0}.
(5,5,5,+,1): Basis Tr((RS + SR)T);
relations Tr((RT)S) + Tr((ST)R) = Tr((RS + SR)T).

(3,3,3,-,3): Basis [[XY]Z];
relations [[XZ]Y] - [[YZ]X] = [[XY]Z];
(Tr(XZ))Y - (Tr(YZ))X = -2[[XY]Z];
$X\circ(Y\circ Z) - Y\circ(X\circ Z) = -\frac{5}{3}[[XY]Z]$,
where $A\circ B = AB + BA - \frac{2}{3}\operatorname{Tr}(AB)I$.

(3,3,3,-,5): Basis $[XY]\circ Z$;
relations $[XZ]\circ Y - [YZ]\circ X = [XY]\circ Z$;
$[X,Y\circ Z] - [Y,X\circ Z] = 3[XY]\circ Z$.

(3,3,3,+,3): Basis (Tr(XY))Z, (Tr(XZ))Y + (Tr(YZ))X;
relations: $(XY+YX)Z + Z(XY+YX) - \frac{4}{3}\operatorname{Tr}(XY)Z = (X\circ Y)\circ Z =$
$= -\frac{1}{3}\operatorname{Tr}(XY)Z + \frac{1}{2}(\operatorname{Tr}(XZ)Y + \operatorname{Tr}(YZ)X)$;
[[XZ]Y] + [[YZ]X] =
$= -\operatorname{Tr}(XY)Z + \frac{1}{2}(\operatorname{Tr}(XZ)Y + \operatorname{Tr}(YZ)X)$;
$X\circ(Y\circ Z) + Y\circ(X\circ Z) =$
$= \operatorname{Tr}(XY)Z + \frac{1}{6}(\operatorname{Tr}(XZ)Y + \operatorname{Tr}(YZ)X)$.

(3,3,3,+,5): Basis $[X\circ Y,Z]$;
relations $[XZ]\circ Y + [YZ]\circ X = [X\circ Y,Z]$;

$$[X, Y\circ Z] + [Y, X\circ Z] = [X\circ Y, Z].$$

(3,3,5,-,3): Basis $[XY]\circ S$;

relations $X\circ[YS] - Y\circ[XS] = 3[XY]\circ S$;

$[Y\circ S, X] - [X\circ S, Y] = [XY]\circ S$.

(3,3,5,-,5): Basis $[[XY]S]$;

relations $[X[YS]] - [Y[XS]] = [[XY]S]$;

$X\circ(Y\circ S) - Y\circ(X\circ S) = [[XY]S]$.

(3,3,5,+,3): Basis $[X\circ Y, S]$;

relations $X\circ[YS] + Y\circ[XS] = [X\circ Y, S]$;

$[X, Y\circ S] + [Y, X\circ S] = [X\circ Y, S]$.

(3,3,5,+,5): Basis $\mathrm{Tr}(XY)S$, $(X\circ Y)\circ S$;

relations $[X[YS]] + [Y[XS]] = 2\,\mathrm{Tr}(XY)S + 3(X\circ Y)\circ S$;

$X\circ(Y\circ S) + Y\circ(X\circ S) = \frac{2}{3}\,\mathrm{Tr}(XY)S - (X\circ Y)\circ S$.

(5,5,3,-3): Basis $[[ST]X]$;

relations $[[XS]T] - [[XT]S] = -[[ST]X]$;

$(X\circ S)\circ T - (X\circ T)\circ S = -[[ST]X]$.

(5,5,3,-,5): Basis $[ST]\circ X$, $[S\circ X, T] - [T\circ X, S]$;

relations $[XS]\circ T - [XT]\circ S = 2[ST]\circ X - ([S\circ X, T] - [T\circ X, S])$.

(5,5,3,+,3): Basis $\mathrm{Tr}(ST)X$, $(S\circ T)\circ X$;

relations $[[XS]T] + [[XT]S] = 2\,\mathrm{Tr}(ST)X + 3(S\circ T)\circ X$;

$(X\circ S)\circ T + (X\circ T)\circ S = \frac{2}{3}\,\mathrm{Tr}(ST)X - (S\circ T)\circ X$.

(5,5,3,+,5): Basis $[S\circ T, X]$;

relations $S\circ[TX] + T\circ[SX] = [S\circ T, X]$;

$[X\circ S, T] + [X\circ T, S] = [S\circ T, X]$.

(5,5,5,-,3): Basis $[RS]\circ T$, $[R, S\circ T] - [S, R\circ T]$;

relations $[RT]\circ S - [ST]\circ R = [R, S\circ T] - [S, R\circ T]$.

(5,5,5,-,5): Basis $[[RS]T]$, $\mathrm{Tr}(RT)S - \mathrm{Tr}(ST)R$;

relations $[[RT]S] - [[ST]R] = [[RS]T]$;

$R\circ(S\circ T) - S\circ(R\circ T) = [[RS]T] + \frac{4}{3}(\mathrm{Tr}(RT)S - \mathrm{Tr}(ST)R)$.

(5,5,5,+,3): Basis $[R\circ S, T]$;

relations $[RT]\circ S + [ST]\circ R = [R\circ S, T]$;

$[R, S\circ T] + [S, R\circ T] = [R\circ S, T]$.

(5,5,5,+,5): Basis $\mathrm{Tr}(RS)T = B_1$, $\mathrm{Tr}(RT)S + \mathrm{Tr}(ST)R = B_2$

$(R\circ S)\circ T = B_3$;

relations $[[TR]S] + [[TS]R] = 2B_1 - 2B_2 + 3B_3$;

$(S\circ T)\circ R + (R\circ T)\circ S = \frac{2}{3}B_1 + \frac{2}{3}B_2 - B_3$.

The claims that the elements are bases and the relations may all be seen by evaluation, in the split case, at suitable "test-triples", as in §5.4. Most of the relations may be seen immediately from the symmetry of Tr(AB), the Jacobi identity, and the fact that each 3 by 3 matrix A annihilates the (characteristic) polynomial

$$p_A(\lambda) = \lambda^3 - Tr(A)\lambda^2 + \frac{1}{2}(Tr(A)^2 - Tr(A^2))\lambda$$
$$- \frac{1}{6}(Tr(A)^3 - 3Tr(A)Tr(A^2) + 2\,Tr(A^3)).$$

§2. Lie Algebras with the Same $\mathcal{s}$-Constituents as in End(V).

Let L be a simple Lie algebra such that:

1) L contains a subalgebra isomorphic to $\mathcal{s}$, also denoted by $\mathcal{s}$.
2) The adjoint representation of $\mathcal{s}$ on L has only irreducible constituents isomorphic to m_1, m_3, m_5.

Thus the $\mathcal{s}$-module structure of L is

$$L \approx \mathcal{D} \oplus (m_3 \otimes A) \oplus (m_5 \otimes B),$$

where $\mathcal{D}$, A, B are trivial $\mathcal{s}$-modules. As in the fundamental work of Tits [T1], and in numerous instances in [Sel], we identify L with the space on the right above and find that the bracket in L has the form:

For $X,Y \in m_3$, a and a' in A, $[X \otimes a, Y \otimes a'] = Tr(XY)d_1(a,a') + [XY] \otimes f_1(a,a') + (X\circ Y) \otimes f_2(a,a')$, where $d_1(a,a') \in \mathcal{D}$, $f_1(a,a') \in A$, $f_2(a,a') \in B$. Each of d_1, f_1, f_2 is bilinear, with f_1 symmetric and d_1, f_2 skew.

For $S \in m_5$, $b \in B$,

$$[X \otimes a, S \otimes b] = (X\circ S) \otimes g_1(a,b) + [XS] \otimes g_2(a,b),$$

where $g_1(a,b) \in A$, $g_2(a,b) \in B$ are bilinear.

For $T \in m_5$, $b' \in B$,

$$[S \otimes b, T \otimes b'] = Tr(ST)d_2(b,b') + [ST] \otimes h_1(b,b') + (S\circ T) \otimes h_2(b,b'),$$

where $d_2(b,b') \in \mathcal{D}$, $h_1(b,b') \in A$, $h_2(b,b') \in B$, all three pairings bilinear, with h_1 symmetric and the others skew.

$\mathcal{D}$ is the centralizer of $\mathcal{s}$ in L, so is a subalgebra, and each of A, B is a right $\mathcal{D}$-module, with action satisfying

$$[X \otimes a, d] = X \otimes ad, \quad [S \otimes b, d] = S \otimes bd$$

for $d \in \mathcal{D}$. All the f_i, g_i, h_i, d_i are $\mathcal{D}$-invariants.

The given injection of $\mathcal{s}$ into L maps $X \in \mathcal{s}$ to $X \otimes c$, where c is a fixed element of A. It follows that

$d_1(c,a) = 0$, $f_2(c,a) = 0$, $g_1(c,b) = 0$, $cd = 0$ for all $a \in A$, $b \in B$, $d \in \mathcal{D}$, and that $f_1(c,a) = a$, $g_2(c,b) = b$.

In terms of these bilinear compositions, we now list consequences of the Jacobi identity in L. These are obtained by expanding the Jacobi identity for typical triples of elements of the forms $X \otimes a$, $S \otimes b$, invoking the relations of §1 and collecting coefficients of the bases of §1 for the various Hom-spaces.

I. $a''d_1(a,a') = \frac{1}{2}(f_1(f_1(a'',a),a') - f_1(f_1(a'',a'),a))$
$+ \frac{1}{2}(g_1(a',f_2(a,a'')) - g_1(a,f_2(a',a''))) + \frac{1}{3}\, g_1(a'',f_2(a,a'))$.

II. $g_2(a'',f_2(a,a')) = \frac{1}{2}(g_2(a,f_2(a',a'')) - g_2(a',f_2(a,a'')))$
$+ \frac{1}{2}(f_2(a,f_1(a',a'')) - f_2(a', f_1(a,a'')))$.

III. $f_2(f_1(a,a'),a'') = \frac{3}{2}(g_2(a,f_2(a',a'')) + g_2(a',f_2(a,a'')))$
$+ \frac{1}{2}(f_2(f_1(a'',a),a') + f_2(f_1(a'',a'),a))$.

IV. $d_1(f_1(a,a'),a'') + d_1(f_1(a',a''),a) + d_1(f_1(a'',a),a') = 0$.

V. $g_1(f_1(a,a'),b) = \frac{3}{2}(g_1(a,g_2(a',b)) + g_1(a',g_2(a,b)))$
$- \frac{1}{2}(f_1(g_1(a,b),a') + f_1(g_1(a',b),a))$.

VI. $h_1(f_2(a,a'),b) = \frac{1}{2}(f_1(g_1(a',b),a) - f_1(g_1(a,b),a'))$
$+ \frac{1}{2}(g_1(a,g_2(a',b)) - g_1(a',g_2(a,b)))$.

VII. $bd_1(a,a') = g_2(a',g_2(a,b)) - g_2(a,g_2(a',b))$
$- \frac{1}{3}(f_2(g_1(a,b),a') - f_2(g_1(a',b),a))$.

VIII. $h_2(f_2(a,a'),b) = \frac{3}{2}(g_2(a,g_2(a',b)) - g_2(a',g_2(a,b)))$
$- \frac{1}{2}(f_2(g_1(a,b),a') - f_2(g_1(a',b),a))$

IX. $g_2(f_1(a,a'),b) = \frac{1}{2}(g_2(a,g_2(a',b)) + g_2(a',g_2(a,b)))$
$- \frac{1}{2}(f_2(g_1(a',b),a) + f_2(g_1(a,b),a'))$

X. $d_2(f_2(a,a'),b) = d_1(g_1(a,b),a') - d_1(g_1(a',b),a)$.

XI. $ad_2(b,b') = h_1(g_2(a,b),b') - h_1(g_2(a,b'),b)$
$+ \frac{1}{3}(g_1(g_1(a,b),b') - g_1(g_1(a,b'),b))$.

XII. $g_1(a,h_2(b,b')) = \frac{1}{2}(g_1(g_1(a,b'),b) - g_1(g_1(a,b),b'))$
$+ \frac{3}{2}(h_1(g_2(a,b),b') - h_1(g_2(a,b'),b))$.

XIII. $f_1(h_1(b,b'),a) = \frac{1}{2}(g_1(g_1(a,b),b') + (g_1(g_1(a,b'),b))$
$+ \frac{1}{2}(h_1(g_2(a,b),b') + h_1(g_2(a,b'),b))$.

XIV. $f_2(a,h_1(b,b')) = h_2(g_2(a,b),b') + h_2(g_2(a,b'),b)$

XV. $g_2(g_1(a,b),b') + g_2(g_1(a,b'),b)$

$$= h_2(g_2(a,b),b') + h_2(g_2(a,b'),b).$$

XVI. $g_2(a,h_2(b,b')) = \frac{1}{2}(g_2(g_1(a,b),b') - g_2(g_1(a,b'),b))$

$$+ \frac{1}{2}(h_2(g_2(a,b),b') - h_2(g_2(a,b'),b)).$$

XVII. $d_1(h_1(b,b'),a) + d_2(g_2(a,b),b') + d_2(g_2(a,b'),b) = 0.$

XVIII. $d_2(h_2(b,b'),b'') + d_2(h_2(b',b''),b) + d_2(h_2(b'',b),b') = 0.$

XIX. $g_1(h_1(b,b'),b'') = h_1(h_2(b',b''),b) + h_1(h_2(b,b''),b').$

XX. $b''d_2(b,b') = g_2(h_1(b'',b),b') - g_2(h_1(b'',b'),b)$

$$+ \frac{1}{3}(h_2(h_2(b'',b),b') - h_2(h_2(b'',b'),b)).$$

XXI. $h_2(h_2(b,b'),b'') = \frac{1}{2}(h_2(h_2(b'',b),b') - h_2(h_2(b'',b'),b))$

$$+ \frac{3}{2}(g_2(h_1(b'',b'),b) - g_2(h_1(b'',b),b')).$$

XXII. $g_2(h_1(b,b'),b'') = \frac{1}{2}(g_2(h_1(b'',b),b') + g_2(h_1(b'',b'),b))$

$$- \frac{1}{2}(h_2(h_2(b',b''),b) + h_2(h_2(b,b''),b').$$

The identities I-XXII, together with the derivation-properties of the mappings $d_1(a,a')$ and $d_2(b,b')$ and anticommutativity, are equivalent to the assertion that our product $\mathcal{D} + (m_3 \otimes A) + (m_5 \otimes B)$ defines a structure of Lie algebra on this space.

§3. Allison's Algebras.

On the space $A \oplus B$ one defines a product by

$$(a,b)(a',b') = (f_1(a,a') + g_1(a,b') - g_1(a',b) + h_1(b,b'), \qquad (1)$$
$$f_2(a,a') + g_2(a,b') + g_2(a',b) + h_2(b,b')),$$

and a conjugation $\overline{(a,b)} = (a,-b)$. Then $A \oplus B$ has the structure of (non-associative) F-algebra in which $(a,b) \to \overline{(a,b)}$ is an involutorial antiautomorphism and $(c,0)$ acts as unit element. The fixed elements of $(a,b) \to \overline{(a,b)}$ are evidently those of A and the skew elements are those of B.

With the definition (1) and involution as above, $A \oplus B$ is a "structurable algebra" in the sense of Allison [A3]. We shall use the term <u>Allison algebra</u> to refer to this structure(*), which is defined as follows:

(*) Some objections to this terminology have been raised by the attributee, on the grounds that it may obscure the importance of the work of I. L. Kantor [Ka]. The relation between the algebras of Kantor and those of Allison is made clear in [A3]. The author finds it somewhat easier to work in Allison's terms, but is quite ready to serve justice by using the term "Kantor algebra" for the structure studied by that author.

An Allison algebra over F is a (finite-dimensional) vector space X with bilinear product xy, unit element c, and involution $x \to \bar{x}$ such that, for all $x,y \in X$, and all $b \in X$ with $\bar{b} = -b$, one has

$$(xy)b - x(yb) = -((xb)y - x(by)) = (bx)y - b(xy). \tag{2}$$

In addition, one defines, for $x,y \in X$, a map $D_{x,y}:X \to X$ which is to send $z \in X$ to

$$zD_{x,y} = \frac{1}{3}[[x,y] + [\bar{x},\bar{y}],z] + [z,y,x] - [z,\bar{x},\bar{y}].$$

(Here $[x,y] = xy - yx$, $[x,y,z] = (xy)z - x(yz)$.) Then one further requires that for $a \in X$, $\bar{a} = a$, and if $\overline{a'} = a'$, $\overline{a''} = a''$,

$$D_{a,a^2} = 0, \quad \text{and} \tag{3}$$

$$[a,a',a''] - [a'',a,a'] = [a',a,a''] - [a'',a',a]. \tag{3'}$$

(The above is not exactly the definition as given by Allison, but it is proved by him to be an equivalent definition - see Theorem 3 of [A3].)

To see that our $X = A \oplus B$ is an Allison algebra, we must verify the identities (2), (3), (3'). By multilinearity it suffices to verify (2) when

(a) $x = b'$, $y = b'' \in B$;
(b) $x = a \in A$, $y = b' \in B$;
(c) $x = b' \in B$, $y = a \in A$;
(d) $x = a$, $y = a' \in A$.

In each case, the associators $[x,y,b]$, $[x,b,y]$, $[b,x,y]$ have components in A and B that may be computed from the definitions, and the alternative laws (2) are implied by certain of our identities I-XXII, as follows:

a) Components in A: XIX
   Components in B: XXII

b) Components in A: XII and XIII
   Components in B: XIV, XV, XVI

c) As in b) above.

d) Components in A: V, VI
   Components in B: VIII, IX.

(3') is the same as II.

To verify the identity (3), one first verifies that, as operators on X stabilizing both A and B,

$$D_{(a,b),(a',b')} = -2(d_1(a,a') + d_2(b,b')). \tag{4}$$

Once (4) is verified, (3) follows at once from IV.

Now $bD_{a,a'} = -2b\ d_1(a,a')$ is implied by VI, VII, VIII. That $a''D_{a,a'} = -2a''\ d_1(a,a')$ is implied by I and II, so we have $D_{a,a'} = -2d_1(a,a')$, from which IV already yields (3). The rest of (4) is implied as follows: That $D_{a,b} = 0 = D_{b,a}$ follows from (2). That $b''D_{b,b'} = -2b''d_2(b,b')$ follows from XIX, XX, XXI and XXII, and $a\ D_{b,b'} = -2a\ d_2(b,b')$ from XI, XII, XIII, XIV, XV, XVI.

Conversely, if X is an Allison algebra with involution $x \to \bar{x}$, unit element c, with A and B respectively the fixed and skew elements for the involution, we may use the relations above to define elements and operators f_i, g_i, h_i, d_i (i = 1,2): (i = 1,2): For example,

$$g_1(a,b) = \frac{1}{2}(ab - ba) \in A; \quad d_2(b,b') = -\frac{1}{2} D_{b,b'}, \text{ a derivation of } X$$

commuting with the involution, thus stabilizing A and B. Then our identities I-XXII may be derived from Allison's, as follows:

Allison's Lemma 6 shows that $D_{a,b} = 0 = D_{b,a}$, and his Corollary 9 shows that

$$D_{xy,z} + D_{yz,x} + D_{zx,y} = 0.$$

Applying these with $x = b$, $y = b'$, $z = b''$ yields $D_{h_2(b,b'),b''} + D_{h_2(b',b''),b} + D_{h_2(b'',b),b'} = 0$, which is equivalent to XVIII.

Inspection of the components in A in the "alternative" identities $[b,b',b''] + [b'',b',b] = 0$ and $[b,b',b''] + [b',b,b''] = 0$ yields XIX. The components in B in the latter relation yield XXII, and XXI is derived from XXII by skew-symmetrizing on b' and b''. Then XX follows from these and the definition of $D_{b,b'} = -2d_2(b,b')$. Thus XVIII-XXII follow from Allison's axioms.

Our relation XVII follows from Allison's

$$D_{ab,b'} + D_{bb',a} + D_{b'a,b} = 0.$$

Computing $a\ d_2(b,b') = -\frac{1}{2} a\ D_{b,b'}$ according to the formula of Allison, and symmetrizing on b,b' gives XIII. Considering the A-components in the relation $[a,b,b'] + [b,a,b'] = 0$ yields XII, and substitution in the formula for $aD_{b,b'}$ of Allison gives XI. The B-components of the alternative relation that gave XII, when skew-symmetrized on b, b', give XVI. Symmetrizing on b, b' and combining with the relation on B-components of

$$[a,b,b'] + [a,b',b] = 0$$

gives XV, which may then be used in the result of symmetrizing on b, b' at the beginning of this sentence to give XIV. Thus all our relations of index $\geq$ XI are established.

The relation X follows from $D_{aa',b} + D_{a'b,a} + D_{ba,a'} = 0$. Symmetrizing on a,a' in the A-component of

$[a,b,a'] + [b,a,a'] = 0$

yields V, and skew-symmetrizing gives VI. Symmetrizing on a,a' in the B-component of this relation gives IX, and skew-symmetrizing gives VIII. Substituting from VIII in the calculation of Allison's $bD_{a,a'}$ yields VII, so that only I-IV remain to be verified.

From $D_{aa',a''} + D_{a'a'',a} + D_{a''a,a'} = 0$ we have IV as before. Allison's identity (7) (our (3)'), is

$$[a,a',a''] - [a'',a,a'] = [a',a,a''] - [a'',a',a],$$

which we have seen is II. Symmetrizing II on a', a'' gives III. Finally, the definition of $a''D_{a,a'}$ by Allison yields I.

That each $D_{x,y}$ is a derivation of Allison's algebra X, commuting with the involution and defined in terms of the operations of the algebra, has as consequence the relation $[D_{x,y},D_{w,z}] = D_{xD_{w,z},y} + D_{x,yD_{w,z}}$ for all $x,y,w,z \in X$, and shows that, with our definitions, $g = \mathrm{Der}(X,-) \oplus (m_3 \otimes A) \oplus (m_5 \otimes B)$ is a Lie algebra. With $c \in A$ as the unit element of X, the map $X \to X \otimes c$ of $s = so(V)$ into g is a monomorphism of Lie algebras, relative to which the adjoint action on g of the subalgebra $so(V) \otimes c$ is equivalent to that of s, where m_3, m_5 are the canonical modules and $\mathrm{Der}(X, -)$, A, B are trivial s-modules. Thus the Lie algebras L under consideration in this section are those obtained from Allison's algebras by the construction above, with the further specification that if L is simple, the trivial s-summand D is spanned by all $d_1(a,a')$ and all $d_2(b,b')$, and is faithfully represented on X.

One sees easily that a proper ideal of X, stable under the involution, say $X_0 = A_0 + B_0$, yields a proper ideal

$$D_{X,X_0} + (m_3 \otimes A_0) + (m_5 \otimes B_0)$$

of L. Thus if L is simple, so is $(X, -)$, and the converse holds as well if

$$L = D_{X,X} + (m_3 \otimes A) + (m_5 \otimes B).$$

It follows as in [A4] (Proposition 8) that L is central simple (i.e., absolutely simple) if and only if $(X, -)$ is. Under the assumption of simplicity for $(X, -)$, $D_{X,X} = \mathrm{Der}(X,-)$ ([A4], Corollary 7.)

Now suppose the 3-dimensional algebra s is split, and let $v_{\pm 1}, v_0$ be a basis for V with $(v_i,v_j) = \delta_{i,-j}$. For $u,v \in V$, define $S_{u,v} \in so(V)$ by

$$w\, S_{u,v} = (w,u)v - (w,v)u,$$

and $T_{u,v} \in m_5$ by

$$wT_{u,v} = (w,u)v + (w,v)u - \frac{2}{3}(u,v)I.$$

Then $E = S_{v_0,v_1}$, $H = 2S_{v_{-1},v_1}$, $F = 2S_{v_0,v_{-1}}$ are a basis for $\mathfrak{so}(V)$ satisfying the canonical relations

$$[EH] = 2E, \ [EF] = H, \ [FH] = -2F.$$

The elements $E \otimes c, H \otimes c, F \otimes c \in L$ thus span a 3-dimensional split simple subalgebra, and the scalar multiples by elements of F (here L is assumed to be central simple) of the element $H \otimes c$ form a one-dimensional split toral subalgebra. We assume from now on that this one-dimensional split subalgebra is maximal among split toral subalgebras of L.

Then it follows (see [A5], Theorem 3.1) that $(X, -)$ is a division algebra, in the sense that, for every $x \neq 0$ in X there is an element $\hat{x} \in X$ such that for all $y \in X$,

$$(x\,\bar{\hat{x}})y + (y\,\bar{\hat{x}})x - (y\,\bar{x})\hat{x} = y,$$

or, equivalently,

$$(\hat{x}\,\bar{x})y + (y\,\bar{x})\hat{x} - (y\,\bar{\hat{x}})x = y.$$

Then $\hat{x}$ is uniquely determined and $\hat{\hat{x}} = x$.

Fix $0 \neq b_0 \in B$. Then from $D_{b_0,b_0} = 0$ and $[b_0,b_0] = 0$, we have

$$[T_{v_1,v_1} \otimes b_0, T_{v_{-1},v_{-1}} \otimes b_0] = [T_{v_1,v_1}, T_{v_{-1},v_{-1}}] \otimes b_0^2 = -2H \otimes b_0^2 .$$

If moreover $b_0^2 = -\frac{\mu}{4}c$, where $0 \neq \mu \in F$, then

$$[T_{v_1,v_1} \otimes b_0, T_{v_{-1},v_{-1}} \otimes \mu^{-1} b_0] = \frac{1}{2}(H \otimes c),$$

and $e = T_{v_1,v_1} \otimes b_0$, $h = \frac{1}{2}(H \otimes c)$, $f = T_{v_{-1},v_{-1}} \otimes \mu^{-1}b_0$ span a 3-dimensional split simple subalgebra with relations as for E, H, F before.

§4. Some Results of Allison.

Allison has shown ([A4], Theorem 10) that any simple Lie algebra over F containing an element $x \neq 0$ with ad x nilpotent contains a 3-dimensional split $\mathfrak{s}$ satisfying the conditions of §2, so resulting from the construction of the last section based on a simple Allison algebra. The eigenvalues of $\mathrm{ad}(H \otimes c)$, with notations as in §3, are then 0, ±2, ±4. When the simple Lie algebra is central simple, all cases where $F\cdot(H \otimes c)$ is not a maximal split toral subalgebra have been dealt with, for structure in [Se1], and for representations in [Se3] and in Chapters 4 and 6. Those cases where $F\cdot(H \otimes c)$ is a maximal toral subalgebra and where ad(H ⊗ c) has only three eigenvalues are treated in [Se1] and [Se3], and those where ad(H ⊗ c) has five eigenvalues, all of multiplicity greater than one, are seen from [Se3], Chapter V, to have been treated, as to representations, in the earlier chapters of this work. The remaining case, in the setting of our construction from Allison algebras, is that where $\mathcal{X}$ is a division algebra and $\mathcal{B}$ has dimension one.

Allison has shown that the Lie algebra $\mathfrak{g}$ constructed as in §2 from one of his algebras is central simple if and only if the Allison algebra is central simple [A4], and he has classified the central simple Allison algebras in [A3]. In [A4] he gives information about the associated Lie algebras $\mathfrak{g}$, listing there his central simple algebras as well. From the classification one sees that the central simple Allison algebras relevant to our remaining cases (dim. $\mathcal{B}$ = 1, dim. $\mathfrak{g}$ = 28,52,78,133 or 248) are in the classes (e), (f) of Theorem 11 of [A4].

The class (e) results as follows: One has a Jordan algebra $\mathcal{J}$ with unit 1 over F, with a linear ("trace') form $t: \mathcal{J} \to F$ such that $t(1) = 3$ and such that, for each $x \in \mathcal{J}$,

$$x^3 - t(x)x^2 + \frac{1}{2}(t(x)^2 - t(x^2))x - \frac{1}{6}(t(x)^3 - 3t(x)t(x^2) + 2t(x^3))1 = 0.$$

The symmetric bilinear form $t(x\cdot y)$ on $\mathcal{J}$ satisfies $t((x\cdot y)\cdot z) = t(x\cdot(y\cdot z))$ for all x,y,z. Set

$$x \times y = x\cdot y - \frac{1}{2}t(x)y - \frac{1}{2}t(y)x + \frac{1}{2}(t(x)t(y) - t(x\cdot y))1,$$

so that $y \times x = x \times y$, $1 \times x = -\frac{1}{2}x + \frac{1}{2}t(x)1$.

Let $n(x) = \frac{1}{6}(t(x)^3 - 3t(x)t(x^2) + 2t(x^3))$, a cubic form on $\mathcal{J}$. By polarizing and normalizing as in [J5] (p. 360), we obtain a symmetric trilinear form (x,y,z) on $\mathcal{J}$ with $(x,x,x) = 3n(x)$, and with $(x,y,z) = t((x \times y)\cdot z) = t((y \times z)\cdot x)$. Let $\mathcal{X}$ be the space of 2 by 2 matrices

$$\begin{pmatrix} \alpha & a \\ b & \beta \end{pmatrix}, \text{ with } \alpha,\ \beta \in F \text{ and } a,b \in \mathcal{J}.$$

Fix $\theta \neq 0$ in F, and define a product in $\mathcal{X}$ by

$$\begin{pmatrix} \alpha & a \\ b & \beta \end{pmatrix} \begin{pmatrix} \gamma & c \\ d & \delta \end{pmatrix} = \begin{pmatrix} \alpha\gamma + \theta t(a\cdot d) & \alpha c + \delta a + 2\theta b \times d \\ \gamma b + \beta d + 2a \times c & \beta\delta + \theta t(b\cdot c) \end{pmatrix}.$$

(The discrepancies here with [A4] are only notational; the bilinear form resulting from n(x) in [A4] by "double directional differentiation of log n(x)" is $t(x\cdot y)$, while the element $x^{\#}$ of [A4] is our $x \times x$. Meanwhile Allison's "$x \times x$" would be $(2x)^{\#} - x^{\#} - x^{\#} = 2x^{\#}$.) The involution in $\mathcal{X}$ sends

$$\begin{pmatrix} \alpha & a \\ b & \beta \end{pmatrix} \text{ to } \begin{pmatrix} \beta & a \\ b & \alpha \end{pmatrix}.$$

The centralizer of $S_{v_{-1},v_1} \otimes \begin{pmatrix}1 & 0\\ 0 & 1\end{pmatrix}$ in the resulting Lie algebra $\mathfrak{g}$ contains

$z = (T_{v_{-1},v_1} - T_{v_0,v_0}) \otimes \begin{pmatrix}1 & 0\\ 0 & -1\end{pmatrix}$.

All derivations of X, as involutorial algebra, must annihilate $\begin{pmatrix}1 & 0\\ 0 & 1\end{pmatrix}$ and must map $\mathcal{B} = F\begin{pmatrix}1 & 0\\ 0 & -1\end{pmatrix}$ to itself. From $\begin{pmatrix}1 & 0\\ 0 & -1\end{pmatrix}^2 = \begin{pmatrix}1 & 0\\ 0 & 1\end{pmatrix}$, we see that all such derivations must annihilate $\mathcal{B}$, and thus that z centralizes $\mathrm{Der}(X,-)$ in $\mathfrak{g}$. All $S_{v_0,v_{\pm1}} \otimes \begin{pmatrix}0 & a\\ 0 & 0\end{pmatrix}$ and all $S_{v_0,v_{\pm1}} \otimes \begin{pmatrix}0 & 0\\ b & 0\end{pmatrix}$ are eigenvectors for ad z, with associated integral eigenvalues (±1). Both $T_{v_1,v_1} \otimes \mathcal{B}$ and $T_{v_{-1},v_{-1}} \otimes \mathcal{B}$ centralize z, and $S_{v_0,v_i} \otimes \begin{pmatrix}1 & 0\\ 0 & 1\end{pmatrix} \pm T_{v_0,v_i} \otimes \begin{pmatrix}1 & 0\\ 0 & -1\end{pmatrix}$, $i = \pm1$ belong to eigenvalue ±3 of ad z. It follows that <u>the case</u> (e) <u>cannot occur if</u> $t_0 = F\cdot(H \otimes c)$ <u>is to be a maximal split torus in</u> $\mathfrak{g}$.

In the case labeled (f) by Allison, J is as above, but is a Jordan algebra over a quadratic extension field E of F. Thus $t(x)$ takes on values in E. If $\alpha \to \bar{\alpha}$ denotes the nontrivial automorphism of E/F, there is a $\bar{\alpha}$-semilinear bijection $w: J \to J$ such that $t(a\cdot b^w) = \overline{t(b\cdot a^w)}$ for all $a,b \in J$, and a fixed non-zero element θ of E such that

$$n(a^w) = N_{E/F}(\theta)\overline{n(a)}$$

for all $a \in J$. The underlying vector space of X is $E \times J$, with F-bilinear product

$$(\alpha,a)(\beta,b) = (\alpha\beta + t(a\cdot b^w),\ \alpha b + \bar{\beta} a + \tfrac{2}{\theta} a^w \times b^w)$$

and involution $(\alpha,a) \to (\bar{\alpha},a)$. Thus $\mathcal{B} = F\zeta \times 0$, where $0 \neq \zeta \in E$, $\bar{\zeta} = -\zeta$.

[It will possibly be remarked that the context of the Jordan algebras J here appears to be less general than that considered by Allison. We indicate that such is not the case, insofar as our purposes are concerned, by appeal to work of McCrimmon [Mc1], [Mc2] (see also [Sp]). Allison deals with the Jordan algebra "of an admissible non-degenerate cubic form with basepoint". If N is the form and c the basepoint, one deduces that the underlying space carries a structure of Jordan algebra with unit element c, in which each x satisfies

$$x^3 - T(x)x^2 + S(x)x - N(x)c = 0. \tag{5}$$

where $S(x)$, $T(x)$ are quadratic resp. linear forms derived from N and c. There is a quadratic operation $x \to x^{\#}$ such that $x^{\#} = x^2 - T(x)x + S(x)c$ with $c^{\#} = c$ and $T(x^{\#}\cdot x) = 3N(x)$ for all x. It is axiomatic that $N(c) = 1$, so that $T(c) = 3$. One effect of the assumption of non-degeneracy is to make the Jordan algebra J <u>semi-simple</u>.

Writing "1" for "c" above, we first suppose that for each $x \in J$, the elements

x^2, x, 1 are linearly dependent. Then either $J = F1$, or J is a commutative associative separable F-algebra of dimension 2 (either $F \times F$ or a quadratic field extension K of F), or J is the Jordan algebra of a quadratic form in a vector space V of dimension at least 2. If $J = F1$, then $T(1) = 3$, $S(1) = T(1^{\#}) = 3$, so for $\xi \in F$, $T(\xi 1) = 3\xi$, $S(\xi 1) = 3\xi^2 = \frac{1}{2}(T(\xi 1)^2)$, $N(\xi 1) = \xi^3 = \frac{1}{6}(T(\xi 1)^3 - 3T(\xi 1)T((\xi 1)^2) + 2T((\xi 1)^3)$, and our conditions apply.

In the remaining cases, each $x \in J$ satisfies a relation

$$x^2 - \tau(x)x + \frac{1}{2}(\tau(x)^2 - \tau(x^2))1 = 0,$$

where τ is a linear form with $\tau(1) = 2$. In [Mc1] (pp. 506-7), these cases are shown only to arise if J has dimension 2, the case $F \times F$ or a quadratic field extension. In this case the central simple Lie algebra to be associated with the construction would have dimension 21, and its representation theory is therefore subsumed under our treatments, in [Se3] and in Chapters 3, 4, of algebras of type B_3 and C_3 respectively. Thus we assume J contains an element x such that 1, x, x^2 are linearly independent.

From the fact that the Jacobian at 1 of the polynomial mapping $x \to x^2$ of J is the invertible linear mapping $y \to 2y$ it follows that there are also $x \in J$ with 1, x^2, x^4 linearly independent. Multiplying (5) by x and using (5) to substitute for x^3 in the result gives

$$x^4 = (T(x)^2 - S(x))x^2 + (N(x) - S(x)T(x))x + T(x)N(x)1,$$

which we solve for $(N(x) - S(x)T(x))x$. Let $f(x) = N(x) - S(x)T(x)$. Then we can write $f(x)x^3$ and $f(x)x$ as linear combinations of x^4, x^2 and 1. Now write (5) as

$$x^3 = T(x)x^2 - S(x)x + N(x)1$$

and square, then multiply by $f(x)$ and substitute from the above for $f(x)x^3$ and $f(x)x$ to get

$$f(x)\ (x^6 - (T(x)^2 - 2S(x))x^4 + a(x)x^2 + b(x)1) = 0.$$

Because the polynomial f is not zero, we must have $T(x^2) = T(x)^2 - 2S(x)$ for x in a Zariski-dense set, so for all x, or $S(x) = \frac{1}{2}(T(x)^2 - T(x^2))$. Applying T to the relation (5) yields

$$3N(x) = T(x^3) - T(x)T(x^2) + S(x)T(x),$$

from which it now follows that

$$N(x) = \frac{1}{6}(T(x)^3 - 3T(x)T(x^2) + 2T(x^3)),$$

and thus that the setting is as we have originally described.]

From polarizing the relation $n(x^w) = N_{E/F}(\theta)\,\overline{n(x)}$ we obtain $(x^w,y^w,z^w) = N_{E/F}(\theta)\,\overline{(x,y,z)}$ for all $x,y,z \in J$, or $t((x^w \times y^w)^w \cdot z) = t((x^w \times y^w)^w \cdot z^w) = N_{E/F}(\theta)\,\overline{t((x \times y)\cdot z)}$, from which it follows that

$$(x^w \times y^w)^w = N_{E/F}(\theta)\, x \times y \tag{6}$$

for all $x,y \in J$. If $u = 1^w$, then $n(u) = N_{E/F}(\theta)$, u is invertible in J with $u^{-1} = n(u)^{-1}\, u \times u$, and $u^{-1^w} = n(u)^{-1}(1^w \times 1^w)^w = 1 \times 1 = 1$.

One further relation will be useful. The U-operator of Jacobson in a Jordan algebra is defined by setting

$$xU_y = 2y\cdot(y\cdot x) - y^2\cdot x.$$

Our relation is

$$w^2 = U_u, \tag{7}$$

where $u^{-1^w} = 1$, $1^w = u$ as above.

This may be seen by noting first that $(1 \times x^w)^w = (u^{-1^w} \times x^w)^w = N_{E/F}(\theta)\, u^{-1} \times x$ for all x, and that $n(u) = N_{E/F}(\theta)$, $u \times u = n(u)u^{-1}$, so that $(1 \times x^w)^w = (u \times u) \times x$. Now $1 \times x^w = -\frac{1}{2}x^w + \frac{1}{2}t(x^w)1 = -\frac{1}{2}x^w + \frac{1}{2}\overline{t(u\cdot x)}1$, so that $(1 \times x^w)^w = -\frac{1}{2}x^{w^2} + \frac{1}{2}t(u\cdot x)u$, or

$$x^{w^2} = t(u\cdot x)u - 2(u \times u) \times x. \tag{8}$$

Expanding $(u \times u) \times x$ according to the definition; using the associativity of the form $t(x\cdot y)$, and polarizing the fundamental cubic relation in J shows that the right hand side of (8) is equal to $x\,U_u$, as asserted.

§5. The Decomposable Case.

The presence of the generic cubic equation assures that the Jordan algebra J has degree at most three, in the sense of Jacobson. The analysis of the last section, eliminating commutative associative algebras of degree two and Jordan algebras of quadratic forms, shows that the degree of J cannot be two. Thus either $J = E$ is of degree one, or J is simple of degree 3, or $J \approx E \times E \times E$, or $J \approx E \times M$, where M is a simple Jordan algebra of degree 2. We consider here the situation when J decomposes (the last two cases indicated).

Here the unit element 1 of J is the sum of two proper orthogonal idempotents e, $1-e$. For each idempotent e, the generic cubic gives that $e(1-t(e) + \frac{1}{2}(t(e)^2 - t(e)))$ is a scalar multiple of 1, so that $t(e)^2 - 3t(e) + 2 = 0$. Replacing e by $1-e$ if necessary, we may assume $t(e) = 1$, $t(1-e) = 2$. Here e and $1-e$ are unit elements for their respective ideals, so that $y \in J\cdot e$ implies $y = y\cdot e$. Polarizing the generic cubic for the coefficient of λ when $x = \lambda y + e$, we find $y = t(y)e$, so that $J\cdot e$ has dimension one. Moreover, $e \times e = 0$, so $e^W \times e^W = 0$ as well, and $t(e\cdot e^W) = t(e^W\cdot e) = \beta \in F$. Thus $e^W = \beta e + z$, $z \in J\cdot(1-e) = \{x \in J | x\cdot e = 0\}$.

For $\lambda \in E$, $x \in J\cdot(1-e)$, expanding the cubic for $x + \lambda e$ and extracting the coefficient of λ gives

$$-x^2 + t(x)x - \frac{1}{2}(t(x)^2 - t(x^2))(1-e) = 0.$$

Thus

$$x \times x = \frac{1}{2}(t(x)^2 - t(x^2))e,$$

so that $z \times z$ above is in $J\cdot e$. Also,

$$e \times z = -\frac{1}{z} z + \frac{1}{2} t(z)(1-e).$$

Now both components, that in $J\cdot e$ and that in $J\cdot(1-e)$, of $e^W \times e^W = 2\beta e \times z + z \times z$, must be zero. From the above this means $t(z)^2 = t(z^2)$ and $\beta t(z)(1-e) = \beta z$. If $\beta \neq 0$, we have $z = t(z)(1-e)$, $t(z) = 2t(z) = 0$, and $e^W = \beta e$.

If $\beta = 0$, then $0 = z \times z = z^2 - t(z)z$, because $t(z)^2 = t(z^2)$, and either $z^2 = 0$ or we may replace z by $t(z)^{-1}z$ to obtain an idempotent in $J\cdot(1-e)$, say e', with $t(e') = 1$, $e^W = \varkappa e'$, $0 \neq \varkappa \in E$.

If $J\cdot(1-e)$ has no proper idempotents, then either $J\cdot(1-e)$ is a quadratic extension field K of E or is the Jordan algebra of a quadratic form over E such that the quadratic form does not represent 1. In neither case does $J\cdot(1-e)$ contain non-trivial elements of square zero, nor idempotents of trace one. Thus $e^W = \beta e$ in this case.

Otherwise, $J\cdot(1-e) \approx E \times E$, or is the Jordan algebra of a quadratic form in at least two dimensions over E, that form representing 1. In the latter case, if V is the quadratic space and $(v|w)$ the associated bilinear form,

$$J\cdot(1\cdot e) \approx E \times V, \text{ with}$$

$$(\xi,v)\cdot(\eta,w) = (\xi\eta + (v|w),\ \xi w + \eta v).$$

Thus $(\xi,v)^2 = (\xi,v)$ means that either $(\xi,v) = (1,0)$ or $(\xi,v) = (\frac{1}{2},v)$ with $(v|v) = \frac{1}{4}$. Thus we may fix $e_1 = (\frac{1}{2},v)$, $e_2 = (-\frac{1}{2},v)$. If $w \in V$, $(v|w) = 0$, then $e_1\cdot(0,w) = \frac{1}{2}(0,w) = e_2\cdot(0,w)$, so

$$J\cdot(1-e) = Ee_1 + Ee_2 + W, \text{ where}$$

$$W = \{x \in J\cdot(1-e) | x\cdot e_1 = \frac{1}{2} x = x\cdot e_2\}.$$

For $x \in W$, $\frac{1}{4}t(x) = t((x\cdot e_1)\cdot e_1) = t(x\cdot(e_1\cdot e_1)) = \frac{1}{2}t(x)$, so that $t(x) = 0$. Thus

$x \times e_1 = 0 = x \times e_2$.

Moreover, $(\xi,v)^2 = (0,0)$ if and only if $\xi = 0 = (v|v)$, and in this case there is $w \in V$, $(v|w) = 1$, $(w|w) = 0$.

For $x \in J\cdot(1-e)$ we see from the above that $e \times x \neq 0$ unless x is a multiple of $1-e$, and then $e \times x = \frac{1}{2}x$. If $t(x) = 0$, $e \times x = -\frac{1}{2}x$. Thus $e(0) = \{x \in J | x \times e = 0\} = Je = Ee$, of E-dimension <u>one</u>. Now suppose $e^w = \varkappa e_1$ or $\varkappa e_2$, say $e^w = \varkappa e_1$. Then $x \in e(0)$ if and only if $x^w \in e^w(0) = e_1(0)$. But $e_1(0)$ has dimension at least two in these cases (containing both e_1 and W). Thus either $e^w = \beta e$ or $e^w = (0,v)$ with $(v|v) = 0$. In the latter case, $t(e^w) = 0$, $t(1-e) = 2$ gives $(1-e) \times e^w = 0 = e^w \times e^w$, and again violates the fact that $e^w(0)$ must have dimension one. That is $e^w = \beta e$ in all cases except that where $J(1-e) \cong E \times E$.

Here $J(1-e) = Ee_1 + Ee_2$, e_i orthogonal idempotents with $t(e_i) = 1$, $J\cdot e_i = Ee_i$, and we may replace e by e_i in the above to obtain an idempotent e' with $t(e') = 1$, $J\cdot e' = Ee'$, $e'^w = \beta e'$ unless, with relabeling if necessary, $e^w = \varkappa_1 e_1$, $e_1^w = \varkappa_2 e_2$, $e_2^w = \varkappa_3 e$. But $w^2 = U_u$ stabilizes each Ee, Ee_i, and this is a contradiction. We have shown

<u>Lemma</u> 7.1. <u>If the Jordan algebra</u> J <u>is not simple, there is an idempotent</u> $e \in J$ <u>with</u> $t(e) = 1$, $J\cdot e = Ee$ <u>and</u> $e^w = \beta e$, <u>where</u> $\beta \in F$.

With e as in the lemma, let $J_0 = J\cdot(1-e) = \{x \in J | x\cdot e = 0\} = \{x \in J | t(x\cdot e) = 0\}$. The last of these characterizations shows that $J_0^w = J_0$.

Allison's presentation of his algebras is based on the operators $V_{x,y}$ on X, for $x,y \in X$, defined to map $z \in X$ to $(x\bar{y})z + (z\bar{y})x - (z\bar{x})y$. With $\zeta \in E$, $\bar{\zeta} = -\zeta$, consider the map $z \to \delta_1 z$ of X to $X = E \times J$ defined as $-\frac{1}{3}V_{(1,0),(\zeta,0)}$ on $E \times 0$ and as $V_{(1,0),(\zeta,0)}$ on $0 \times J$. It is directly checked from the definitions that

$$\delta_1(\alpha, a) = (\alpha\zeta, \zeta a), \text{ so that}$$

$\delta_1^2 z = \zeta^2 z$ for all $z \in X$.

Next define $\delta_2 z$ to be the effect on z of $V_{(1,0),(0,\zeta e)}$, with e as in the lemma. Then one finds

$$\delta_2(\alpha,0) = (0, \bar{\alpha}\zeta e); \quad \alpha \in E;$$

$$\delta_2(0,\mu e) = (\beta\bar{\mu}\zeta,0), \quad \mu \in E;$$

for $a \in J_0$,

$$\delta_2(0,a) = (0, \frac{-2}{\theta}\zeta e^w \times a^w).$$

Now $t((e^w \times a^w)\cdot e^w) = t((e^w \times e^w)\cdot a^w) = 0$, so $e^w \times a^w \in J_0$. Thus $\delta_2^2(0,a) =$

$$(0, \frac{-4\zeta^2}{\theta\bar\theta}(e^w \times a^w)^w \times e^w) = (0, -4\zeta^2 \beta(e \times a) \times e)$$

$$= (0, -\zeta^2\beta a), \quad \text{because} \quad (e \times a) \times e = \frac{1}{4} a.$$

From the values given, it is easily seen that $\delta_2^2(\alpha,\mu e) = -\beta\zeta^2(\alpha,\mu e)$, so that $\delta_2^2 z = -\beta\zeta^2 z$ for all $z \in X$. The values given are sufficient to show that $\delta_1\delta_2 = -\delta_2\delta_1$, as endomorphisms of X. Thus $1, \delta_1, \delta_2, \delta_1\delta_2$ are a basis for a quaternion algebra Q, that denoted $(\zeta^2, -\beta\zeta^2)$ in the notation of [O'M], §57, and X is a left Q-module. From [O'M], 57:9 and 57:10, Q is a division algebra if and only if β is not a norm in $N_{E/F}(E)$. We verify that $\beta \notin N_{E/F}(E)$, by using Allison's result that X is a division algebra.

Namely suppose $\beta = \xi\bar\xi$, $\xi \in E$. Let $x = (\xi,e) \in X$. Because X is a division algebra there is $y \in X$ with $V_{y,x}$ = identity. Let $y = (\alpha, \lambda e + a_0)$, where $\alpha, \lambda \in E$, $a_0 \in J_0$. We then have in particular

$$(1,0)V_{y,x} = (1,0), \quad \text{or}$$

$$(1,0) = y\bar{x} + \bar{x}y - \bar{y}x$$

$$= (2\,\alpha\bar\xi - \bar\alpha\xi + \bar\lambda\beta, (\alpha + \xi\lambda)e + \xi a_0 + \frac{2}{\theta} a_0^w \times e^w).$$

From the coefficient of e in the second component, $\alpha = -\xi\lambda$, so that $-2\xi\bar\xi\lambda + \xi\bar\xi\bar\lambda + \bar\lambda\beta = 1$, or $2\beta(\lambda - \bar\lambda) = 1$. This is only possible if $\beta = -\bar\beta$, contradicting $\beta \in F$. Thus Q is a quaternionic division algebra, and X a left Q-vector space. In particular, the F-dimension of X, and that of J_0, are divisible by 4.

Let $\delta \to \bar\delta$ be the usual involution in Q, so that $\bar\delta_i = -\delta_i$, $i = 1,2$.

We identify the Q-subspace $0 \times J_0$ with J_0 by $(0,a) \leftrightarrow a$; then $q(a,b) = \dfrac{tr(ab^w)-tr(ba^w)}{2\zeta} = \dfrac{tr(ab^w)-\overline{tr(ab^w)}}{2\zeta}$ defines a skew-symmetric F-valued F-bilinear form on J_0. We set

$$\langle a,b\rangle = q(a,b) + \delta_1^{-1} q(\delta_1 a,b) + \delta_2^{-1} q(\delta_2 a,b) + (\delta_1\delta_2)^{-1} q(\delta_1\delta_2 a,b),$$

a Q-valued antihermitian Q-form on J_0, readily seen to be nondegenerate from the nondegeneracy of $tr(ab)$. Let Y be a 3-dimensional Q-space with basis y_{-1}, y_0, y_1 and antihermitian Q-form satisfying $\langle y_{-1},y_1\rangle = 1 = -\langle y_1,y_{-1}\rangle$; $\langle y_{-1},y_{-1}\rangle = 0 = \langle y_0,y_{-1}\rangle = \langle y_0,y_1\rangle = \langle y_1,y_1\rangle$; $\langle y_0,y_0\rangle = -\delta_0^{-1}$. Form the orthogonal direct sum

$$V = Y \oplus J_0,$$

with nondegenerate antihermitian form $\langle v,w\rangle$. We shall exhibit an F-isomorphism of Lie algebras between g and the Lie algebra of Q-endomorphisms of V skew with respect to the form $\langle v,w\rangle$. This will show that the algebra g of this case is not new, its modules having been treated in earlier chapters.

For $y, v, w \in V$, the map $y \to \langle y,v\rangle w + \langle y,w\rangle v$ is in the Lie algebra $K(V)$ of skew endomorphisms of V. We denote this map, as before, by $S_{v,w}$. Then $S_{v,w} = S_{w,v}$, $S_{v,\lambda w} = S_{\bar{\lambda} v,w}$, and $[S_{v,w}, S_{v',w'}] = S_{vS_{v',w'},w} + S_{v,wS_{v',w'}}$. (Here, as usual, V is regarded as right module for $\mathrm{End}_{\mathcal{Q}}(V)$.) Moreover, $K(V)$ is spanned over F by the $S_{v,w}$, and in particular by the $S_{a,b}$ $(a,b \in J_0)$; the $S_{\lambda y_0,y_1}$ $(\lambda \in \mathcal{Q})$; the $S_{\lambda y_0,y_{-1}}$ $(\lambda \in \mathcal{Q})$; the $S_{\lambda y_{-1},y_1}$ $(\lambda \in \mathcal{Q})$; the $S_{\lambda y_i,y_i}$ $(-1 \le i \le 1,\ \lambda \in F)$; and the $S_{y_i,a}$ $(-1 \le i \le 1,\ a \in J_0$. It will be observed in the sequel that the image of our map $g \to K(V)$ contains all these, so that g is mapped onto $K(V)$. The map will clearly be non-zero, and will be seen to be a morphism of Lie algebras. By the simplicity of g it will be an isomorphism.

We begin by setting

$$\rho(S_{v_0,v_1} \otimes (1,0)) = S_{y_0,y_1},$$

$$\rho(S_{v_0,v_{-1}} \otimes (1,0)) = -S_{\delta_1 y_0,y_{-1}},$$

$$\rho(S_{v_{-1},v_1} \otimes (1,0)) = -S_{y_{-1},y_1}.$$

(Note that $S_{\varkappa v,v} = S_{v,\varkappa v} = S_{v,\bar{\varkappa} v}$ shows that $S_{\varkappa v,v} = 0$ whenever $\bar{\varkappa} = -\varkappa$.) Here our split 3-dimensional algebra $\mathfrak{s}$ is identified with the span of the three elements for which ρ is defined, and we extend ρ to $\mathfrak{s}$ by F-linearity. Using the parenthetical remark, one sees at once that $\rho\colon \mathfrak{s} \to K(V)$ is a Lie morphism.

Next we set

$$\rho(T_{v_1,v_1} \otimes (\zeta,0)) = -\,S_{y_1,y_1}, \quad \text{and define}$$

$\rho((T_{v_1,v_1} \otimes (\zeta,0))\,(\mathrm{ad}(S_{v_0,v_1} \otimes (1,0)))^k$ to be equal to $(-S_{y_1,y_1})(\mathrm{ad}\,S_{-\delta_1 y_0,y_{-1}})^k$ for all $k > 0$, so that

$$\rho(T_{v_0,v_1} \otimes (\zeta,0)) = S_{\delta_1 y_0,y_1},$$

$$\rho((T_{v_{-1},v_1} - T_{v_0,v_0}) \otimes (\zeta,0)) = -S_{\delta_1 y_{-1},y_1} - \zeta^2\, S_{y_0,y_0},$$

$$\rho(T_{v_0,v_{-1}} \otimes (\zeta,0)) = -\zeta^2\, S_{y_{-1},y_0},$$

$$\rho(T_{v_{-1},v_{-1}} \otimes (\zeta,0)) = \zeta^2\, S_{y_{-1},y_{-1}}.$$

If $m_5 \subset g$ denotes the span of the five elements on which ρ is here defined, m_5 is an irreducible $\mathfrak{s}$-submodule in the adjoint action of $\mathfrak{s}$ on g, and we extend ρ to m_5 by linearity. With the definitions above, one sees that $[\rho(m), \rho(s)] = \rho([ms])$ for all $m \in m_5$, $s \in \mathfrak{s}$. (It actually suffices to verify this for our five basic elements m, and for $s = S_{v_0,v_1} \otimes (1,0)$.) Then if $m_0 = T_{v_1,v_1} \otimes (\zeta,0)$, we find that $[m_0,m] \in \mathfrak{s}$ for each of our basis elements m of m_5, and that

$\rho([m_0,m]) = [\rho(m_0),\rho(m)]$ for these elements m. Thus $\{n \in \mathcal{m}_5 \mid$ for all $m \in \mathcal{m}$, $[n,m] \in \mathcal{s}$ and $\rho([n,m]) = [\rho(n),\rho(m)]\}$ is a non-zero $\mathcal{s}$-submodule of $\mathcal{m}_5$, so is $\mathcal{m}_5$. It follows that $\mathcal{s} + \mathcal{m}_5$ is a Lie subalgebra of $\mathcal{g}$ and that ρ is a Lie morphism of $\mathcal{s} + \mathcal{m}_5$ into $K(V)$.

Next set $\rho(S_{v_0,v_1} \otimes (0,e)) = \zeta^{-2}\, S_{\delta_1\delta_2 y_0,y_1}$. Then the condition that ρ be an $\mathcal{s}$-morphism forces us to choose

$$\rho(S_{v_{-1},v_1} \otimes (0,e)) = -\zeta^{-2} S_{\delta_1\delta_2 y_{-1},y_1},$$

$$\rho(S_{v_0,v_{-1}} \otimes (0,e)) = -S_{\delta_1 y_0,y_{-1}}.$$

The requirement that ρ preserve brackets with $\mathcal{m}_5$ forces

$$\rho(S_{v_0,v_1} \otimes (0, \zeta e)) = -S_{\delta_2 y_0,y_1},$$

$$\rho(S_{v_{-1},v_1} \otimes (0, \zeta e)) = S_{\delta_2 y_{-1},y_1},$$

$$\rho(S_{v_0,v_{-1}} \otimes (0, \zeta e)) = S_{\delta_1\delta_2 y_0,y_{-1}},$$

and one checks as before that this defines a mapping $\rho: \mathcal{s} + \mathcal{m}_5 + \mathcal{s} \otimes (0,Ee) \to K(V)$ preserving brackets with $\mathcal{s} + \mathcal{m}_5$. Now $D_{(0,e),(0,\zeta e)} \in \mathcal{g}$ is not in the space on which ρ has been defined; we define

$$\rho(D_{(0,e),(0,\zeta e)}) = \frac{2}{3}\beta\zeta^2(S_{y_0,y_0} - 2\, S_{\delta_1 y_{-1},y_1})$$

and extend ρ by F-linearity to $\mathcal{s} + \mathcal{m}_5 + (\mathcal{s} \otimes (0,Ee)) + FD_{(0,e),(0,\zeta e)}$, a subalgebra of $\mathcal{g}$. By calculations, reduced to essentials as indicated above, ρ is a Lie morphism of this subalgebra to $K(V)$.

For $a \in J_0$, we now define

$$\rho(S_{v_0,v_1} \otimes (0,a)) = S_{a,y_1};$$

$$\rho(S_{v_{-1},v_1} \otimes (0,a)) = -S_{\delta_1 a,y_0};$$

$$\rho(S_{v_0,v_1} \otimes (0,a)) = S_{\delta_1 a,y_{-1}}.$$

Using the fact that $D_{(\varkappa,0),(0,a)} = 0 = D_{(0,\varkappa e),(0,a)}$ for all $\varkappa \in E$, we may verify at once that the above definition is consistent with all brackets with elements of $\mathcal{s} + \mathcal{m}_5 + \mathcal{s} \otimes (0,Ee)$, and therefore preserves brackets with $D_{(0,e),(0,\zeta e)}$ as well.

Now there is an F-linear mapping ψ from $\Lambda_F^2(E) \oplus \Lambda_F^2(J_0)$ onto $\mathrm{Der}(X,-)$ mapping $e \wedge \zeta e$ to $D_{(0,e),(0,\zeta e)}$ and $a \wedge b\,(a,b \in J_0)$ to $D_{(0,a),(0,b)}$. Likewise there is an F-linear mapping $\sigma : \Lambda_F^2(E) \oplus \Lambda_F^2(J_0)$ to $K(V)$ sending $e \wedge \zeta e$ to $\frac{2}{3}\beta\zeta^2(S_{y_0,y_0} - 2\, S_{\delta_1 y_{-1},y_1}) = \rho(D_{(0,e),(0,\zeta e)})$, $\zeta e \wedge e$ to the negative of this quantity, and $a \wedge b$ to

$$\frac{1}{3}\zeta^2\, q(a,b)S_{y_0,y_0} - \frac{2}{3}\, q(a,b)S_{\delta_1 y_{-1},y_1} + S_{\delta_1 a,b}\,. \tag{9}$$

(It will be noted that $S_{\delta_1 b,a} = S_{a,\delta_1 b} = -S_{\delta_1 a,b}$.) In fact $\operatorname{Ker}\psi \subseteq \operatorname{Ker}\sigma$, so that there is a well-defined F-linear mapping $\rho:\operatorname{Der}(X,-) \to K(V)$ agreeing with our previous ρ on $D_{(0,e),(0,\zeta e)}$ and sending $D_{(0,a),(0,b)}$ to (9) above.

To indicate why $\operatorname{Ker}\psi \subseteq \operatorname{Ker}\sigma$, let $\alpha_0 \in F$, $a_1,\dots,a_n$; $b_1,\dots,b_n$ elements of J_0 and suppose

$$0 = \psi(\alpha_0 e \wedge \zeta e + \sum_{i=1}^{n} a_i \wedge b_i) = \alpha_0\, D_{(0,e),(0,\zeta e)} + \sum_{i=1}^{n} D_{(0,a_i),(0,b_i)}. \tag{10}$$

The derivation on the right in (10) sends $(0,e)$ to $-\frac{8}{3}\alpha_0(0,\ \beta\zeta e) - (\frac{4}{3}\sum_{i=1}^{n} q(a_i,b_i))(0,\ \zeta e)$. Thus we have

$$2\,\alpha_0\beta + \sum_{i=1}^{n} q(a_i,b_i) = 0. \tag{11}$$

If $c \in J_0$, we have $(0,c)D_{(0,e),(0,\zeta e)} = \frac{4}{3}(0,\ \beta\zeta a)$, while $(0,c)D_{(0,a),(0,b)}$ (for $a,b \in J_0$) is

$$\begin{aligned}(0,\ &\frac{2}{3}\, q(a,b)\ \zeta\ c + tr(c\cdot b^W)a - tr(c\cdot a^W)b - tr(b)tr(c)a^W\\ &+ tr(b\cdot c)a^W + tr(a)tr(c)b^W - tr(a\cdot c)b^W + [tr(b)tr(c)tr(a^W)\\ &- tr(b\cdot c)tr(a^W) - tr(a)tr(c)tr(b^W) + tr(a\cdot c)tr(b^W)](1-e)).\end{aligned}$$

From (10) we see that for all $c \in J_0$, we must have

$$\begin{aligned}0 = &\sum_{i=1}^{n} (tr(c\cdot b_i^W)a_i - tr(c\cdot a_i^W)b_i)\\ &+ \sum_{i=1}^{n} (tr(b_i\cdot c)a_i^W - tr(a_i\cdot c)\cdot b_i^W - tr(b_i)tr(c)a_i^W + tr(a_i)tr(c)b_i^W)\\ &+ [\sum_{i=1}^{n} (tr(b_i)tr(c)tr(a_i^W) - tr(b_i\cdot c)tr(a_i^W) - tr(a_i)tr(c)tr(b_i^W)\\ &\qquad + tr(a_i\cdot c)tr(b_i^W))](1-e).\end{aligned} \tag{12}$$

$$\begin{aligned}\text{Now}\ \ \langle c,\delta_1 a\rangle\ \ b = &\ q(c,\ \zeta a)b + q(\delta_1 c,\ \zeta a)\delta_1^{-1}\, b\\ &+ q(\delta_2,c,\ \zeta a)\delta_2^{-1}b + q(\delta_1\delta_2 c,\ \zeta a)\delta_2^{-1}\ \delta_1^{-1}\ b.\end{aligned} \tag{13}$$

The first two terms here combine to

$$(\frac{-\zeta\ tr(c\cdot a^W) - \zeta\ tr(c\cdot a^W)}{2\zeta})\ b + (\frac{-\zeta^2\ tr(c\cdot a^W) + \zeta^2\ \overline{tr(c\cdot a^W)}}{2\zeta})\ \zeta^{-1}\ b = -tr(c\cdot a^W)b.$$

Also $\delta_2^{-1}b = -\beta^{-1}\zeta^{-2}\delta_2 b = \beta^{-1}\zeta^{-1}\cdot\frac{2}{\theta}e^w \times b^w$

$$= \frac{2\zeta^{-1}}{\theta} e \times b^w = -\frac{\zeta^{-1}}{\theta} b^w + \frac{\zeta^{-1}}{\theta} tr(b^w)(1-e);$$

$$\delta_2^{-1}\delta_1^{-1} b = -\beta^{-1}\zeta^{-4}\delta_2 \zeta b = -\beta^{-1}\zeta^{-2}\frac{2}{\theta} e^w \times b^w$$

$$= \frac{\zeta^{-2}}{\theta} b^w - \frac{\zeta^{-2}}{\theta} tr(b^w)(1-e);$$

$$q(\delta_2 c, \zeta a) = \frac{tr((-\frac{2\zeta}{\theta} e^w \times c^w)\cdot -\zeta a^w) - \overline{tr(\quad)}}{2\zeta}$$

and $\frac{2\zeta^2}{\theta} tr((e^w \times c^w)\cdot a^w) = 2\zeta^2\bar{\theta}\,\overline{tr((e \times c)\cdot a)}$

$$= -\zeta^2\bar{\theta}\,\overline{tr(c\cdot a)} + \zeta^2\bar{\theta}\,\overline{tr(c)tr(a)};$$

$$q(\delta_1\delta_2 c, \zeta a) = \frac{tr((\frac{-2\zeta^2}{\theta} e^w \times c^w)\cdot - \zeta a^w) - \overline{tr(\quad)}}{\zeta 2}, \text{ and}$$

$$\frac{2\zeta^3}{\theta} tr((e^w \times c^w)\cdot a^w) = -\zeta^3\bar{\theta}\,\overline{tr(c\cdot a)} + \zeta^3\bar{\theta}\,\overline{tr(c)}\,\overline{tr(a)}.$$

Now substitution in the last two terms of (13) shows that these combine to

$$-tr(c\cdot a)b^w + tr(c)tr(a)b^w$$
$$+ tr(c\cdot a)tr(b^w)(1-e) - tr(c)tr(a)tr(b^w)(1-e).$$

That is, the relation (12) expresses exactly the relation

$$c\sum_{i=1}^{n} S_{\delta_1 a_i, b_i} = 0. \tag{14}$$

Thus the fact that our element is in Ker ψ yields the two relations (11) and (14), the latter for all $c \in J_0$. It follows that $\sum_{i=1}^{n} S_{\delta_1 a_i, b_i} = 0$, so that σ sends $\alpha_0 e \wedge \zeta e + \Sigma\, a_i \wedge b_i$ to

$$\frac{1}{3}\zeta^2(2\,\alpha_0\beta + \Sigma\, q(a_i,b_i))(S_{y_0,y_0} - 2S_{\delta_1,y_{-1},y_1}),$$

and this is zero by (11). Our choice of σ was made so that

$$[\rho(S_{v_{-1},v_1} \otimes (0,a)), \rho(S_{v_{-1},v_1} \otimes (0,b))] = \rho([S_{v_{-1},v_1} \otimes (0,a), S_{v_{-1},v_1} \otimes (0,b)]),$$

and one also sees that it has the property of being consistent with all other brackets among the pairs of elements of $\mathcal{S} \cup (m_5 \otimes (F\zeta,0)) \cup (\mathcal{S} \otimes (0,J))$. Using the Jacobi identity, we see that the mapping ρ, as now defined on all of g, is an F-Lie morphism of g to $K(V)$, whose image contains a set of linear generators for $K(V)$. This proves the

Theorem 7.2. When the Jordan algebra J is not simple (over E), the Lie algebra g associated with the Allison algebra $(X,-) = (E \times J, -, w)$ is isomorphic to a Lie algebra $K(V)$, where V is a vector space over a quaternionic division algebra Q over F and where $K(V)$ is the set of Q-endomorphisms of V skew with respect to a nondegenerate skew-hermitian form on V (with respect to the standard involution in $\mathcal{Q}$.)

§6. The Case of a Special Simple Jordan Algebra.

Here we assume the Jordan algebra J is simple and special. That is, there is an injective E-linear mapping $\varphi: J \to R$, where R is an associative E-algebra with unit 1_R, such that $\varphi(1_J) = 1_R$ and

$$\varphi(a\cdot b) = \frac{1}{2}(\varphi(a)\,\varphi(b) + \varphi(b)\varphi(a)) \quad \text{for all} \quad a,b \in J.$$

We may further assume R is the subalgebra of R generated by $\varphi(J)$, in which case R is of finite dimension over E. By [J5], p. 209, the cubic cases under consideration admit only the following possibilities:

i) J is a cubic field extension of E. Then φ is a field-isomorphism of J onto $R \approx J$. We identify J with R in this case.

ii) J is the Jordan algebra of linear transformations T of a 3-dimensional E-module V, self-adjoint with respect to a nondegenerate symmetric bilinear form on V. Then $R \approx \mathrm{End}_E(V)$, which we view as acting on the right on V, and φ may be taken to be the identity mapping of J into R.

iii) J is the Jordan algebra of all elements of a central simple associative E-algebra R of degree 3, with $a\cdot b = \frac{1}{2}(ab + ba)$. Here there are two possibilities for φ: the identity mapping of J into R, and the identity mapping of J into the opposite algebra R^{op}.

iv) There is a quadratic field extension K of E and a central simple associative K-algebra R of degree 3, with involution * of second kind, fixing E, and J is the Jordan algebra of *-fixed elements of R. Our φ may again be taken to be the inclusion $J \hookrightarrow R$.

v) There is a central simple associative E-algebra R of degree 6 over E, with involution * of first kind and symplectic type, and J is the Jordan algebra of *-fixed elements of R, with φ the inclusion $J \to R$.

The sense in which φ is uniquely determined in the above is that if S, ψ is another "associative specialization" as in our first paragraph (with R and S assumed simple), there is an E-isomorphism of associative algebras $\Upsilon: R \to S$ (in iii), one must add: "or $\Upsilon: R^{op} \to S$") such that $\Upsilon(a) = \Upsilon(\varphi(a))$ for all $a \in J$. Moreover, the involution * in iv), v) of R, as well as the adjoint in ii), are determined by J, in that S must also be involutorial, $\psi(J)$ the self-adjoint elements of S, and Υ an isomorphism of involutorial algebras.

More precisely, the meaning of this is that if V' is a faithful irreducible special right J-module, so that V' is a finite-dimensional E-space, ψ an injective E-linear map into $\mathrm{End}_E(V')$, acting on the right on V', such that $\psi(1_J) = I_{V'}$,

$\psi(a\cdot b) = \frac{1}{2}(\psi(a)\psi(b) + \psi(b)\psi(a))$, the action being irreducible, then:

In case ii), there is an E-linear isomorphism $s:V \to V'$ such that $\psi(a) = s^{-1}\varphi(a)s$ (right operations!) for all $a \in J$.

In case iii), either $R \simeq \mathrm{End}_E(V)$, where V is a 3-dimensional E-space, or $R \simeq \mathrm{End}_D(D) \simeq D$, where D is a central division algebra of degree 3 over E. In the former subcase, either $\psi(a) = s^{-1}\varphi(a)s$ for all $a \in J$, where $s:V \to V'$ is an E-isomorphism as in case ii), or $\varphi(a)$ is regarded as an element of $\mathrm{End}_E(V^*) \simeq R^{op}$, and $s:V^* \to V'$ as an isomorphism with $\psi(a) = s^{-1}\varphi(a)s$. In the latter subcase, V' identifies with D' or with D'^{op}, where D' is an E-division algebra isomorphic to D, and there is an isomorphism of division algebras $s:D \to D'$ or $s:D^{op} \to D'$ such that $\psi(a) = s^{-1}\varphi(a)s$, with appropriate identifications.

In case iv), either $R = \mathrm{End}_K(V)$, V of dimension 3 over K, in which case V' is also a 3-dimensional K-space and s a K-isomorphism $V \to V'$ with $\psi(a) = s^{-1}\varphi(a)s$ for all $a \in J$, or $R = D$, D a central simple algebra of degree 3 over K, and V' is a K-vector space K-isomorphic to D, thus carrying a structure of involutorial division algebra D' isomorphic to that of D. V' is regarded as right $\mathrm{End}_{D'}(V')$-module, and the identification with D' means simply that there is an involutorial isomorphism of K-algebras $s:D \to D'$ such that $\psi(a) = s(\varphi(a))$ for all $a \in J$.

In case v), either $R = \mathrm{End}_E(V)$, V a 6-dimensional E-space, and V' is a second 6-dimensional E-space, $s:V \to V'$ an E-isomorphism such that $\psi(a) = s^{-1}\varphi(a)s$ for all $a \in J$, <u>or</u> $R = \mathrm{End}_{\mathcal{Q}}(V)$, V a 3-dimensional $\mathcal{Q}$-space where $\mathcal{Q}$ is a quaternionic division algebra over E; here V' is a 3-dimensional $\mathcal{Q}'$-space where $\mathcal{Q}'$ is a quaternionic division algebra isomorphic to $\mathcal{Q}$. Thus we may regard V' as (left) $\mathcal{Q}$-space: Then there is a $\mathcal{Q}$-isomorphism $s:V \to V'$ such that $\psi(a) = s^{-1}\varphi(a)s$ for all $a \in J$.

We use this information to make a sharper identification of the mapping $w:J \to J$. With $1^w = u \in J$, we have seen that $n(u) = \theta\bar{\theta}$, $t(ua) = t(a^w)$ for all $a \in J$, and $w^2 = U_u$: $aU_u = uau$, when J is identified with $\varphi(J) \subseteq R$. Moreover $(a^w \times b^w)^w = \theta\bar{\theta}\, a \times b$, for all $a,b \in J$.

Now let $\bar{R}$ be the additive group R, made into a ring with product $x \mathbin{\bar{\cdot}} y = xu^{-1}y$, so with unit element u, and into an E-algebra with scalar multiplication $\lambda \mathbin{\bar{\cdot}} x = \bar{\lambda}x$. For $a \in J$, we have

$$(u^{-1}a)^3 - t(u^{-1}a)(u^{-1}a)^2 + s(u^{-1}a)(u^{-1}a) - n(u^{-1}a)1 = 0 \tag{15}$$

where $s(x) = \frac{1}{2}(t(x)^2 - t(x^2))$, $n(x) = \frac{1}{6}(t(x)^3 - 3t(x)t(x^2) + 2t(x^3))$ as before. This is evident except in cases iv), v), because all elements of R satisfy such an equation. In the remaining cases, the involution $x \to x^*$ of R having J as fixed elements may be replaced by the involution $x \to u^{-1}x^*u$, having $u^{-1}J$ as fixed elements, as in [Se5], §2.4. When (15) is multiplied on the left by u, we obtain the following relation in $\bar{R}$:

$$a^{\bar{3}} - \overline{t(u^{-1}a)} \mathbin{\bar{\cdot}} a^{\bar{2}} + \overline{s(u^{-1}a)} \mathbin{\bar{\cdot}} a - \overline{n(u^{-1}a)} \cdot 1_{\bar{R}} = 0. \tag{16}$$

Now $n(u^{-1}a) = n(u^{-1})n(a)$. When n is the (reduced) norm of a central simple algebra, or the norm of a field extension, this is clear. In the remaining case v), this follows generally from Theorem 6, p. 242 of [J5], or may also be verified by direct computation in the split case. Thus the norm of a as element of $\bar{R}$ is given by

$$n_{\bar{R}}(a) = \overline{n(u^{-1}a)} = \overline{n(u)}^{-1}\,\overline{n(a)} = (\theta\,\bar{\theta})^{-1}\,\overline{n(a)}.$$

Replacing a by a^w, we have

$$n_{\bar{R}}(a^w) = (\theta\bar{\theta})^{-1}\,\overline{n(a^w)} = (\theta\,\bar{\theta})^{-1}\,\theta\,\bar{\theta}\,n(a) = n(a).$$

Thus w is an E-linear mapping from J, as Jordan subalgebra of R, to J, as Jordan subalgebra of $\bar{R}$, mapping 1_R to $u = 1_{\bar{R}}$, and preserving the norm. From [J5], p. 244 (Theorem 7), it follows that w is an isomorphism of Jordan algebras. From Jacobson's determination of associative specializations it follows that either there is an isomorphism $\psi : R \to \bar{R}$ of associative E-algebras such that for all $a \in J$, $a^w = \psi(a)$, or we are in case iii) and there is an anti-isomorphism $\psi : R \to \bar{R}$ such that $a^w = \psi(a)$ for all $a \in J$. That is, in the former case there is a "conjugate-linear" mapping $\psi: R \to R$ mapping 1 to u, with $\psi(xy) = \psi(x)u^{-1}\,\psi(y)$ for all $x,y \in R$, and such that $a^w = \psi(a)$ for all $a \in J$, while in the latter case the only change is that $\psi(xy) = \psi(y)u^{-1}\,\psi(x)$.

Moreover, $\psi(\psi(x)) = u\,x\,u$ for all $x \in R$. This follows from the formulas above, the fact that $\psi^2(a) = a^{w^2} = u\,a\,u$ for all $a \in J$, so that $\psi^2(u^{-1}) = u = \psi(1)$ and $\psi(u^{-1}) = 1$, and the fact that $R = J + [JJ]$.

Let V be the special J-module denoted by "V" in cases ii)-v), or by "D" in cases iii), iv), whichever of these is acted on by R as endomorphisms. We make a second structure $\bar{V}$ on V, as follows:

1) The underlying additive group of $\bar{V}$ is that of V; in particular, the sets V and $\bar{V}$ are identified.

2) $\bar{V}$ is an E-vector space, with $\lambda \mathbin{\bar{\cdot}} v = \bar{\lambda}\,v$ (multiplication by scalars in V) for $\lambda \in E$, $v \in \bar{V}$.

3) A pairing $(v,x) \to v \mathbin{\bar{\cdot}} x : \bar{V} \times \bar{R} \to \bar{V}$ is defined by $v \mathbin{\bar{\cdot}} x = vu^{-1}x$. Then $\bar{V}$ is an irreducible right $\bar{R}$-module.

Now we combine this observation with the fact that $\bar{R}, w$ is a (simple) "associative specialization" of J. In particular, the considerations developed for general V' in the cases ii) - v) now apply to $\bar{V}$. They will be involved in this context in §8.2, where the study of $\mathcal{g}$-modules will yield further constraints on R, and hence on J.

§7. Questions of Existence: Examples.

Allison and Faulkner [AF] have given a general process for constructing central simple Allison algebras from Jordan algebras of degree 4, with norm criteria as to when the Allison algebra is a division algebra. In Theorem 7.1 of that paper and the subsequent example 7.2 one obtains central Allison division algebras over fields $k = F(\xi)$, ξ transcendental over F, constructed from central Jordan division algebras of degree 4 over F. These Jordan division algebras are either:

a) the symmetric elements, with respect to an involution of first kind and orthogonal type in a central associative division algebra of degree 4 over F;

b) the totality of elements of such an associative division algebra;

c) the symmetric elements, with respect to an involution of second kind, of a division algebra, central of degree 4 over a quadratic extension of F;

d) the symmetric elements of a central associative division algebra of degree 8 over F, with respect to an involution of first kind and symplectic type.

By Theorem 6.6 of [AF], the dimension of the space of skew elements in the associated Allison algebra $(X,-)$ is 1, so $X \cong E \times J$ as above, where E is a quadratic extension of k and J is a cubic Jordan algebra over E. In the respective cases a)-d), the algebra X as constructed in [AF] has k-dimension 20, 32, 32, 56, and a split form of the Jordan algebra J is recovered. Upon splitting, one sees that the associated Lie algebra g is a k-form of the split algebra of respective type E_6, E_7, E_7, E_8.

In particular, we take $F = \mathbb{Q}$, so that $k = \mathbb{Q}(\xi)$ admits central simple Lie algebras g that are of relative rank one, non-reduced, with a root-space of dimension one, and of type E_6, E_7, E_8. Thus the context of this chapter includes forms of these exceptional types.

Briefly, the construction of [AF] is as follows: Let B be a Jordan algebra of degree 4, in particular an algebra $A \otimes_F F(\xi)$ over $k = F(\xi)$, where A is one of a)-d). Let B_0 be the set of elements of B of trace zero, so that $B = B_0 \oplus k\, 1_B$. Define $\theta: B \to B$ by $b^\theta = -b$ if $b \in B_0$, $1^\theta = 1$. Let $X = B \times B$, with product

$$(a,b)(a',b') = (a \cdot a' + \xi(b \cdot b^\theta)^\theta,\ a^\theta \cdot b' + (b^\theta \cdot a^\theta)^\theta),$$

and with $\overline{(a,b)} = (a, -b^\theta)$. Then $(X,-)$ is an Allison algebra, denoted "CD(B,n,ξ)" in [AF], and has the properties cited above in cases a)-d).

One is not certain that the construction above exhausts all Allison division algebras - if this were known to be the case, the realization $X = B \times B$ would surely be preferable to that involving the "semi-isotopy" w.

CHAPTER 8: CONSTRUCTION OF MODULES FOR THE SUPER-EXCEPTIONAL ALGEBRAS OF RANK ONE

The algebras of the title above are those considered in the bulk of Chapter 7. Here the Allison algebra is $X = E \times J$, where J is either one of the special simple Jordan algebras over E of §7.6 or is an exceptional simple Jordan algebra over E. A one-dimensional maximal F-split torus in the Lie algebra g is

$$t_0 = S_{v_{-1},v_1} \otimes (F,0),$$

with centralizer

$$g_0 = \mathrm{Der}(X,-) + t_0 + S_{v_{-1},v_1} \otimes (0,J) + (T_{v_{-1},v_1} - T_{v_0,v_0}) \otimes (F\zeta,\ 0),$$

and $\mathrm{Der}(X,-) = D_{X,X}$.

§1. Construction of Certain g_0-Modules.

We shall see that $[g_0g_0]$ is a simple Lie algebra over F, absolutely simple except when J is a cubic field extension of E, and with $g_0 = [g_0g_0] + t_0$. Thus $[g_0g_0]$ is a simple module for its own (right) adjoint action, and prescribed scalar action of t_0 extends this structure to that of simple g_0-module. Likewise, any one-dimensional F-space Fv becomes a g_0-module with $[g_0g_0]$ acting trivially and t_0 by any preassigned scalar function. No more need be said about these modules until they are invoked in our summary.

A more interesting module is X, the Allison algebra. Here we define a map $\rho\colon [g_0g_0] \to \mathrm{End}_F(X)$, with X regarded as right $\mathrm{End}_F(X)$-module, by

$$(\alpha,a)\ \rho((T_{v_{-1},v_1} - T_{v_0,v_0}) \otimes (\lambda\zeta,0)) = (3\alpha\lambda\zeta,\lambda\zeta a),$$

$$(\alpha,a)\ \rho(S_{v_{-1},v_1} \otimes (0,b)) = (\alpha,a)(0,b),$$

$$(\alpha,a)\ \rho(D_{(0,b),(0,c)}) = (\alpha,a)D_{(0,b),(0,c)},$$

for $a,b,c \in J$; $\alpha,\ \lambda \in F$. Because $\mathrm{Der}(X,-) = D_{X,X}$ and $D_{(1,0),X} = 0 = D_{(\zeta,0),X}$, the last condition amounts to $(\alpha,a)\rho(D) = (\alpha,a)D$ for all $D \in \mathrm{Der}(X,-)$, and thus is evidently well-defined. From

$$[(T_{v_{-1},v_1} - T_{v_0,v_0}) \otimes (\zeta,0),\ S_{v_{-1},v_1} \otimes (0,b)]$$
$$= 2\ S_{v_{-1},v_1} \otimes (0,\zeta b),$$ we find writing T for $T_{v_{-1},v_1} - T_{v_0,v_0}$,

$$(\alpha,a)[\rho(T\otimes(\zeta,0)),\ \rho(S_{v_{-1},v_1}\otimes(0,b))]$$

$$= (3\alpha\zeta,\ \zeta a)(0,b) - ((\alpha,a)(0,b))\ \rho(T\otimes(\zeta,0))$$

$$= (\zeta t(ab^w) - 3\zeta t(ab^w),\ 3\alpha\zeta b - \alpha\zeta b - \frac{2\zeta}{\theta} a^w \times b^w - \frac{2\zeta}{\theta} a^w \times b^w)$$

$= (\alpha,a)\ \rho([T\otimes(\zeta,0),\ S\otimes(0,b)])$.

From $[S_{v_{-1},v_1}\otimes(0,b),\ S_{v_{-1},v_1}\otimes(0,c)] =$

$$= S_{v_{-1},v_1}\circ S_{v_{-1},v_1}\otimes\left(\frac{t(bc^w)-t(cb^w)}{2},\ 0\right)$$

$$+ \mathrm{Tr}(S_{v_{-1},v_1}^2)\ d_1((0,b),(0,c))$$

$$= \frac{1}{3} T\otimes(t(bc^w) - t(cb^w),0)$$

$$+ D_{(0,c),(0,b)},$$

we see that $\rho([S_{v_{-1},v_1}\otimes(0,b),\ S_{v_{-1},v_1}\otimes(0,c)])$ sends (α,a) to

$$((t(bc^w) - t(cb^w))\alpha,\ \tfrac{1}{3}(t(bc^w))a)$$

$$+ (\alpha,a)D_{(0,c),(0,b)}.$$

Now $(\alpha,a)D_{(0,c),(0,b)} = \frac{2}{3}[(0,c),(0,b)](\alpha,a)$ $+\frac{1}{3}(\alpha,a)[(0,c),(0,b)] + ((\alpha,a)(0,b))(0,c) - ((\alpha,a)(0,c))(0,b)$, from the definition of an inner derivation, and

$[(0,c),(0,b)] = (t(cb^w),0) = (t(cb^w) - \overline{t(cb^w)},0)$ means that the first two terms combine to

$$((t(cb^w) - t(bc^w))\alpha,\ \tfrac{1}{3}(t(cb^w) - t(bc^w))a).$$

It follows at once that

$$\rho([S_{v_{-1},v_1}\otimes(0,b),\ S_{v_{-1},v_1}\otimes(0,c)]) =$$

$$[\rho(S_{v_{-1},v_1}\otimes(0,b)),\ \rho(S_{v_{-1},v_1}\otimes(0,c))].$$

That $\rho([T\otimes(\zeta,0),D_{(0,b),(0,c)}]) = [\rho(T\otimes(\zeta,0)),\ \rho(D_{(0,b),(0,c)})]$ now follows from the above and the Jacobi identity. To complete showing that ρ is a representation, it suffices only to notice that

$$(\alpha,a)\ \rho([S_{v_{-1},v_1}\otimes(0,d),\ D_{(0,b),(0,c)}])$$

$$= (\alpha,a)\ \rho(S_{v_{-1},v_1}\otimes(0,d)D_{(0,b),(0,c)})$$

$$= (\alpha,a)\cdot((0,d)D_{(0,b),(0,c)})$$

$$= ((\alpha,a)\cdot(0,d))D_{(0,b),(0,c)} - (\alpha,a)D_{(0,b),(0,c)}\cdot(0,d)$$
$$= (\alpha,a)[\rho(S_{v_{-1},v_1} \otimes (0,d)),\ \rho(D_{(0,b),(0,c)})].$$

In any non-zero $[\mathcal{g}_0\mathcal{g}_0]$-submodule $\mathcal{Y}$ of $\mathcal{X}$, the action of $T \otimes (\zeta,0)$ shows that each component is an E-subspace of E resp. $\mathcal{J}$, contained in $\mathcal{Y}$. If the E-component is non-zero, it is E, and acting with the $\rho(S_{v_{-1},v_1} \otimes (0,b))$ then shows $\mathcal{Y} = \mathcal{X}$. If the E-component were zero, operating with $\rho(S_{v_{-1},v_1} \otimes (0,b))$ would show that the $\mathcal{J}$-component must be in the radical of the trace form on $\mathcal{J}$, so would be zero as well. Thus $\mathcal{X}$ is an <u>irreducible</u> $[\mathcal{g}_0\mathcal{g}_0]$-module, and any scalar function on t_0 makes $\mathcal{X}$ into an irreducible $\mathcal{g}_0$-module.

Next we consider more closely the case i) of §7.6, where $\mathcal{J}$ is a cubic field extension of E. Let $1^w = u$, $n(u) = \theta\bar{\theta}$ as before. We distinguish two cases:

a) $u \in E$. Then from $t(u) = t(1\ 1^w) = \overline{t(1^w 1)} = \overline{t(u)}$, we have $3\bar{u} = 3u$, and $u \in F$.

For $x \in \mathcal{J}$, set $x^\sigma = x^w u^{-1} = u^{-1}x^w = (u^{-1}x)^w$. If $x \in E$, we have $x^\sigma = (x1)^\sigma = \overline{(x1)^w u^{-1}} = \bar{x}\,u\,u^{-1} = \bar{x}$, and in general for $x,y,z \in \mathcal{J}$,

$$t(x^\sigma) = t(x^w u^{-1}) = t(xu^{-1^w}) = \overline{t(x1)} = \overline{t(x)},$$
$$t(x^\sigma y^\sigma) = t(x^w \underline{y^w u^{-2}} = u^{-2}\ t(x^w y^w)$$
$$= u^{-2}\ \underline{t(x^{w^2} y)} = u^{-2}\ \overline{t(xU_u)y)}$$
$$= u^{-2}\ \underline{t(u^2 xy)} = t(xy),$$
$$t(x^\sigma y^\sigma z^\sigma) = u^{-3} t(x^w y^w z^w)$$
$$= u^{-3} t((x^w\ \ y^w)z^w) + \frac{1}{2}u^{-3}\ t(x^w)t(y^w z^w)$$
$$+ \frac{1}{2}u^{-3}\ t(y^w)t(x^w z^w) - \frac{1}{2}u^{-3} t(x^w)t(y^w)t(z^w)$$
$$+ \frac{1}{2}u^{-3}\ t(x^w y^w)t(z^w).$$

The first term of this last expression is equal to

$$u^{-3}\ \theta\bar{\theta}\ \overline{t((x \times y)z)} = n(u)^{-1}\ \theta\bar{\theta}\ \overline{t((x \times y)z)} = \overline{t((x \times y)\cdot z)}.$$

Together with what we have shown before, this yields $t(x^\sigma y^\sigma z^\sigma) = \overline{t(xyz)}$. Thus for all x,y,z

$$t(x^\sigma y^\sigma z^\sigma) = \overline{t(xyz)} = \overline{t((xy)z)} = t((zy)^\sigma z^\sigma).$$

From the non-degeneracy of the trace form it follows that σ <u>is an automorphism of $\mathcal{J}$ extending</u> $x \to \bar{x}$ <u>on</u> E. Furthermore,

$$y^{\sigma^2} = y^{w^2} u^{-2} = uyuu^{-2} = y, \quad \text{so that}$$

σ has period two. Thus the fixed field D of σ has codimension 2 in $\mathcal{J}$, and $\mathcal{J}$

<u>has the structure of tensor product</u>

$J \cong E \otimes_F D$, <u>with</u> w <u>being given as the mapping sending</u> $\xi \otimes d$ <u>to</u> $\bar{\xi} \otimes du$. We shall obtain a similar result in

Case b). $u \notin E$. Then $J = E(u)$. We have $u = 1^w$, $u^w = 1^{w^2} = 1U_u = u^2$, $u^{2^w} = u^{ww} = u^3$, and <u>in general</u> $u^{j^w} = u^{j+1}$. Thus $t(u) = t(11^w) = \overline{t(1^w 1)} = \overline{t(u)}$, $t(u^2) = t(u^w) = t(1u^w) = \overline{t(1^w u)} = \overline{t(u^2)}$, $t(u^3) = t(uu^w) = \overline{t(u^3)}$, and the minimum polynomial of u over E has coefficients in F. Thus $F(u) = D$ <u>is a cubic subfield of</u> J, <u>and</u> $J = E \otimes D$. Let σ be the automorphism of J mapping $\xi \otimes d$ to $\bar{\xi} \otimes d$. Then w sends $\xi \otimes d$ to $\bar{\xi} \otimes du$ as before. We have proved

<u>Proposition</u> 8.1. <u>Suppose the Jordan algebra</u> J <u>of</u> §7.6 <u>is as in</u> i), <u>i.e., a cubic extension field of</u> E. <u>Then the automorphism</u> $\lambda \to \bar{\lambda}$ <u>of</u> E <u>has an extension to an automorphism</u> σ <u>of period</u> 2 <u>of</u> J. <u>The fixed field</u> D <u>of</u> σ <u>contains an element</u> u <u>with</u> $n(u) = \theta\bar{\theta}$, <u>and</u> $J \cong E \otimes_F D$, <u>with</u> $(\xi \otimes d)^w = \bar{\xi} \otimes du$ <u>for all</u> $\xi \in E$, $d \in D$.

Now let M be the same vector space as J, over F. We introduce a structure of $[g_0 g_0]$-module on $M \times M$, one that will be a prototype for all our subsequent constructions. The (right) representation ρ will have, for $x,y \in M$, $a \in J$,

$(x,y)\,\rho(S_{v_{-1},v_1} \otimes (0,a)) = (ya^w, xa)$, and for $\lambda \in F$,

$$(x,y)\,\rho(T \otimes (\lambda\zeta,0)) = (\lambda\zeta x, -\lambda\zeta y).$$

From $[T \otimes (\zeta,0), S_{v_{-1},v_1} \otimes (0,a)] = 2S_{v_{-1},v_1} \otimes (0,\zeta a)$, we see that these definitions are consistent with respect to the homomorphism-property of ρ. For ρ to be a $[g_0,g_0]$-representation, we must define ρ on $\mathrm{Der}(X,-)$ so that

$$\rho([S_{v_{-1},v_1} \otimes (0,a),\ S_{v_{-1},v_1} \otimes (0,b)])$$
$$= [\rho(S_{v_{-1},v_1} \otimes (0,a)),\ \rho(S_{v_{-1},v_1} \otimes (0,b))],$$

that is, we must have

$$(x,y)\,\rho(D_{(0,b),(0,a)}) = (xab^w, ya^w b) - (xba^w, yb^w a)$$
$$- \frac{1}{3}((t(ab^w) - t(ba^w))x,\ (t(ba^w) - t(ab^w))y) \qquad (1)$$
$$= ((ab^w - ba^w - \tfrac{1}{3}t(ab^w - ba^w))x,\ (ba^w - ab^w - \tfrac{1}{3}t(ba^w - ab^w))y).$$

To see that there is a well-defined action of $\mathrm{Der}(X,-) = D_{X,X} = D_{(0,J),(0,J)}$ on $J \times J$ satisfying (1) it suffices to show that if $\Sigma D_{(0,b_i),(0,a_i)} = 0$, then $\Sigma(a_i b_i^w - b_i a_i^w - \frac{1}{3}\,\mathrm{tr}(a_i b_i^w - b_i a_i^w)1) = 0$ in J.

All that is necessary for this is to note that $D_{(0,b),(0,a)}$ sends $(0, 1_J)$ to

$$2(0,\ ba^w - ab^w - \frac{1}{3}\, t(ba^w - ab^w)1_J).$$

Thus we have a well-defined F-linear mapping $\rho : [g_0, g_0] \to \mathrm{End}_F(M \times M)$. To see that $M \times M$ is a $[g_0, g_0]$-module, it suffices to check the action of ρ on brackets of the form

$$[S_{v_{-1}, v_1} \otimes (0,c),\ D_{(0,b),(0,a)}]. \tag{2}$$

(Note that $[T \otimes (\zeta,0), D] = 0$ for all $D \in \mathrm{Der}(X,-)$.)

For once we know that ρ preserves brackets (2), and if $D \in \mathrm{Der}(X,-)$, we shall have

$$\begin{aligned}
&\rho([D_{(0,c_1),(0,c_2)}, D])\\
&\quad = \rho([[S_{v_{-1},v_1} \otimes (0,c_2),\ S_{v_{-1},v_1} \otimes (0,c_1)]D])\\
&\quad = \rho([S_{v_{-1},v_1} \otimes (0,c_2)D,\ S_{v_{-1},v_1} \otimes (0,c_1)])\\
&\quad + \rho([S_{v_{-1},v_1} \otimes (0,c_2),\ S_{v_{-1},v_1} \otimes (0,c_1)D]\\
&\quad = [\rho(S_{v_{-1},v_1} \otimes (0,c_2)D),\ \rho(S_{v_{-1},v_1} \otimes (0,c_1))]\\
&\quad + [\rho(S_{v_{-1},v_1} \otimes (0,c_2)),\ \rho(S_{v_{-1},v_1} \otimes (0,c_1)D)]\\
&\quad = [[\rho(S_{v_{-1},v_1} \otimes (0,c_2)),\ \rho(D)]\ \rho(S_{v_{-1},v_1} \otimes (0,c_1))]\\
&\quad + [\rho(S_{v_{-1},v_1} \otimes (0,c_2)),\ [\rho(S_{v_{-1},v_1} \otimes (0,c_1)),\ \rho(D)]]\\
&\quad = [\rho(D_{(0,c_1),\ (0,c_2)}),\ \rho(D)]
\end{aligned}$$

by the Jacobi identity and $[\rho(T \otimes (\zeta,0)),\ \rho(D)] = 0$.

Now $[\rho(S_{v_{-1},v_1} \otimes (0,c)),\ \rho(D_{(0,b),(0,a)})]$ sends (x,y) to

$$\begin{aligned}
&2(yc^w(ab^w - ba^w - \frac{1}{3}\, t(ab^w - ba^w)1),\\
&\qquad xc(ba^w - ab^w - \frac{1}{3}\, t(ba^w - ab^w)1).
\end{aligned} \tag{3}$$

Meanwhile

$$[S_{v_{-1},v_1} \otimes (0,c),\ D_{(0,b),(0,a)}] = S_{v_{-1},v_1} \otimes (0,c)D_{(0,b),(0,a)},$$

and $(0,c)D_{(0,b),(0,a)} =$

$$\begin{aligned}
&(0,\ \frac{1}{3}\, (t(ba^w) - t(ab^w))c + t(ca^w)b - t(ca^w)a\\
&\qquad + 4(c \times a) \times b^w - 4(c \times b) \times a^w).
\end{aligned} \tag{4}$$

Now we expand $(c \times a) \times b^w$, $(c \times b) \times a^w$ according to their definitions and use the polarized form of the basic relation

$$a^3 - t(a)a^2 + \frac{1}{2}(t(a)^2 - t(a^2))a - \frac{1}{6}(t(a)^3 - 3t(a)t(a^2) + 2t(a^3))1_J = 0$$

to substitute for the terms cab^w and cba^w and ultimately to deduce that

$$(0,c)D_{(0,b),(0,a)} = (0,\ 2c(ba^w - ab^w) - \frac{2}{3}\,t(ba^w - ab^w)c).$$

Thus $\rho([S_{v_{-1},v_1} \otimes (0,c),\ D_{(0,b),(0,a)}])$ sends (x,y) to

$$\begin{aligned}&(2y(c(ba^w) - c(ab^w) - \tfrac{1}{3}\,t(ba^w - ab^w)c)^w,\\ &\qquad 2x(c(ba^w) - c(ab^w) - \tfrac{1}{3}\,t(ba^w - ab^w)c)).\end{aligned} \tag{5}$$

To verify the equality of (3) and (5), it remains only to show that

$$(c(ba^w - ab^w))^w = c^w(ab^w - ba^w),$$

by the semilinearity of w. Here we may assume $c = \gamma \otimes c_0$, $b = \beta \otimes b_0$, $a = \alpha \otimes a_0$, with $\alpha, \beta, \gamma \in E$, $a_0, b_0, c_0 \in D$ as above; then

$$cba^w - cab^w = (\gamma\beta\bar{\alpha} - \gamma\alpha\bar{\beta}) \otimes a_0b_0c_0u,$$

$$\begin{aligned}(cba^w - cab^w)^w &= (\bar{\gamma}\,\bar{\beta}\alpha - \bar{\gamma}\,\bar{\alpha}\,\beta) \otimes a_0b_0c_0u^2\\ &= c^wab^w - c^wba^w,\end{aligned}$$

and our ρ is a representation as claimed.

From the fact that X is a faithful irreducible $\mathfrak{g}_0$-module it follows that $[\mathfrak{g}_0\ \mathfrak{g}_0]$ is semisimple, and thus that $M \times M$ is a completely reducible $[\mathfrak{g}_0\ \mathfrak{g}_0]$-module. We next consider the centralizer of $\rho([\mathfrak{g}_0\mathfrak{g}_0])$ in $\mathrm{End}_F(M \times M)$.

For fixed $c \in J$, the map $(x,y) \to (cx,cy)$ evidently centralizes the action of $[\mathfrak{g}_0\mathfrak{g}_0]$ on $M \times M$, so we may regard J as a subfield of $\mathrm{End}_{[\mathfrak{g}_0\mathfrak{g}_0]}(M \times M)$. If ψ is a general $[\mathfrak{g}_0\mathfrak{g}_0]$-endomorphism of $M \times M$, let

$$\psi((x,0)) = (f_1(x),f_2(x)),\ \psi((0,x)) = (g_1(x),g_2(x)),$$

where f_i, g_i are F-linear maps $M \to M$. If $f_1(1) = c$, the fact that ψ commutes with the action of $S_{v_{-1},v_1} \otimes (0,x)$ on $(1,0)$ yields $g_2(x) = cx$. The same test applied to $(0,1)$ gives $f_1(x^w) = cx^w$, so $f_1(y) = cy$. Thus ψ is the sum of a map $(x,y) \to (cx,cy)$ and a map $\varphi\colon (x,y) \to (g(y),f(x))$ in $\mathrm{End}_{[\mathfrak{g}_0\mathfrak{g}_0]}(M \times M)$.

The commutativity of φ with the action of $S_{v_{-1},v_1} \otimes (0,a)$ gives

$$f(x)a^w = g(xa),\ f(ya^w) = g(y)a$$

for all $x,y,a \in J$. In particular if $f(1) = b$ we have $g(a) = ba^w$, $f(a^w) = bua$, or

$$\varphi(x,y) = (by^w,\ bux^{w^{-1}}) = (by^w,\ bu^{-1}x^w)$$

for all $x,y \in J$. Using the information $J = E \otimes_F D$, $(\xi \otimes d)^w = \bar{\xi} \otimes du$ as determined above, it follows that the above $\varphi \in \mathrm{End}_{[g_0 g_0]}(M \times M)$ for each $b \in J$. Thus $\mathrm{End}_{[g_0 g_0]}(M \times M)$ consists of the mappings

$$(x,y) \to (cx + by^w,\ cy + bu^{-1}x^w)$$

for all fixed pairs $(b,c) \in J \times J$, and is isomorphic as F-vector space to $J \times J$.

The map as above for $c = 0$, $b = 1$ has square sending (x,y) to $((u^{-1}x^w)^w,\ u^{-1}y^{w^2}) = (ux,uy)$. If we denote this map by φ_1, we thus have $\varphi_1^2 = u \in D$. Thus $\varphi_1^{-1} = u^{-1}\varphi_1$, so for $c \in J$, $\varphi_1 c \varphi_1^{-1}$ (now operating on the left) sends (x,y) to

$$((c(u^{-1}(u^{-1}x)^w))^w,\ u^{-1}(c(u^{-1}y)^w)^w)$$
$$= (c^\sigma x,\ c^\sigma y),$$

where σ is the automorphism of J agreeing with $\xi \to \bar{\xi}$ on E and fixing D.

Every element of $\mathrm{End}_{[g_0 g_0]}(M \times M)$ is uniquely expressible in the form $c + b\varphi_1$, with $c,b \in J$. Thus $\mathrm{End}_{[g_0 g_0]}(M \times M)$ is a quaternion algebra over D, with basis $1, \zeta, \varphi_1, \zeta\varphi_1$. In the notation of §57 of [O'M], this is the quaternion algebra $\left(\frac{\zeta^2,u}{D}\right)$. _The_ g_0_-module_ $M \times M$ _is irreducible, of_ F-_dimension_ 12, _if and only if_ $\left(\frac{\zeta^2,u}{D}\right)$ _is a division algebra. If this quaternion algebra is not a division algebra, i.e., if_ $u \in N_{J/D}(J)$, _then_ $M \times M$ _is the direct sum of two isomorphic irreducible_ g_0_-modules of_ F-_dimension_ 6.

When $S_{v_{-1},v_1} \otimes (1,0)$ is taken to act as the identity on $M \times M$, we shall see that the irreducible g_0-module so obtained is the keystone to the rational representation theory of the larger Lie algebra g.

§2. Special Absolutely Irreducible Jordan Algebras.

a) _The homogeneous case._

The Jordan algebras J of this section are those of cases ii), iii), v) of §7.6, where in case iii) $J = \mathrm{End}_E(V)$, V a 3-dimensional E-vector space, and in case v), J is the elements of $\mathrm{End}_E(V)$, V a 6-dimensional E-space, that are symmetric with respect to a symplectic involution. In case iii) the special J-module denoted in §7.6 as $\bar{V}$ was either isomorphic to V or to the dual V^*. For us, the "homogeneous case" will refer to that where $\bar{V}$ is isomorphic to V. We let M denote the special absolutely irreducible J-module V in these cases, and we follow the lead of §1 above in defining a structure of $[g_0 g_0]$-module on $M \times M$. That is, we define a map $\rho: S_{v_{-1},v_1} \otimes (0,J) \to \mathrm{End}_F(M \times M)$ as in §1, by

$$(x,y)\,\rho(S_{v_{-1},v_1} \otimes (0,a)) = (ya^w,\ xa),$$

and we set

$$(x,y)\,\rho(T \otimes (\lambda\zeta,0)) = (\lambda\zeta x,\ -\lambda\zeta y), \quad \text{for } \lambda \in F.$$

It is still the case that $D_{(E,0),X} = 0$, and that $\mathrm{Der}(X,-)$ annihilates $(E,0)$ with $(0,c)D_{(0,b),(0,a)}$ being given by the formula (4) of §1. In particular, $c = 1 = 1_J$ gives

$$(0,1)D_{(0,b),(0,a)} = 2(0,\ b\cdot a^w - a\cdot b^w - \tfrac{1}{3}(t(b\cdot a^w) - t(a\cdot b^w))1),$$

where $a\cdot b = \frac{1}{2}(ab + ba)$. Thus if $\sum_i D_{(0,b_i),(0,a_i)} = 0$ in $\mathrm{Der}(X,-)$, we have

$$\sum_i (b_i\cdot a_i^w - a_i \cdot b_i^w) = \tfrac{1}{3}\sum_i (t(b_i\cdot a_i^w) - t(a_i\cdot b_i^w))1.$$

For $a,b,c \in J$, expanding $(4(c \times a) \times b$ according to definitions and substituting for the term $(c\cdot a)\cdot b = \frac{1}{4}(cab + acb + bca + bac)$ from the polarized generic cubic yields $4(c \times a) \times b = -abc - cba + t(ab)c + t(bc)a$, all products here being the associative product in $\mathrm{End}(V)$, where the extension of t to $\mathrm{End}(V)$ is defined below. Thus

$$(0,c)D_{(0,b),(0,a)} = (0,\ (ba^w - ab^w)c + c(a^wb - b^wa) + \tfrac{2}{3}\,(t(ab^w - ba^w))c),$$

and if $\sum_i D_{(0,b_i),(0,a_i)} = 0$, we have

$$0 = \sum_i \{(b_ia_i^w - a_ib_i^w)c + c(a_i^wb_i^w - b_i^wa_i) + \tfrac{2}{3}\,t(a_ib_i^w - b_ia_i^w)c\} \tag{7}$$

for all $c \in J$. For $c = 1$, we recover our relation (6), in associative form

$$\sum_i b_ia_i^w + a_i^wb_i - a_ib_i^w - b_i^wa_i = \tfrac{2}{3}\sum_i t(b_ia_i^w - a_ib_i^w)1,$$

$$\text{or}\quad \sum_i a_i^wb_i - b_i^wa_i = \sum_i (a_ib_i^w - b_ia_i^w + \tfrac{2}{3}\,t(b_ia_i^w - a_ib_i^w)1).$$

Substituting in (7) gives

$$0 = \sum_i ((b_ia_i^w - a_ib_i^w)c - c(b_ia_i^w - a_ib_i^w))$$

for all $c \in J$. Now the absolute irreducibility of the action of J on V yields

$$\sum_i (b_ia_i^w - a_ib_i^w) = \xi \in E.$$

Now our E-linear mapping $t{:}J \to E$ is in each case under consideration the restriction to J of an E-linear mapping $t\colon \mathrm{End}_E(V) \to E$. In cases ii), iii), t is the ordinary trace, while in case v), t is one-half the ordinary trace. A similar observation holds in the remaining cases labeled iii), v), with the "ordinary" trace replaced by the reduced trace. In cases iv), it is the K-valued K-linear trace or reduced trace that extends t to the enveloping associative algebra of J.

Now $t(\xi) = t(\xi 1_J) = 3\xi$, and

$$t(\sum_i b_i a_i^w - a_i b_i^w) = \overline{t(\sum a_i b_i^w - b_i a_i^w))} = -3\bar{\xi},$$

so $\bar{\xi} = \lambda\xi$, $\lambda \in F$. Moreover

$$\sum_i (a_i^w b_i - b_i^w a_i) = -\bar{\xi} + \frac{2}{3} t(\xi)1 = \xi, \quad \text{as well.}$$

Thus, for all $x,y \in M$ we have

$$0 = \sum_i x(a_i b_i^w - b_i a_i^w) - \frac{1}{3} t(a_i b_i^w - b_i a_i^w)x \quad \text{and}$$

$$0 = \sum_i y(a_i^w b_i - b_i^w a_i) - \frac{1}{3} t(b_i a_i^w - a_i b_i^w)y.$$

Accordingly <u>there is a well-defined linear map</u> $\rho: Der(X,-) \to End_F(M \times M)$ <u>such that</u> $\rho(D_{(0,b),(0,a)})$ <u>sends</u> (x,y) <u>to</u>

$$(xab^w - xba^w,\ ya^w b - yb^w a) \tag{8}$$
$$-\frac{1}{3}((t(ab^w) - t(ba^w))x,\ (t(ba^w) - t(ab^w))y).$$

We now have a linear map $\rho : [g_0, g_0] \to End_F(M \times M)$, which we claim is a (right) representation of $[g_0 g_0]$. As in §1, verification of this claim reduces to showing

$$\rho(S_{v_{-1},v_1} \otimes (0,c)D_{(0,b),(0,a)}) = [\rho(S_{v_{-1},v_1} \otimes (0,c)),\ \rho(D_{(0,b),(0,a)})].$$

The former sends (x,y) to

$$(y(ba^w c + ca^w b - ab^w c - cb^w a + \frac{2}{3} t(ab^w - ba^w)c)^w,$$
$$x(ba^w c + ca^w b - ab^w c - cb^w a + \frac{2}{3} t(ab^w - ba^w)c)),$$

and the latter, to

$$-\ (ya^w bc^w - yb^w ac^w,\ xab^w c - xba^w c)$$
$$+\ \frac{1}{3}((t(ba^w) - t(ab^w))yc^w,\ (t(ab^w) - t(ba^w))xc)$$
$$+\ (yc^w ab^w - yc^w ba^w,\ xca^w b - xcb^w a)$$
$$-\ \frac{1}{3}((t(ab^w) - t(ba^w))yc^w,\ (t(ba^w) - t(ab^w))xc).$$

Thus to show ρ is a representation, it suffices to show

$$(ba^w c + ca^w b - ab^w c - cb^w a)^w = b^w ac^w - a^w bc^w + c^w ab^w - c^w ba^w. \tag{9}$$

In the cases in question, "homogeneity" means that there is a semilinear mapping $s:M \to M$ such that for all $a \in J$, $x \in M$,

$$x\, a^w = xs^{-1}u\ a\ s \quad \text{(see §7.6)}.$$

Thus $x(a^w)^w = x(aU_u) = x\ u\ a\ u = xs^{-1}u(s^{-1}u\ a\ s)s = x\ u^2s^{-2}as^2$, because u commutes with s. Thus $a\ u = us^{-2}as^2$, and the E-endomorphism $s^{-2}u$ of M centralizes the action of J. Because J acts absolutely irreducibly on M, we have $s^2 = \Upsilon u$, for $\Upsilon \in E$. Because s commutes with u, s commutes with Υ, and so $\Upsilon \in F$.

Now $(ba^wc)^w = s^{-1}u(ba^wc)s = s^{-1}u\ b\ s^{-1}u\ a\ s\ c\ s$. Substituting $\Upsilon^{-1}u^{-1}s$ for the internal factor s^{-1} gives

$$\begin{aligned}(ba^wc)^w &= s^{-1}u\ b\ s\ \Upsilon^{-1}\ a\ s\ c\ s \\ &= (s^{-1}u\ b\ s)a(s\ \Upsilon^{-1}cs) = b^wac^w\end{aligned}$$

because $s^{-1}u = \Upsilon^{-1}s$. This proves (9), and the following

<u>Proposition</u> 8.2. <u>The</u> F-<u>space</u> $M \times M$ <u>has a unique structure of right</u> $[\mathcal{G}_0\mathcal{G}_0]$-<u>module in which</u>

$$(x,y)(T \otimes (\zeta,0)) = (\zeta\, x,\ -\zeta\, y),$$

$$(x,y)(S_{v_{-1},v_1} \otimes (0,a)) = (ya^w,\ xa)$$

<u>for all</u> $x,y \in M$, <u>all</u> $a \in J$.

As before, we know that $[\mathcal{G}_0\mathcal{G}_0]$ is semisimple. Indeed in §5 it will be shown that $[\mathcal{G}_0\mathcal{G}_0]$ is absolutely simple of type C_3, A_5 or D_6, respectively. Thus $M \times M$ is completely reducible, of F-dimension 12, 12 or 24, respectively. We next consider the centralizer of the action of the action of $[\mathcal{G}_0\mathcal{G}_0]$ on $M \times M$.

As in §1, each centralizing transformation is the sum of a map $(x,y) \to (f(x),f(y))$ where f centralizes the action of J on M, so has $f(x) = \lambda x$, $\lambda \in E$, and of a map $(x,y) \to (g_1(y),\ f_2(x))$, where

$$g_1(xa) = f_2(x)a^w, \quad \text{so} \quad g_1(x) = f_2(x)u, \quad \text{and}$$

where $g_1(y)a = f_2(ya^w)$, so $f_2(y)ua = f_2(ya^w)$, $f_2(xa)u = f_2(x)a^w$, all $x,y \in M$; and all $a \in J$. Now set $h(x) = f_2(xs)$. Then $h(\zeta\, x) = f_2((\zeta x)s)$

$$= -f_2((xs)\cdot(\zeta 1)) = -f_2(xs)(\zeta 1)^w\ u^{-1} = \zeta\, f_2(xs)$$

$= \zeta\, h(x)$, so $h:M \to M$ is E-linear. Moreover, for $a \in J$, $h(xa) = f_2(xas) = f_2(xu^{-1}s\ a^w) = f_2(xu^{-1}s)ua = f_2(xs)a = h(x)a$, using $f_2(y)ua = f_2(ya^w)$ with $a = 1$ to give $f_2(y)u = f_2(yu)$. It follows that there is $\mu \in E$ with $h(x) = \mu x$ for all $x \in M$, so that $f_2(x) = \mu(xs^{-1})$ for all x, and $g_1(y) = \mu(ys^{-1})u$.

Conversely, for fixed $\mu \in E$, the map $\varphi_\mu : (x,y) \to (\mu(ys^{-1})u, \mu(xs^{-1}))$ centralizes the action of $T \otimes (\zeta,0)$ and of $S_{v_{-1},v_1} \otimes (0,a)$. (For the latter, note that

$$(\mu(xas^{-1}u), \mu(ya^w s^{-1})) = (\mu(xs^{-1})a^w, \mu(ys^{-1}ua))$$

follows because $xs^{-1}a^w = xs^{-2}u\ a\ s = x\ a\ \gamma^{-1}s = x\ a\ s^{-1}u$.) Thus the centralizer of the action of $[g_0 g_0]$ on $M \times M$ consists of the maps

$$(x,y) \to (\lambda\ x + \mu(ys^{-1})u,\ \lambda y + \mu(xs^{-1})$$

for $\lambda, \mu \in E$, and has F-dimension 4. For $\lambda = 0$, $\mu = 1$, the square of the map $\varphi_1 : (x,y) \to (ys^{-1}u, xs^{-1})$ sends (x,y) to $(xs^{-2}u, ys^{-2}u) = \gamma^{-1}(x,y)$. We have, for $\lambda \in E$,

$$\varphi_\lambda((x,y)) = (\lambda(ys^{-1}u), \lambda(xs^{-1})) = \lambda\varphi_1((x,y))$$
$$= ((\bar{\lambda}y)s^{-1}u, (\bar{\lambda}x)s^{-1}) = \varphi_1\ \bar{\lambda}((x,y)).$$

It follows that our centralizer is the F-quaternion algebra $(\zeta^2, \gamma^{-1}) = \left(\frac{\zeta^2, \gamma^{-1}}{F}\right)$. The $[g_0 g_0]$-module $M \times M$ is irreducible if $\gamma \notin N_{E/F}(E)$, and otherwise is the sum of two isomorphic irreducible submodules.

Now in cases ii) and iii), we fix a basis e_1, e_2, e_3 for M over E. Then s is represented by a 3 by 3 E-matrix $(\sigma) = (\sigma_{ij})$: $e_i s = \sum_j \sigma_{ij} e_j$, and s^2 by $(\bar{\sigma})(\sigma)$. Thus $\det(s^2) = N_{E/F}(\det(\sigma)) = \det(\gamma u) = \gamma^3 \det(u) = \gamma^3 n(u) = \gamma^3 N_{E/F}(\theta) = \gamma N_{E/F}(\gamma\theta)$, and $\gamma \in N_{E/F}(E)$. We conclude:

<u>Proposition</u> 8.3. <u>In the present context, when</u> $[M:E] = 3$, <u>the</u> $[g_0 g_0]$<u>-module</u> $M \times M$ <u>is the direct sum of two isomorphic irreducible</u> 6-<u>dimensional modules</u>.

Now a central simple Lie algebra of type C_3 can only have a 6-dimensional irreducible module if that Lie algebra is split, and likewise for a central simple Lie algebras of type A_5. For in each case, 6 is the minimum dimension for non-trivial irreducible representations in the split case. If the original algebra is not split it has an irreducible representation (one of the "defining" ones) over F, of dimension greater than 6, but whose irreducible constituents upon splitting include that of dimension 6 which is assumed above. But this is impossible, as in §2.3. (Or one may remark that, by dimensions $[g_0 g_0]$ must be $sp(W)$ or $s\ell(W)$ where W is our 6-dimensional F-space.) Thus we have proved:

<u>Proposition</u> 8.4. <u>If</u> J <u>is a special absolutely irreducible Jordan algebra over</u> E, $[J:E] = 15$, <u>or</u> $[J:E] = 9$ <u>and we are in the non-homogeneous case</u>.

<u>Corollary</u>. <u>There is no super-exceptional central simple Lie algebra of rank one and dimension</u> 52.

For in this case we must have $[J:E] = 6$, and J must be the symmetric elements of $M_3(E)$ with respect to an involution of orthogonal type. Thus J is absolutely irreducible over E, and E^3 is the only special irreducible module. Thus the hypotheses of Proposition 8.4 would be satisfied, but clearly not the conclusion.

b) The non-homogeneous case.

In this case, $J = \mathrm{End}_E(V)$, V a 3-dimensional space as above, the operation being $\frac{1}{2}(ab + ba) = a\cdot b$ (right action on V). The dual space $V' = \mathrm{Hom}_E(V,E)$ is then a left $\mathrm{End}_E(V)$-module, with $< av'|v >= < v'|va >$ for all $a \in \mathrm{End}_E(V)$, $v \in V$, $v' \in V'$, so is also a special (absolutely) irreducible J-module. From §7.6, we see that the assumption of non-homogeneity, i.e., that the new structure of J-module on V introduced there is not isomorphic to the original structure on V, implies that it is isomorphic to V'. Thus there is an invertible semilinear map $s:V' \to V$ such that for all $a \in J$, $v \in V$,

$$v\ a^w = (au(vs^{-1}))s.$$

Likewise, there is an invertible semilinear map $r: V \to V'$ such that for all $a \in J$, $v' \in V'$,

$$a^w v' = ((v'r^{-1})ua)r.$$

(For otherwise there would be such a map $r:V' \to V'$ with

$$a^w v' = (au(v'r^{-1}))r \quad \text{for all} \quad a,$$

but then $q = s^{-1}rs:V \to V$ would satisfy

$$v\ a^w = ((v\ q^{-1})ua)q \quad \text{for all} \quad v \in V,\ a \in J, \quad \text{and}$$

we would be in the homogeneous case.)

Now
$$\begin{aligned} v\ a^{(w^2)} &= (a^w u(vs^{-1}))s \\ &= (((u(vs^{-1}))r^{-1})ua)rs. \end{aligned}$$

With $a = 1$ we find $vu = (u(vs^{-1}))s$, $uv' = v'r^{-1}ur$, so the above is

$$v(rs)^{-1}\ u^2\ a(rs).$$

From $a^{(w^2)} = u\ a\ u$, we find as before that

$$(rs)^{-1}u = \gamma\ I \in \mathrm{End}_E(V), \quad \gamma \in E.$$

As in a), we now proceed to define a structure of right $[\mathfrak{g}_0\mathfrak{g}_0]$-module on $M \times M$, where $M = V$, by

$$(x,y)(T \otimes (\zeta,0)) = (\zeta x, -\zeta y),$$

$$(x,y)(S_{v_{-1},v_1} \otimes (0,a)) = (ya^w, xa),$$

$$(x,y)D_{(0,b),(0,a)} = (xab^w - xba^w,\ ya^wb - yb^wa)$$

$$- \frac{1}{3}(t(ab^w - ba^w)x,\ t(ba^w - ab^w)y).$$

That there is a well-defined F-linear mapping $\rho\colon [\mathfrak{g}_0\mathfrak{g}_0] \to \mathrm{End}_F(M \times M)$ giving this action follows as before, by showing that if $a_i, b_i \in J$ satisfy (7), then

$$\sum_i (b_i a_i^w - a_i b_i^w - \frac{1}{3} t(b_i a_i^w - a_i b_i^w) 1_J)$$

$$= \sum_i (a_i^w b_i - b_i^w a_i - \frac{1}{3} t(a_i^w b_i - b_i^w a_i) 1_J)$$

centralizes the action of J on V, so is scalar (in E). Because its trace is zero, the displayed element must be zero, and this implies as before that ρ is well-defined.

To see that ρ is a (right) representation of $[\mathfrak{g}_0\mathfrak{g}_0]$ it is enough as before to verify that

$$\rho(S_{v_{-1},v_1} \otimes (0,c)D_{(0,b),(0,a)})$$

$$= [\rho(S_{b_{-1},v_1} \otimes (0,c)),\ \rho(D_{(0,b),(0,a)})],$$

and for this to verify (9), or that

$$(ab^wc + cb^wa)^w = a^wbc + c^wba^w, \tag{10}$$

for all $a,b,c \in J$.

For $v \in V = M$,

$$v(ab^wc + cb^wa)^w = ((ab^wc + cb^wa)u(vs^{-1}))s,$$

and

$$(a\,b^wc\,u(vs^{-1}))s = (a((cu(vs^{-1}))r^{-1}ub)r)s,$$

while

$$vc^wba^w = (au((vc^w)b)s^{-1})s \tag{11}$$

$$= (au((cu(vs^{-1}))sb)s^{-1})s.$$

Now $v(rs)^{-1}u = \gamma\, v$ for all $v \in M$, or $vs^{-1}r^{-1}u = \gamma v$, or $(u(vs^{-1}))r^{-1} = \gamma v$, or $u(vs^{-1}) = (\gamma\, v)r$. Applying this to (11) with v replaced by $(cu(vs^{-1}))s)b$, we see that

$$vc^wba^w = (a(\gamma((cu(vs^{-1}))sb)r)s \tag{12}$$

$$= (a(\gamma(cu(vs^{-1}))s)b)r)s.$$

Replacing v by $v's$ in $u(vs^{-1}) = (\gamma v)r$ gives $uv' = (\gamma(v's))r$, or $(uv')r^{-1} = \gamma(v's)$, or $(v'r^{-1})u = \gamma(v's)$. Applying to $v' = cu(vs^{-1})$, we have $vc^w ba^w = a((cu(vs^{-1}))r^{-1}ub)r)s$, which is the term developed above in $v(ab^w c + cb^w a)^w$. By symmetry, $va^w bc^w$ is the other term, and our verification that $M \times M$ is a $[g_0 g_0]$-module is complete.

To describe the $[g_0 g_0]$-endomorphisms of $M \times M$, we note as before that these are sums of scalar multiples by E and of maps $(x,y) \to (g(y), f(x))$, where

$$g(y)a = f(ya^w),\ g(xa) = f(x)a^w$$

for all $x, y \in M$, $a \in J$, so that $a = 1$ gives

$$g(x) = f(xu) = f(x)u \quad \text{for all } x.$$

Moreover f and g are semilinear $V \to V$.

Now define an E-linear map $h: V \to V'$ by $h(x) = (f(x)u)s^{-1} = f(xu)s^{-1} = u(f(x)s^{-1})$. Then $h(xa) = u(f(xa)s^{-1}) = (f(xa)u)s^{-1} = g(xa)s^{-1} = (f(x)a^w)s^{-1} = au(f(x)s^{-1}) = ah(x)$, and h is a J-homomorphism $V \to V'$. Because V and V' are non-isomorphic irreducible J-modules, $h = 0$. Thus the centralizer of the action of $[g_0 g_0]$ on $M \times M$ is E, and we have proved

Proposition 8.5. *In the non-homogeneous case, with $J = M_3(E)$, the module $M \times M$ for $[g_0 g_0]$ is irreducible, with centralizer E. Thus it splits upon extension of the base field into two non-isomorphic absolutely irreducible modules, each of dimension 6.*

§3. The Module $M \times M$ for Other Special J.

In all the other cases, we may follow the same procedure as in §2 to make $M \times M = V \times V$ into a module for $[g_0 g_0]$. Here the cases are as follows:

iii) J is a central simple associative division algebra A over E, with operation $\frac{1}{2}(ab + ba)$, and $V = M = A$, $[A:E] = 9$, with $x \cdot a = xa$ for $x \in M$, $a \in J$. There is a non-isomorphic module V', identified with A as vector space, with $v' \cdot a = av'$ for all $a \in J$, $v' \in V'$. The map t is the reduced trace on A. Here we have in the homogeneous case a semilinear invertible map $s: M \to M$ such that for all $v \in M$, $a \in J$, $v \cdot a^w = (((vs^{-1}) \cdot u) \cdot a)s$, $((vs^{-1}) \cdot u)s = v \cdot u$, $s^2 = du$, $d \in \mathrm{End}_J(M) \approx A^{op}$. In the non-homogeneous case, we have $s: V' \to V$ and $r: V \to V'$, invertible semilinear maps, such that for all $v \in V$, $v' \in V'$, $a \in J$,

$$v \cdot a^w = (a \cdot u \cdot (vs^{-1}))s \tag{13}$$
$$a^w v' = ((v'r^{-1})ua)r.$$

Moreover, $(vu)s^{-1} = u(vs^{-1})$, $(uv')r^{-1} = (v'r^{-1})u$ for all $v \in V$, $v' \in V'$, and

$rs = du$ for $d \in \mathrm{End}_J(V) \approx A^{op}$ as above.

To see that our mapping ρ is well-defined in this case, note that the relation (7) of §2 tells us that

$$\sum_i (b_i\bar{a}_i^w - a_ib_i^w - \frac{1}{3} t(b_ia_i^w - a_ib_i^w)1)$$
$$= \sum_i (b_i^wa_i - a_i^wb_i - \frac{1}{3} t(b_i^wa_i - a_i^wb_i)1)$$

must be central in A, hence a scalar in E. No further comment is then required.

In showing that ρ is a representation, we had to verify (9). In the homogeneous case, this comes down to showing that

$$s^{-1}u\ b\ s^{-1}u\ a\ s\ c\ s = (s^{-1}u\ b\ s)\ a(s^{-1}u\ c\ s).$$

Here $s^2 = du$ gives $s\ d^{-1}u^{-1} = s(du)^{-1} = s^{-1}$, $s^{-1}u = sd^{-1} = d^{-1}s$ and the left-hand side above is

$$s^{-1}u\ b\ s\ d^{-1}a\ s\ c\ s = (s^{-1}u\ b\ s)a\ d^{-1}s\ c\ s$$
$$= (s^{-1}u\ b\ s)a(s^{-1}u\ c\ s), \quad \text{as required.}$$

The calculation in the non-homogeneous case also follows the pattern of §2. In the non-homogeneous case it follows as in §2 that $\mathrm{End}_{[g_0g_0]}(M \times M)$ consists of the maps $(x,y) \to (dx,dy)$, $d \in \mathrm{End}_J(M)$, so $\mathrm{End}_{[g_0g_0]}(M \times M) \approx A^{op}$, a division algebra with center E, a quadratic extension of F. Accordingly, $M \times M$ is an irreducible $[g_0g_0]$-module, such that upon passage to a splitting field it has two isomorphism classes of (absolutely) irreducible constituents, each of multiplicity three and dimension 6.

In the homogeneous case, we see as before that $\mathrm{End}_{[g_0g_0]}(M \times M)$ consists of the maps

$$(x,y) \to (d_1x + d_2(ys^{-1})u,\ d_1y + d_2(xs^{-1})) \tag{14}$$

for d_1, d_2 arbitrary in $\mathrm{End}_J(M)$,

so is an F-algebra of dimension 36. Those elements of this centralizer that centralize $(x,y) \to (\zeta x, \zeta y)$ have the form $(x,y) \to (dx,dy)$, from which it follows that the center of $\mathrm{End}_{[g_0g_0]}(M \times M)$ consists of maps $(x,y) \to (\lambda x, \lambda y)$, $\lambda \in E$, and only such as commute with $(x,y) \to ((ys^{-1})u, xs^{-1})$, i.e., with $\lambda \in F$. That is, $C = \mathrm{End}_{[g_0g_0]}(M \times M)$ _is a central simple algebra_.

If C is split, then the image of $M \times M$ under a primitive idempotent in C is a 6-dimensional $[g_0g_0]$-module, and $[g_0g_0]$ is in any case a form of A_5, of dimension 35. Thus $[g_0g_0]$ would have a faithful 6-dimensional module, and would have to be split. If $C \approx M_3(Q)$, Q a quaternionic division algebra, then $[g_0g_0]$ acts as Q-endomorphisms on a 3-dimensional left Q-space, so may be identified with a Lie subalgebra of $M_3(Q)$, which must be the derived algebra. Again $[g_0 g_0]$ must contain nilpotent elements, and likewise if $C \approx M_2(D)$, D a cubic central division algebra. Thus C must be a central division algebra over E. In the sense of [Ab], the index of C is 6, and this must be its order in the Brauer group over F, with $C \approx Q \otimes D$ where Q is a quaternionic division algebra and D a cubic central division algebra. In particular, $M \times M$ must be an irreducible $[g_0 g_0]$-module.

iv) J is the set of hermitian elements of a central simple associative algebra A over K, $[A:K] = 9$, $[K:E] = 2$, with respect to an involution of second kind fixing E. Here $V = M$ is the unique irreducible right A-module, and is the unique special irreducible J-module. We have $[M:E] = 6$ or $[M:E] = 18$, according as A is split or division. The map t is the K-valued ordinary or reduced trace on A, whose values on J lie in E. The argument of the last subsection for the homogeneous case applies here without change to show that $M \times M$, with the action of $[g_0 g_0]$ as before, is a (right) $[g_0 g_0]$-module, and that $\mathrm{End}_{[g_0 g_0]}(M \times M)$ consists of the maps

$$(x,y) \to (d_1 x + d_2(ys^{-1})u,\ d_1 y + d_2(xs^{-1})) \tag{15}$$

for $d_1, d_2 \in \mathrm{End}_A(M) = K$ or A^{op}, according as A is split or not. (Here $s: M \to M$ is as before.)

As before, the center of $\mathrm{End}_{[g_0 g_0]}(M \times M)$ consists of maps $(x,y) \to (\lambda x, \lambda y)$, where $\lambda \in K$. Now conjugation by s induces an automorphism of A, hence an automorphism of K, and the relation $s^2 = du$ shows that this automorphism of K is of order 2, with fixed field $K_0 \supset F$ such that $K = E \otimes_F K_0$. In order for $(x,y) \to (\lambda x, \lambda y)$ to commute with $(x,y) \to ((ys^{-1})u, xs^{-1})u, xs^{-1})$, we must have $\lambda \in K_0$, and this characterizes our center. Thus the center of $\mathrm{End}_{[g_0 g_0]}(M \times M)$ is the field K_0, quadratic over F, the semisimple algebra $C = \mathrm{End}_{[g_0 g_0]}(M \times M)$ is simple with center K_0, and $M \times M$ is a $[g_0 g_0]$-module, which has non-isomorphic irreducible constituents (upon extension of the base field).

When $[M:E] = 6$, $[C:K_0] = 4$, and $M \times M$ has two isomorphic irreducible $[g_0 g_0]$-summands, each of dimension 12, when $C \approx M_2(K_0)$; otherwise C is a central division algebra over K_0, and $M \times M$ is irreducible. When $[M:E] = 18$, $[C:K_0] = 36$, and $M \times M$ is the sum of isomorphic $[g_0 g_0]$-irreducible submodules of dimensions 12, 24, 36, or 72, according as C is of order 1,2,3, or 6 in the Brauer group over K_0.

v) J is the set of hermitian elements of a central simple associative algebra A over E, $[A:E] = 36$, $[J:E] = 15$, with respect to an involution of first kind. By the considerations of §2, we may assume that $A \approx M_3(Q) \approx \mathrm{End}_Q(V)$, where V is a 3-dimensional left vector space over the quaternionic division algebra Q over E, and J is the self-adjoint Q-endomorphisms of V with respect to a nondegenerate form, hermitian with respect to the canonical involution in Q. Here we take $M = V$, the unique special irreducible J-module, and $M \times M$ is a $[g_0 g_0]$-module as *in the homogeneous cases above.* Also as above, $\mathrm{End}_{[g_0g_0]}(M \times M)$ *is a central simple F-algebra, so the* $[g_0g_0]$*-module* $M \times M$ *has all of its irreducible constituents (even upon splitting) isomorphic.*

We have $[M:E] = 12$, $\mathrm{End}_{[g_0g_0]}(M \times M)$ is isomorphic as vector space to $Q \times Q$, of F-dimension 16, and thus as algebra is either $M_4(F)$, $M_2(P)$ or D, where P is a quaternionic division algebra over F and D is a central division algebra over F, $[D:F] = 16$. In the case of $M_4(F)$, $M \times M$ has $[g_0g_0]$-submodules of dimension $\frac{1}{4}[M \times M:F] = 12$ for $[g_0g_0]$, and we shall see in §5 that $[g_0g_0]$ is central simple of type D_6. In this case $[g_0g_0]$ must be the skew transformations with respect to an anisotropic symmetric bilinear form (over F) on such a 12-dimensional module.

When the endomorphism algebra is $M_2(P)$, $M \times M$ has irreducible submodules of dimension 24, and otherwise $M \times M$ is irreducible of dimension 48. *Thus the* $[g_0g_0]$*-module* $M \times M$ *is either irreducible, or is the sum of* 2 *isomorphic irreducible submodules of dimension* 24, *or the sum of* 4 *such submodules of dimension* 12.

§4. Splitting-The Allison Algebras.

Let $(X,-)$ be an Allison algebra of type (f): $X = E \times J$, a division algebra with J a simple Jordan algebra over E. Let L be a finite extension field of E in which the parameter θ in the definition of $(X,-)$ is a cube, say $\theta = \varkappa^3$, $\varkappa \in L$. We may assume L to be Galois over F, and we may assume the nontrivial automorphism, call it σ, of E/F extended to L, so $(\varkappa^\sigma)^3 = \theta^\sigma = \bar{\theta}$.

Form the vector space $L \otimes_E J$, having the structure of L-Jordan algebra. The pairing $(\lambda,a) \to \lambda^\sigma \otimes a^w$ of $L \times J$ to $L \otimes_E J$ is biadditive and E-balanced, so defines an additive map $L \otimes_E J \to L \otimes_E J$ sending $\lambda \otimes a$ to $\lambda^\sigma \otimes a^w$, and this is a σ-semilinear bijection of L-vector spaces.

Now form the E-vector space

$$Y = E \times E \times (\varkappa \otimes J) \times (\varkappa^{-1} \otimes J),$$

the last two factors regarded as E-subspaces of $L \otimes_E J$. For $(\mu,\nu,\ \varkappa \otimes x,\ \varkappa^{-1} \otimes y) \in Y$ $(\mu,\ \nu \in E;\ x,y \in J)$, set

$$\overline{(\mu,\nu,\ \varkappa \otimes x,\ \varkappa^{-1} \otimes y)} = (\nu,\mu,\ \varkappa \otimes x,\ \varkappa^{-1} \otimes y),$$

and define a product by

$$\begin{aligned}
&(\mu,\nu,\ \varkappa \otimes x,\ \varkappa^{-1} \otimes y) \cdot (\mu',\ \nu',\ \varkappa \otimes x',\ \varkappa^{-1} \otimes y') \\
&= (\mu\mu' + t(x \cdot y'),\ \nu\nu' + t(x' \cdot y),\ \varkappa \otimes (\mu x' + \nu' x + \tfrac{2}{\theta}\, y \times y'), \\
&\qquad \varkappa^{-1} \otimes (\mu' y + \nu y' + 2\,\theta\, x \times x')).
\end{aligned} \tag{16}$$

Then $\mathcal{Y}$ becomes a non-associative E-algebra in which $Y \to \bar{Y}$ is an involution (E-linear). Set

$$(\mu,\nu,\ \varkappa \otimes x,\ \varkappa^{-1} \otimes y)^* = (\nu^\sigma,\mu^\sigma,\ \varkappa \otimes y^{w^{-1}},\ \varkappa^{-1} \otimes x^w),$$

a σ-semilinear bijection of $\mathcal{Y}$. One verifies easily that $Y \to Y^*$ commutes with the involution $Y \to \bar{Y}$, fixes the unit element $(1,1,0,0)$, and is an F-automorphism of $\mathcal{Y}$ with respect to the product (16). (For the last, note that

$$t(x^{w^{-1}} \cdot y^w) = t(x \cdot y)^\sigma,$$

$$(x \times x')^{w^{-1}} = \left(\frac{1}{\theta\theta^\sigma}(x^w \times x'^w)^w\right)^{w^{-1}} = \frac{1}{\theta\theta^\sigma}\ x^w \times x'^w,$$

$$(y \times y')^w = (y^{w^{-1}w} \times y'^{w^{-1}w})^w = \theta\theta^\sigma\ y^{w^{-1}} \times y'^{w^{-1}}).$$

Moreover, $*$ has period two. Thus the $*$-fixed elements of $\mathcal{Y}$ constitute an F-form of the involutorial E-algebra $(\mathcal{Y},-)$. These $*$-fixed elements have the form

$$(\mu,\mu^\sigma,\ \varkappa \otimes x,\ \varkappa^{-1} \otimes x^w), \tag{17}$$

so are isomorphic as F-module to $E \times \mathcal{J}$, the element (17) corresponding to (μ,x). The involution $Y \to \bar{Y}$ sends the element (17) to that corresponding to (μ^σ,x). A straightforward verification shows that under this correspondence the $*$-fixed elements of $\mathcal{Y}$ are isomorphic, as involutorial algebra over F, to the Allison algebra $(\mathcal{X},-)$.

Thus $(\mathcal{X},-)$ is an F-form of the E-algebra $(\mathcal{Y},-)$, so $(\mathcal{X},-)_E = (E \otimes_F \mathcal{X},\ \mathrm{id} \otimes -)$ is an involutorial E-algebra isomorphic to $(\mathcal{Y},-)$. We next consider $\mathcal{Y}_L = L \otimes_E \mathcal{Y}$ as involutorial L-algebra.

As L-module, $\mathcal{Y}_L = L \times L \times (L \otimes \varkappa\mathcal{J}) \times (L \otimes \varkappa^{-1}\mathcal{J})$, where we write $\varkappa\mathcal{J}$, $\varkappa^{-1}\mathcal{J}$ for $\varkappa \otimes \mathcal{J}$, $\varkappa^{-1} \otimes \mathcal{J}$. For $\lambda \in L$, $x \in \mathcal{J}$, the map $(\lambda,\ \varkappa x) \to \lambda\varkappa \otimes x \in L \otimes_E \mathcal{J}$ is an E-bilinear pairing, so there is an L-linear bijection $L \otimes_E \varkappa\mathcal{J} \to L \otimes_E \mathcal{J}$ mapping $\lambda \otimes \varkappa x$ to $\lambda\varkappa \otimes x$, and a similar bijection $L \otimes \varkappa^{-1}\mathcal{J} \to L \otimes \mathcal{J}$ mapping $\lambda \otimes \varkappa^{-1}y$ to $\lambda\varkappa^{-1} \otimes y$. Thus there is an L-linear bijection $\mathcal{Y} \to \mathcal{Z}$, where $\mathcal{Z} = L \times L \times (L \otimes_E \mathcal{J}) \times (L \otimes_E \mathcal{J})$, mapping $(\lambda,\mu,\ \lambda' \otimes \varkappa x,\ \mu' \otimes \varkappa^{-1}y)$ to $(\lambda,\mu,\ \lambda'\varkappa \otimes x,\ \mu'\varkappa^{-1} \otimes y)$.

We transport the structure of Allison L-algebra from $\mathcal{Y}_L$ to $\mathcal{Z}$ by this mapping, obtaining on $\mathcal{Z}$, $\overline{(\lambda,\mu,\ u,v)} = (\mu,\lambda,\ u,v)$ and

$$(\lambda,\mu,\ u,v)\cdot(\lambda',\mu',\ u',v') = (\lambda\lambda' + t(u\cdot v'),\mu\mu' + t(v\cdot u'),$$

$$\lambda u' + \mu' u + 2v \times v',\ \lambda' v + \mu v' + 2u \times u').$$

[By biadditivity, it suffices to check these when $u = \alpha \otimes x$, $v = \beta \otimes y$, $u' = \alpha' \otimes x'$, $v' = \beta' \otimes y'$ for $\alpha,\alpha',\beta,\beta' \in L$: $x,x',y,y' \in \mathcal{J}$; the product in question must then be what corresponds to the product in $\mathcal{Y}_L$ of $(\lambda,\mu,\ \alpha\varkappa^{-1} \otimes \varkappa x, \beta\varkappa \otimes \varkappa^{-1}y)$ and $(\lambda',\mu',\alpha'\varkappa^{-1} \otimes \varkappa x',\beta'\varkappa \otimes \varkappa^{-1}y')$, namely

$$(\lambda\lambda' + \alpha\beta' \otimes t(x\cdot y'),\ \mu\mu' + \beta\alpha' \otimes t(x'\cdot y),$$

$$\lambda\alpha' \varkappa^{-1} \otimes \varkappa x' + \mu' \alpha\varkappa^{-1} \otimes \varkappa x + \beta\beta' \varkappa^2 \otimes \frac{2}{\theta} \varkappa y \times y',$$

$$\lambda'\beta\varkappa \otimes \varkappa^{-1} y + \mu\beta'\varkappa \otimes \varkappa^{-1} y' + \alpha\alpha'\varkappa^{-2} \otimes 2\,\theta\varkappa\, x \times x').$$

The fact that $\varkappa^3 = \theta \in E$ enables us to pass θ, θ^{-1} from one tensor factor to the other and to complete the proof.]

§5. Splitting-the Lie Algebras.

We assume L to be a (finite-galois as needed) extension field of F, containing E, over which the Allison algebra $(X,-)_L$ is isomorphic to $L \times L \times J_L \times J_L$ as in §8.4. Further, we assume L is large enough so that J_L is split. In the cases i) - vi) as numbered in Chapter 7, this means:

i): $J_L \cong Lf_1 \oplus Lf_2 \oplus Lf_3$, where the f_i are non-zero orthogonal idempotents.

ii): (This case is vacuous for our purposes by §3; however, a few details deferred there are needed to establish this fact, and are given at the end of this section.)

iii), iv): $J_L \cong M_3(L)$, with Jordan product $a \cdot b = \frac{1}{2}(ab + ba)$.

v) J_L is the Jordan algebra of self-adjoint L-linear transformations of L^6 with respect to a non-degenerate alternate bilinear form (u,v). We take a canonical basis $u_{-3}, u_{-2}, u_{-1}, u_1, u_2, u_3$ for L^6, with $(u_i, u_j) = \delta_{i,-j}\,\mathrm{sgn}(j)$.

vi) J_L is the split exceptional Jordan algebra over L. We identify J_L with 3 by 3 hermitian matrices over the split octonions as in [Sel].

We give a sharper description of the derivations of $(X,-)_L = L \times L \times J_L \times J_L$, recalling that these are to commute with the involution. Each derivation D annihilates the subspace $L \times L$ and stabilizes the subspace $J_L \times J_L$. One verifies that the image of $(0,0,1,0)$ has the form $(0,0,c,0)$ with $t(c) = 0$, and such a D sends $(0,0,0,1)$ to $(0,0,0,-c)$. Direct calculation shows that a derivation with this effect on $(0,0,1,0)$ and $(0,0,0,1)$ is

$$\frac{1}{2} D_{(0,0,1,0),(0,0,0,c)}, \tag{18}$$

which sends (α,β,a,b) to $(0,0,a \cdot c, -b \cdot c)$ whenever $c \in J_L$, $t(c) = 0$.

Thus every derivation is the sum of an inner derivation of the form (18) and a derivation D annihilating both $(0,0,1,0)$ and $(0,0,0,1)$. Now one verifies that this latter D must be of the form $(\alpha,\beta,a,b) \to (0,0,a\Delta,b\Delta)$, where $\Delta: J_L \to J_L$ is a derivation of Jordan algebras. It follows that the derivations of $(X,-)_L$ are L-isomorphic to the space of pairs (c,Δ), where $c \in J_L$, $t(c) = 0$, and where $\Delta \in \mathrm{Der}(J_L)$.

In case i), J_L has no nontrivial derivations, and a space of dimension 2 of elements of trace 0, with basis $f_1 - f_2$ and $f_2 - f_3$. Thus $\mathrm{Der}(X,-)$ is 2-dimensional and commutative, and

$$h = \mathrm{Der}(X,-) + F(T \otimes (\zeta,0)) + F\cdot(S_{v_{-1},v_1} \otimes (1,0))$$

is a 4-dimensional commutative subalgebra of $\mathfrak{g}$. Upon field extension to L, h_L has basis:

$$t_i = \frac{1}{2} D_{(0,0,1,0),(0,0,0,f_i-f_{i+1})},\ i = 1,2:$$

$$t_3 = T \otimes (1,-1,0,0) = (T_{v_{-1},v_1} - T_{v_0,v_0}) \otimes (1,-1,0,0);$$

$$t_4 = S_{v_{-1},v_1} \otimes (1,1,0,0).$$

The (right) adjoint representation of h_L on $\mathfrak{g}_L$ decomposes $\mathfrak{g}_L$ into root-spaces and shows that h_L is a splitting Cartan subalgebra of $\mathfrak{g}_L$. The roots are a system of type D_4. Bases for root-spaces corresponding to simple roots, based on lexicographic ordering by (rational) values at t_1, t_2, t_3, t_4, are as follows:

$S_{v_{-1},v_1} \otimes (0,0,0,f_2)$, for a root γ_1;

$S_{v_0,v_1} \otimes (1,1,0,0) + T_{v_0,v_1} \otimes (1,-1,0,0)$, for a root γ_2;

$S_{v_{-1},v_1} \otimes (0,0,0,f_1)$, for a root γ_3;

$S_{v_{-1},v_1} \otimes (0,0,0,f_3)$, for a root γ_4.

Here γ_1, γ_3, γ_4 vanish at $S_{v_{-1},v_1} \otimes (1,0)$, identified with $S_{v_{-1},v_1} \otimes (1,1,0,0) \in h_L$, while γ_2 has the value 1 here.

The fundamental weights $\omega_1,\ldots,\omega_4$ for $\mathfrak{g}_L$ relative to h_L are: $\omega_2 = \gamma_1 + 2\gamma_2 + \gamma_3 + \gamma_4$, $\omega_i = \gamma_i + \gamma_2 + \frac{1}{2}\gamma_j + \frac{1}{2}\gamma_k$, $i = 1,3,4$, where $\{2,i,j,k\} = \{1,2,3,4\}$. If α is the simple root for $\mathfrak{g}$ relative to $t_0 = F\cdot(S_{v_{-1},v_1} \otimes (1,0))$ with value 1 at the displayed element, then we have $\omega_2|_{t_0} = 2\alpha$, $\omega_i|_{t_0} = \alpha$ for $i = 1,3,4$. According to our general theory, in order to construct all F-irreducible $\mathfrak{g}$-modules, we must construct all λ-admissible $\mathfrak{g}_0$-modules for $\lambda = \alpha$, 2α.

In particular, let us consider an irreducible $\mathfrak{g}_L$-module of highest t_0-weight α. The highest h_L-weight of such a module must be one of the ω_i $(i \neq 2)$, and

$$h_1 = \frac{2}{3} t_1 - \frac{2}{3} t_2 + \frac{1}{3} t_3 - 2t_4;$$

$$h_3 = -\frac{4}{3} t_1 - \frac{2}{3} t_2 + \frac{1}{3} t_3 - 2t_4;$$

$$h_4 = \frac{2}{3} t_1 + \frac{4}{3} t_2 + \frac{1}{3} t_3 - 2t_4$$

are a basis for $[\mathfrak{g}_0\mathfrak{g}_0]_L \cap h_L$, a splitting Cartan subalgebra of $[\mathfrak{g}_0\mathfrak{g}_0]_L$, consisting of coroots for the restrictions to this space of γ_1, γ_3, γ_4, respectively. Here $[\mathfrak{g}_0\mathfrak{g}_0]_L$ is a direct sum of three copies of $\mathfrak{sl}(2,L)$; for $i = 1,3,4$, the highest t_0-weight space of an irreducible $\mathfrak{g}_L$-module of highest weight ω_i is annihilated by the $(\mathfrak{g}_L)_{\pm\gamma_j}$, h_j for $j \in \{1,3,4\}$, $j \neq i$, and is the unique 3-dimensional irreducible module for $(g_L)_{\gamma_i} + (g_L)_{-\gamma_i} + Lh_i \simeq s\ell(2,L)$.

iii), iv): Let e_{ij} be the usual 3 by 3 matrix units over L, now with $1 \leq i, j \leq 3$. These form a basis for J_L, and $\mathrm{Der}(X,-)_L$ has dimension 16. We shall see that $\mathrm{Der}(X,-)_L$ is the sum of two simple Lie algebras of type A_2.

If h is a Cartan subalgebra of g of the form

$$F\cdot S_{v_{-1},v_1} \otimes (1,0) + F\cdot(T \otimes (\zeta,0)) + h_0, \tag{19}$$

where h_0 is a Cartan subalgebra of $\mathrm{Der}(X,-)$, we may assume our extension field L splits h_0. Here it will be convenient to change notation, writing t_1 for the element designated by "t_4" under i) above, and t_2 for the element previously denoted "$-t_3$".

Over the algebraic closure $\bar{L}$ of L, there is a product of factors $\exp(D_i)$, where $D_i \in \mathrm{Der}((X,-))_{\bar{L}}$ are nilpotent, which conjugates $(h_0)_{\bar{L}}$ to $(h')_{\bar{L}}$, where h' is a splitting Cartan subalgebra of $\mathrm{Der}(X,-)_L$, whose basis $t_3,\ldots,t_6$ we give explicitly below. Each such $\exp(D_i)$ fixes all elements of X of the form $(\xi,0)$ $(\xi \in \bar{L})$, so that by conjugation it fixes t_1 and t_2. Moreover, only finitely many elements of $\bar{L}$ are involved in this conjugation, so we may enlarge L as needed to assume the conjugation occurs in g_L. That is, we have an automorphism of g_L fixing t_1 and t_2 and mapping h_L as above to a maximal toral subalgebra of g_L with a prescribed splitting basis. Then the roots also correspond, as well as their values at t_1. In the sequel, we summarize this argument by saying simply that, over L, a Cartan subalgebra of the form (19) may be assumed to have K-basis $t_1,t_2,\ldots,t_n$, where $t_3,\ldots,t_n$ will be explicitly given elements of $\mathrm{Der}(X,-)_L$.

In the case at hand, we take $t_1 = S_{v_{-1},v_1} \otimes (1,1,0,0)$, $t_2 = T \otimes (-1,1,0,0)$, as stated, and $t_3 = D_{(0,0,e_{12},0),(0,0,0,e_{21})}$,

$$t_4 = D_{(0,0,e_{23},0),(0,0,0,e_{32})},$$
$$t_5 = D_{(0,0,e_{11},0),(0,0,0,e_{11})},$$
$$t_6 = D_{(0,0,e_{22},0),(0,0,0,e_{22})},$$

as our assumed basis for $(h_0)_L$. Then with h as in (19), h_L is a splitting Cartan subalgebra for g_L, relative to which a simple system of roots $\gamma_1,\ldots,\gamma_6$ has root-vectors:

$$e_{\gamma_1} = D_{(0,0,e_{12},0),(0,0,0,e_{31})};$$
$$e_{\gamma_2} = D_{(0,0,e_{23},0),(0,0,0,e_{12})};$$
$$e_{\gamma_3} = S_{v_{-1},v_1} \otimes (0,0,e_{31},0);$$
$$e_{\gamma_4} = D_{(0,0,e_{21},0),(0,0,0,e_{13})};$$
$$e_{\gamma_5} = D_{(0,0,e_{13},0),(0,0,0,e_{32})};$$

$$e_{\gamma_6} = S_{v_0,v_1} \otimes (1,1,0,0) + T_{v_0,v_1} \otimes (1,-1,0,0).$$

This system is of type E_6, with diagram

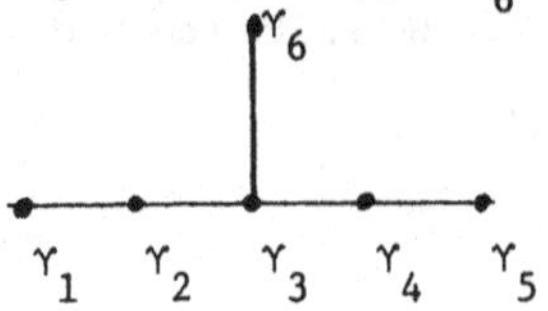

Moreover, $\gamma_i(t_1) = 0$ for $i \neq 6$, while $\gamma_6(t_1) = 1$. Thus if $\mathcal{t}_0 = Ft_1$ is our maximal split torus in $\mathcal{g}$, with α as fundamental root, we have $\gamma_i\big|_{\mathcal{t}_0} = 0$ for $i \neq 6$, $\gamma_6\big|_{\mathcal{t}_0} = \alpha$.

The fundamental weights for $\mathcal{g}_L$ relative to $\mathcal{h}_L$ are seen from [Bo], Planche V (p. 261), to be:

$$\omega_1 = \tfrac{1}{3}(4\gamma_1 + 5\gamma_2 + 6\gamma_3 + 4\gamma_4 + 2\gamma_5 + 3\gamma_6);$$
$$\omega_2 = \tfrac{1}{3}(5\gamma_1 + 10\gamma_2 + 12\gamma_3 + 8\gamma_4 + 4\gamma_5 + 6\gamma_6);$$
$$\omega_3 = 2\gamma_1 + 4\gamma_2 + 6\gamma_3 + 4\gamma_4 + 2\gamma_5 + 3\gamma_6$$
$$\omega_4 = \tfrac{1}{3}(4\gamma_1 + 8\gamma_2 + 12\gamma_3 + 10\gamma_4 + 5\gamma_5 + 6\gamma_6);$$
$$\omega_5 = \tfrac{1}{3}(2\gamma_1 + 4\gamma_2 + 6\gamma_3 + 5\gamma_4 + 4\gamma_3 + 3\gamma_6);$$
$$\omega_6 = \gamma_1 + 2\gamma_2 + 3\gamma_3 + 2\gamma_4 + \gamma_5 + 2\gamma_6 .$$

The respective restrictions to $\mathcal{t}_0$ are: α, 2α, 3α, 2α, α, 2α. Thus it is the admissible $\mathcal{g}_0$-modules corresponding to these $\mathcal{t}_0$-weights that must be determined to complete our constructions.

Clearly $e_{\gamma_1},\dots,e_{\gamma_5}$ are in $(\mathcal{g}_0)_L$, as are the root-vectors corresponding to $-\gamma_i$ $(i \neq 6)$, and inspection of the root-system E_6 shows that these ten root-vectors generate the Lie algebra $[\mathcal{g}_0\mathcal{g}_0]_L$, a simple Lie algebra of type A_5. If $\pi_1,\dots,\pi_5$ are the fundamental weights of $[\mathcal{g}_0\mathcal{g}_0]_L$ associated with the restrictions of $\gamma_1,\dots,\gamma_5$ to $\mathcal{h}_L \cap [\mathcal{g}_0\mathcal{g}_0]_L = \mathcal{c}$, we have $\omega_i\big|_{\mathcal{c}} = \pi_i$, $i \neq 6$; $\omega_6\big|_{\mathcal{c}} = 0$. In particular, if ω is the highest weight of an irreducible representation of $\mathcal{g}_L$ such that $\omega\big|_{\mathcal{t}_0} = \alpha$, then $\omega = \omega_1$ or ω_5 and the representation of $[\mathcal{g}_0\mathcal{g}_0]_L$ on the highest $\mathcal{t}_0$-weight space is that of highest $\mathcal{c}$-weight π_1 or π_5, one of the fundamental 6-dimensional representations of the split A_5.

v) Here we write $S_{u,w}$ for the map $x \to (x,u)w + (x,w)u$ of L^6, where (w,x) is our alternate form. The derivations of J_L are the maps $T \to [T,S]$, where S is a linear combination of the $S_{u,w}$, and $\mathrm{Der}(J_L)$ is thus isomorphic as L-Lie algebra to the set of those S satisfying $(wS,x) + (w,xS) = 0$ for all $w,x \in L^6$. Likewise J_L is spanned by the mappings $T_{u,w}: x \to (x,u)w - (x,w)u$.

Let $t_1, t_2 \in \mathfrak{g}_L$ be as in iii), iv). Let $t_3 \in \mathrm{Der}(X,-)_L$ send (ξ,η,x,y) to $(0,0,x\cdot(T_{u_{-1},u_1} - T_{u_{-2},u_2}), y\cdot(T_{u_{-2},u_2} - T_{u_{-1},u_1}))$, $t_4 : (\xi,\eta,x,y) \to (0,0,x\cdot(T_{u_{-2},u_2} - T_{u_{-3},u_3}), y\cdot(T_{u_{-3},u_3} - T_{u_{-2},u_2}))$, and let t_{4+j}, $1 \le j \le 3$, send (ξ,η,x,y) to $\frac{1}{4}(0,0,[x,S_{u_{-j},u_j}],[y,S_{u_{-j},u_j}])$. Then $t_1,\ldots,t_7$ are a basis for a commutative subalgebra $\mathcal{C}$ of $\mathfrak{g}_L$, whose adjoint action on $\mathfrak{g}_L$ is diagonalizable, and $\mathcal{C}$ is its own centralizer. That is, $\mathcal{C}$ is a splitting Cartan subalgebra of $\mathfrak{g}_L$. In fact, $\mathcal{C} = Lt_1 + Lt_2 + \mathcal{C}_0$, where $\mathcal{C}_0$ is a splitting Cartan subalgebra of $\mathrm{Der}(X,-)_L$, with basis the t_i, $i > 2$.

A set of simple roots $\gamma_1,\ldots,\gamma_7$ relative to $\mathcal{C}$ is of type E_7, with diagram

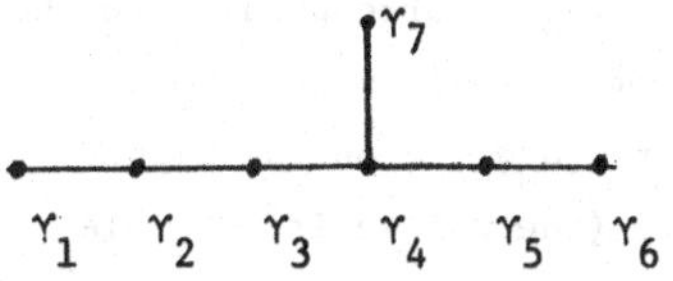

and associated root-vectors

$$e_{\gamma_1} = D_{S_{u_{-2},u_2}}; \quad e_{\gamma_2} = D_{S_{u_2,u_3}} - \frac{1}{2} R_{T_{u_2,u_3}};$$

$$e_{\gamma_3} = D_{S_{u_{-3},u_{-3}}}; \quad e_{\gamma_4} = D_{S_{u_1,u_3}} + \frac{1}{2} R_{T_{u_1,u_3}};$$

$$e_{\gamma_5} = S_{v_{-1},v_1} \otimes (0,0,T_{u_{-1},u_1},0);$$

$$e_{\gamma_6} = S_{v_0,v_1} \otimes (1,1,0,0) - T_{v_0,v_1} \otimes (1,-1,0,0);$$

$$e_{\gamma_7} = D_{S_{u_{-1},u_{-1}}}.$$

The notations here are as follows: For $c \in J_L$, $t(c) = 0$, R_c is the derivation $(\xi,\eta,x,y) \to (0,0,x\cdot c,-y\cdot c)$ of $(X,-)_L$, and $D_{S_{u,w}}$ is the derivation

$$(\xi,\eta,x,y) \to \frac{1}{4}(0,0,[x,S_{u,w}],[y,S_{u,w}]).$$

As before, we may assume $\mathcal{C}$ is of the form $(F\cdot(S_{v_{-1},v_1} \otimes (1,0)) + F\cdot(T \otimes (\zeta,0)) + h_0)_L$, where h_0 is a Cartan subalgebra of $\mathrm{Der}(X,-)$, so is defined over F. For $t_0 = Ft_1$, we have $\gamma_i|_{t_0} = 0$ for $i \ne 6$, $\gamma_6|_{t_0} = \alpha$. From [Bo], Planche VI (p. 265), the fundamental weights for $\mathfrak{g}_L$ are:

$$\omega_1 = \frac{1}{2}(3\gamma_1 + 4\gamma_2 + 5\gamma_3 + 6\gamma_4 + 4\gamma_5 + 2\gamma_6 + 3\gamma_7);$$

$$\omega_2 = 2\gamma_1 + 4\gamma_2 + 5\gamma_3 + 6\gamma_4 + 4\gamma_5 + 2\gamma_6 + 3\gamma_7;$$

$$\omega_3 = \frac{1}{2}(5\gamma_1 + 10\gamma_2 + 15\gamma_3 + 18\gamma_4 + 12\gamma_5 + 6\gamma_6 + 9\gamma_7);$$

$$\omega_4 = 3\gamma_1 + 6\gamma_2 + 9\gamma_3 + 12\gamma_4 + 8\gamma_5 + 4\gamma_6 + 6\gamma_7;$$

$$\omega_5 = 2\gamma_1 + 4\gamma_2 + 6\gamma_3 + 8\gamma_4 + 6\gamma_5 + 3\gamma_6 + 4\gamma_7;$$

$$\omega_6 = \gamma_1 + 2\gamma_2 + 3\gamma_3 + 4\gamma_4 + 3\gamma_5 + 2\gamma_6 + 2\gamma_7;$$

$$\omega_7 = \frac{1}{2}(3\gamma_1 + 6\gamma_2 + 9\gamma_3 + 12\gamma_4 + 8\gamma_5 + 4\gamma_6 + 7\gamma_7).$$

The respective restrictions to $\mathfrak{t}_0$ are: $\alpha, 2\alpha, 3\alpha, 4\alpha, 3\alpha, 2\alpha, 2\alpha$. Thus we must construct all λ-admissible $\mathfrak{g}_0$-modules for $\lambda = \alpha, 2\alpha, 3\alpha, 4\alpha,$ to have completed our constructive program.

The subalgebra $[\mathfrak{g}_0\mathfrak{g}_0]_L$ is generated by the e_{γ_i}, $i \neq 6$, and by the corresponding $e_{-\gamma_i}$, and is a simple Lie algebra of type D_6. The restrictions to $\mathfrak{c} \cap [\mathfrak{g}_0\mathfrak{g}_0]_L$, in terms of the fundamental weights $\pi_1, \ldots, \pi_6$ (corresponding to the restrictions of $\gamma_1, \ldots, \gamma_5, \gamma_7$, respectively), of the ω_i are in order: π_1, π_2, π_3, π_4, π_5, 0, π_6. Thus, for example, the representation of $[\mathfrak{g}_0\mathfrak{g}_0]_L$ on the highest $\mathfrak{t}_0$-weight space of the $\mathfrak{g}_L$-module of highest weight ω_1 (the unique irreducible $\mathfrak{g}_L$-module of highest $\mathfrak{t}_0$-weight α) is the 12-dimensional representation with highest weight π_1.

(vi) Here J_L is the split exceptional Jordan algebra, the derivations of J_L are the split Lie algebra F_4, and the derivations of $(X,-)_L$ identify with the direct sum of the space $R_{(J_0)_L}$ of multiplications by elements of J_L of trace 0, and the space $\mathrm{Der}\, J_L$. If $c_1, c_2 \in (J_0)_L$ and if $D_1, D_2 \in \mathrm{Der}\, J_L$, then $R_{c_i} + D_i$ sends (ξ,η,x,y) to $(0,0,x\cdot c_i + xD_i, -y\cdot c_i + yD_i)$, and

$$[R_{c_1} + D_1, R_{c_2} + D_2] = R_{c_1D_2 - c_2D_1} + (D_{c_1,c_2} + (D_1,D_2]),$$

where D_{c_1,c_2} is the inner derivation of J_L sending x to

$$(c_1\cdot x)\cdot c_2 - (c_2\cdot x)\cdot c_1 .$$

The bracket above is the commutator in $(R_{J_0})_L + \mathrm{Der}\, J_L$, viewed as operators on J_L, that identifies this space with the split exceptional algebra of type E_6 (cf. [J5], [Se6]).

As before, a splitting Cartan subalgebra $\mathfrak{c}$ of $\mathfrak{g}_L$ may be taken to be

$$\mathfrak{c} = L\, t_1 + L\, t_2 + \mathfrak{c}_0,$$

where $\mathfrak{c}_0$ is a splitting Cartan subalgebra of $\mathrm{Der}(X,-)_L$. A fundamental set of roots $\gamma_3, \ldots, \gamma_8$ for $\mathrm{Der}(X,-)_L$ relative to $\mathfrak{c}_0$ has corresponding root-vectors e_{γ_i}, $3 \le i \le 8$, annihilating t_1 and t_2. Thus $\gamma_3, \ldots, \gamma_8$ may be regarded as functions on $\mathfrak{c}$, roots of $\mathfrak{g}_L$ with root-vectors e_{γ_i}. Such a set of root-vectors, with corresponding basis for $\mathfrak{c}_0$, is given in [Se6], pp. 306-308, with a slight correction to be found in [Se1], p. 110.

In our setting, we must further compensate for the fact that, in [Se6], $a\cdot b$ was defined as ab + ba, whereas here we have $a\cdot b = \frac{1}{2}(ab + ba)$. Adapting [Se6] to take this into account, but otherwise using its notations, we have the following basis for $\mathcal{C}_0$:

$$t_{i+4} = (0,0,0,E_{9-i,9-i} - E_{ii}) \in \mathrm{Der}\ J_L,\ 1 \le i \le 4,$$

$$t_{i+2} = R_{w_i},\ w_i = e_{ii} - e_{i+1,i+1},\ i = 1,2.$$

Here the E_{ij} are regarded as 8×8 matrix-units acting on a split octonion algebra $\mathcal{O}$, with a distinguished basis $u_1,\dots,u_8$. The e_{ij} are 3×3 matrix units in 3×3 $\mathcal{O}$-matrices. Here J_L is identified with hermitian 3×3 $\mathcal{O}$-matrices relative to the standard involution in $\mathcal{O}$, so that w_1, w_2 are a basis for the diagonal matrices of trace zero in J_L.

With the basis $t_1,\dots,t_8$ for $\mathcal{C}$ as the source for a lexicographic ordering of the roots of $\mathfrak{g}_L$, we find the simple positive roots to have the following 8-tuples of values at $t_1,\dots,t_8$, with corresponding root-vectors:

γ_1: $(1, -3,0,0,0,0,0,0)$,

$e_{\gamma_1} = S_{v_0,v_1} \otimes (1,1,0,0) - T_{v_0,v_1} \otimes (1,-1,0,0)$,

γ_2: $(0,2,-1,1,0,0,0,0)$,

$e_{\gamma_2} = S_{v_{-1},v_1} \otimes (0,0,e_{22},0)$,

γ_3: $(0,0, \frac{1}{2}, -1, -\frac{1}{4}, -\frac{1}{4}, -\frac{1}{4}, -\frac{1}{4})$,

$e_{\gamma_3} = R_{u_5}(2,3) + (0,0,u_5,0)$ (in the notations of [Se6] for operators on J_L, identified as above with $\mathrm{Der}(X,-)_L$).

γ_4: $(0,0,0,0,0,0, \frac{1}{2}, \frac{1}{2})$,

$e_{\gamma_4} = (0,0,0,E_{35} - E_{46}) \in \mathrm{Der}\ J_L \subset \mathrm{Der}(X,-)_L$.

γ_5: $(0,0,0,0,0, \frac{1}{2}, -\frac{1}{2}, 0)$,

$e_{\gamma_5} = (0,0,0,E_{23} - E_{67}) \in \mathrm{Der}\ J_L$,

γ_6: $(0,0,0,0,0,0, \frac{1}{2}, -\frac{1}{2})$,

$e_{\gamma_6} = (0,0,0,E_{34} - E_{56}) \in \mathrm{Der}\ J_L$,

γ_7: $(0,0,\frac{1}{2},\frac{1}{2}, -\frac{1}{4}, -\frac{1}{4}, -\frac{1}{4}, \frac{1}{4})$

$e_{\gamma_7} = R_{u_5}(1,3) - (0,u_5,0,0) \in \mathrm{Der}(X,-)_L$, as for γ_3.

γ_8: $(0,0,0,0, \frac{1}{2}, -\frac{1}{2}, 0,0)$,

$e_{\gamma_8} = (0,0,0,E_{12} - E_{78}) \in \mathrm{Der}\ J_L$.

The diagram is of type E_8:

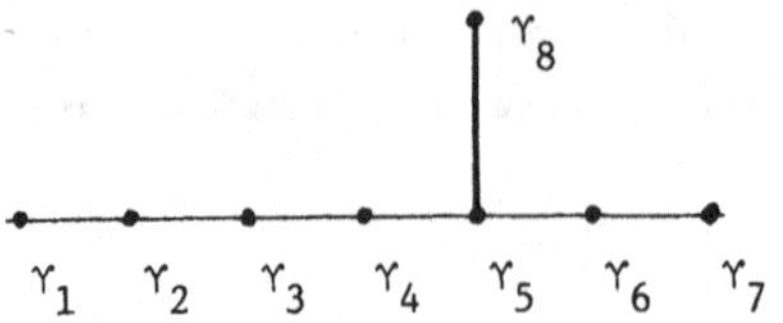

where we may now assume as before that $c_0 = (h_0)_L$, where h_0 is a Cartan subalgebra of $\mathrm{Der}(X,-)$ over F.

One has $\gamma_i(t_1) = 0$ for all $i > 1$, $\gamma_1(t_1) = \alpha(t_1) = 1$. From the expressions in [Bo], Planche VII (p. 269) for the fundamental weights $\omega_1,\ldots,\omega_8$ in terms of the γ_i we find, with $t_0 = Ft_1$ our maximal F-split torus in g, that:

$$\omega_7\big|_{t_0} = 2\,\alpha = \omega_1\big|_{t_0}\,;\ \omega_8\big|_{t_0} = 3\,\alpha = \omega_2\big|_{t_0}\,;$$

$$\omega_6\big|_{t_0} = 4\,\alpha = \omega_3\big|_{t_0}\,;\ \omega_4\big|_{t_0} = 5\,\alpha;\ \omega_5\big|_{t_0} = 6\,\alpha.$$

(The labeling of the diagram differs from that in [Bo].) Thus it is admissible g_0-modules of t_0-weights $i\alpha$, $2 \le i \le 6$, that must be constructed to complete our program.

The root-vectors e_{γ_i}, $i > 1$, and the corresponding $e_{-\gamma_i}$ generate $[g_0g_0]_L$, a subalgebra of type E_7. If β_i, $1 \le i \le 7$, is the restriction of γ_{i+1} to the Cartan subalgebra $c \cap [g_0g_0]_L$ of $[g_0g_0]_L$, then $\beta_1,\ldots,\beta_7$ have corresponding fundamental weights $\pi_1,\ldots,\pi_7$. As before the representation of $[g_0g_0]_L$ on the highest t_0-weight space of an irreducible g_L-module of highest weight ω_j has highest weight 0 if $j = 1$ and π_{j-1} if $j > 1$. Thus, for example, if $j = 1$, this is the trivial representation of $[g_0g_0]_L$ (with t_1 acting as multiplication by 2), and if $j = 7$ it is the adjoint representation of $[g_0g_0]_L$, again with t_1 represented by 2.

(ii) In order to be sure this case is excluded, we need only see that $[g_0g_0]_L$ is simple of type C_3. Here we write e_{ij}, $-1 \le i,j \le 1$, for 3×3 matrix units, regarded as transformations of L^3 with respect to the basis u_{-1},u_0,u_1, and acting on the right. (Thus $(u_i,u_j) = \delta_{i,-j}$ and $u_ie_{jk} = \delta_{ij}u_k$.)

A basis for $(J_L)_0$ consists of $T_{u_{-1},u_1} - T_{u_0,u_0} = e_{-1,-1} - 2e_{00} + e_{11}$; $T_{u_{-1},u_0} = e_{0,-1} + e_{1,0}$; $T_{u_0,u_1} = e_{-1,0} + e_{0,1}$; $\frac{1}{2}T_{u_1,u_1} = e_{-1,1}$; and $\frac{1}{2}T_{u_{-1},u_{-1}} = e_{1,-1}$. The derivations of J_L are realized by forming commutators with the three skew transformations $S_{u_0,u_1} = e_{0,1} - e_{1,0}$; $S_{u_{-1},u_0} = e_{1,0} - e_{0,-1}$; $S_{u_{-1},u_1} = e_{11} - e_{-1,-1}$. Thus $\mathrm{Der}(X,-)_L$ has dimension 8. In terms of inner derivations, a basis for $\mathrm{Der}(X,-)_L$ may be chosen to contain

$t_3 = D_{(0,0,e_{00},0),(0,0,0,e_{00})}$ and

$t_4 = D_{(0,0,e_{-1,1},0),(0,0,0,e_{1,-1})}$,

along with

$e_{\gamma_4} = D_{(0,0,T_{u_0,u_1},0),(0,0,0,T_{u_{-1},u_1})}$ and

$e_{\gamma_3} = D_{(0,0,T_{u_{-1},u_1},0),(0,0,0,T_{u_{-1},u_1} - T_{u_0,u_0})}$.

With t_1 and t_2 as in iii) - vi),

$e_{\gamma_2} = S_{v_{-1},v_1} \otimes (0,0,e_{1,-1},0)$,

$e_{\gamma_1} = S_{v_1,v_0} \otimes (1,1,0,0) - T_{v_1,v_0} \otimes (1,-1,0,0)$,

we have that t_1,t_2,t_3,t_4 are a basis for a Cartan subalgebra $\mathcal{C}$ that splits g_L, relative to which the e_{γ_i}, $1 \le i \le 4$, are root-vectors corresponding to a set of simple roots forming a root-system of type F_4, with diagram

As explained under iii), we may assume t_3, t_4 are a basis for $(\mathcal{C}_0)_L$, where $\mathcal{C}_0$ is a Cartan subalgebra of Der(X,-).

If $\mathcal{t}_0 = Ft_1$ is our maximal split torus in g, we have $\gamma_1|_{\mathcal{t}_0} = \alpha$, $\gamma_i|_{\mathcal{t}_0} = 0$ for $i > 1$, and $[g_0 g_0]_L$ is generated by e_{γ_i}, $i > 1$, along with the corresponding $e_{-\gamma_i}$. It follows that $[g_0 g_0]_L$ is the split simple Lie algebra C_3.

§6. Splitting - the $[g_0 g_0]$-Modules.

We consider here the effect of extension to the splitting field L on the $[g_0 g_0]$-modules of §§1-3, and some comparisons of irreducible constituents of these $[g_0 g_0]_L$-modules with the highest $\mathcal{t}_0$-weight spaces of irreducible g_L-modules considered in §5.

For the trivial module, nothing needs to be said. The adjoint module $[g_0 g_0]_L$ is simple except when J is a cubic extension field of E, in which case $[g_0 g_0]_L$ has been seen in §5 to decompose into three nonisomorphic 3-dimensional irreducible modules for the algebra $[g_0 g_0]_L$ of type $A_1 \times A_1 \times A_1$. When this is the case, these $[g_0 g_0]_L$-modules are isomorphic to the highest $\mathcal{t}_0$-weight spaces of the irreducible g_L-modules of highest weights $2\omega_1$, $2\omega_3$, $2\omega_4$.

In cases iii), iv), we see from 5 that the trivial $[g_0 g_0]$-module, with t_1

acting as multiplication by 2, is an F-form of the $(g_0)_L$-module which is the highest t_0-weight space of the g_L-module of highest weight ω_6, namely the adjoint g_L-module. Meanwhile $[g_0g_0]_L$, the adjoint $[g_0g_0]_L$-module, is the highest t_0-weight space for the g_L-module of highest weight $\omega_1 + \omega_5$ (or of highest weight $\omega_1 + \omega_5 + k\omega_6$, any k).

In case v), it follows similarly that the trivial $[g_0g_0]$-module, with t_1 acting as twice identity, is an F-form of the highest t_0-weight space of the adjoint g_L-module, while the adjoint module $[g_0g_0]_L$ is isomorphic, as $[g_0g_0]_L$-module, to the highest t_0-weight space of the g_L-module of highest weight ω_2 (or $\omega_2 + k\omega_7$, for any k).

Finally, in case vi), the trivial $[g_0g_0]$-module, with t_1 acting as twice the identity, is the highest t_0-weight space of the adjoint module g, and is thus an F-form of the highest t_0-weight space of the adjoint module g_L, with highest weight ω_1. The adjoint module $[g_0g_0]_L$ is $[g_0g_0]_L$-isomorphic to the highest t_0-weight space of the g_L-module of highest weight ω_7 (or $\omega_7 + k\omega_1$, for any k).

From the general considerations of Chapter 2 concerning admissible g_0-modules, it follows that we have the following table of admissible modules:

t_0-weight	case	$[g_0g_0]$-module	t_1 represented by	g_L-weight
0	all	trivial	0	0
α	none			(respectively)
2α	all	trivial	2	$\omega_2;\omega_6;\omega_7;\omega_1$
4α	v);vi)	trivial	4	$2\omega_7;2\omega_1$
6α	vi)	trivial	6	$3\omega_1$
2α	i)	adjoint	2	$2\omega_1,2\omega_3,2\omega_4$
2α	iii);iv); v);vi)	adjoint	2	$\omega_1+\omega_5;\omega_2;\omega_7$
4α	v);vi)	adjoint	4	$\omega_2+\omega_7;\omega_7+\omega_1$
6α	vi)	adjoint	6	$\omega_7+2\omega_1$

Next we consider the $[g_0g_0]$-module which is the Allison algebra X, of respective F-dimensions 8, 20, 32, 56, according as the case is i), iii)-iv), v) or vi).

In case i), the vector space X_L has basis $(1,1,0,0)$, $(1,-1,0,0)$, $(0,0,f_i,0)$, $(0,0,0,f_i)$, in our previous notation, where f_1, f_2, f_3 are orthogonal idempotents in D_L, where D is our σ-fixed cubic subfield of J containing u. Thus $u = \delta_1f_1 + \delta_2f_2 + \delta_3f_3$, $\delta_i \neq 0$ in L. Extended to L, our representation ρ of $[g_0g_0]_L$ on X_L represents

$$S_{v_{-1},v_1} \otimes (0,0,0,f_i), \quad 1 \le i \le 3, \tag{20}$$

by $(\mu, \nu, x, y) \to (t(x\cdot f_i), 0, 2yxf_i, \nu f_i)$ and so all three of these annihilate the common space with $x = 0 = y$, $\mu = 0$. Thus X_L is an irreducible $[g_0g_0]_L$-module with highest weight-vector $(1,0,0,0)$. The elements (20) are fundamental root-vectors belonging to positive roots of $[g_0g_0]_L$, with the

$$S_{v_{-1},v_1} \otimes (0,0,f_i,0)$$

belonging to the corresponding negative roots. Now $(S_{v_{-1},v_1} \otimes (0,0,f_i,0))$ sends $(1,0,0,0) \in X_L$ to $(0,0,f_i,0) \in X_L$, and this in turn to zero. It follows that the highest weight of X_L as $[g_0g_0]_L$-module is $\frac{1}{2}(\gamma_1 + \gamma_3 + \gamma_4)$, as function on $[g_0g_0]_L \cap h_L$. The highest weights of g_L with this module as highest t_0-weight space are the $\omega_1 + \omega_3 + \omega_4 + k\omega_2$. In particular, the $[g_0g_0]$-module X becomes a $3\,\alpha$-admissible (absolutely irreducible) g_0-module if we let $t_1 = S_{v_{-1},v_1} \otimes (1,0)$ act as $3\cdot$Identity. Because $\omega_1 + \omega_3 + \omega_4 + k\,\omega_2\big|_{t_0} = (3 + 2k)\alpha$, X can only be λ-admissible for those λ of this form, with $k \geq 0$.

Next consider the cases iii), iv), where X_L has as basis the $(0,0,e_{ij},0)$, $(0,0,0,e_{ij})$, $1 \leq i,j \leq 3$, as well as $(1,1,0,0)$ and $(1,-1,0,0)$. The elements of X_L annihilated by all $\rho(e_{\gamma_i})$, $1 \leq i \leq 5$, are seen to be of the form $(0,\beta,0,0)$, $\beta \in L$. Thus there is a one-dimensional highest weight space for the representation of $[g_0g_0]_L$ on X_L, and therefore X_L is an irreducible $[g_0g_0]_L$-module of dimension 20. There is only one irreducible module of dimension 20 for a split algebra of type A_5, namely that of highest weight π_3, in the labeling of §5. Only g_L-modules of highest weight $\omega_3 + k\omega_6$ can have X_L as highest t_0-weight module (for $[g_0g_0]_L$). Thus X is an absolutely irreducible, $3\,\alpha$-admissible g_0-module. The corresponding representation of g is absolutely irreducible of dimension 2925.

When $\dim.X = 32$, we find that $(0,1,0,0) \in X_L$ is a weight vector annihilated by $e_{\gamma_1},\dots,e_{\gamma_5}$, e_{γ_7}, and by all of $\mathrm{Der}(X,-)_L$, but not by

$$e_{-\gamma_5} = S_{v_{-1},v_1} \otimes (0,0,0,T_{u_{-1},u_1}).$$

Thus $(0,1,0,0)$ is the highest weight vector for an irreducible $[g_0g_0]_L$-submodule of X_L with highest weight of the form $m\pi_5$, $m > 0$, in our labeling of §5. Now the $[g_0g_0]_L$-module of highest weight π_5 is a half-spin module, of dimension 32, and for $m > 1$ the dimension of the irreducible $[g_0g_0]_L$-module of highest weight $m\pi_5$ is greater than 32. It follows that X_L is this half-spin module, and that X is an absolutely irreducible $[g_0g_0]$-module.

The $[g_0g_0]_L$-module X_L can only occur as highest t_0-weight space for irreducible g_L-modules of highest weight $\omega_5 + k\omega_6$, with restriction $(3 + 2k)\alpha$ to t_0. The only such restriction in our range is $3\,\alpha$. Thus X is a $3\,\alpha$-admissible, absolutely irreducible, g_0-module. The irreducible g-module with $3\,\alpha$ as highest t_0-weight, and with X as highest t_0-weight space, is absolutely irreducible, of dimension 8645.

In case vi), X_L is a 56-dimensional non-trivial module for the split Lie algebra $[g_0g_0]_L$, of type E_7. Because E_7 has no non-trivial irreducible module of dimension less than 56, and only one of that dimension, with highest weight π_1, our

X_L must be this irreducible module. It can occur as highest $\mathfrak{t}_0$-weight space only for irreducible $\mathfrak{g}_L$-modules of highest weights $\omega_2 + k\omega_1$, with restrictions $(3 + 2k)\alpha$ to $\mathfrak{t}_0$. In particular, the $[\mathfrak{g}_0\mathfrak{g}_0]$-module X, with $\mathfrak{t}_0$ acting by 3α, is a 3α-admissible, absolutely irreducible, $\mathfrak{g}_0$-module; with $\mathfrak{t}_0$ acting by 5α, it is 5α-admissible. The irreducible $\mathfrak{g}$-module with 3α as highest $\mathfrak{t}_0$-weight, and with X as highest $\mathfrak{t}_0$-weight space, is absolutely irreducible, of dimension 30,380.

Finally, we consider the splitting of the modules $M \times M$. Here the information derived in §3 concerning $C = \text{End}_{[\mathfrak{g}_0\mathfrak{g}_0]}(M \times M)$, and the splitting of $[\mathfrak{g}_0\mathfrak{g}_0]_L$ as in §5 yield a great deal of information. We proceed by cases, in descending order:

v) We know that $[\mathfrak{g}_0\mathfrak{g}_0]_L$ is a split algebra of type D_6, so has an irreducible module, unique to within isomorphism, of dimension 12, with highest weight π_1, and no irreducible module of lower dimension. We also know from §2 and §3 that either $[M:E] = 6$, i.e., $[M \times M:F] = 24$ and C is a quaternion algebra over F, or $[M:E] = 12$, $[M \times M:F] = 48$, and C is a central simple algebra of dimension 16 over F. In the former case, two orthogonal idempotents in C_L split $(M \times M)_L$ into non-trivial isomorphic $[\mathfrak{g}_0\mathfrak{g}_0]_L$-submodules, each of dimension 12. In the latter case, four orthogonal idempotents in C_L split $(M \times M)_L$ into four such isomorphic modules. All these have highest weight π_1, the restriction of the weight ω_1 of $\mathfrak{g}_L$, and of $\omega_1 + k\omega_6$ for any k, a weight whose restriction to $\mathfrak{t}_0$ is $(1 + 2k)\alpha$. Accordingly, if we let $\mathfrak{t}_0$ act on $M \times M$ by α, then all $\mathfrak{g}_0$-irreducible constituents of $M \times M$ are isomorphic and are α-admissible. Any of them is the unique α-admissible $\mathfrak{g}_0$-module. With $k = 1$ and $\mathfrak{t}_0$ acting by 3α, each such constituent is a 3α-admissible $\mathfrak{g}_0$-module.

iv) In case iv), we have either $[M:E] = 6$, $[M \times M:F] = 24$, or $[M:E] = 18$, $[M \times M:F] = 72$. In the former case, $[C:F] = 8$, and C_L is the product of two split quaternion algebras, so that $(M \times M)_L$ splits into four $[\mathfrak{g}_0\mathfrak{g}_0]_L$-submodules, isomorphic in pairs. But $[\mathfrak{g}_0\mathfrak{g}_0]_L$ is a split algebra of type A_5, so has all its irreducible modules of dimension at least 6, with exactly two, those of highest weights π_1 and π_5, being of dimension 6. It follows that $(M \times M)_L$ is the sum of two $[\mathfrak{g}_0\mathfrak{g}_0]_L$-modules of highest weight π_1, and two of highest weight π_5. When $[M:E] = 18$, $[C:F] = 72$ and C_L is the product of two matrix algebras $M_6(L)$, so that $(M \times M)_L$ splits into <u>twelve</u> $[\mathfrak{g}_0\mathfrak{g}_0]_L$-submodules, isomorphic in sets of six. Again all must have dimension 6, of highest weights π_1 and π_5. Because the center of C is the field K_0, all irreducible $[\mathfrak{g}_0\mathfrak{g}_0]$-submodules of $M \times M$ are isomorphic, and any one of these yields $[\mathfrak{g}_0\mathfrak{g}_0]_L$-modules of both highest weights π_1 and π_5 upon splitting.

The highest weights of $\mathfrak{g}_L$ with these restrictions are $\omega_1 + k\omega_6$, $\omega_5 + k\omega_6$, with restrictions $(2k + 1)\alpha$ to $\mathfrak{t}_0$. Accordingly it follows that, with $\mathfrak{t}_0$ acting by α, <u>an irreducible $[\mathfrak{g}_0\mathfrak{g}_0]$-submodule of $M \times M$ is the unique α-admissible $\mathfrak{g}_0$-module</u>. (We also can make this into a 3α-admissible $\mathfrak{g}_0$-module, as before.)

iii) In the non-homogeneous cases of iii), we have seen that either $[M:E] = 3$ and

$C \cong E$, or $[M:E] = 9$ and $C \cong A^{op}$, a division algebra with center E. In either case, $M \times M$ is an irreducible $[g_0g_0]$-module, and $M_L \times M_L$ has non-isomorphic $[g_0g_0]_L$-submodules. When $[M:E] = 3$, we have $[M_L \times M_L:L] = 12$, $[g_0g_0]_L$ is of type A_5, and again $M_L \times M_L$ must be the sum of two irreducible 6-dimensional $[g_0g_0]_L$-modules, of respective highest weights π_1 and π_5. When $[M:E] = 9$, C_L has six orthogonal idempotents, $[M_L \times M_L:L] = 36$, and as in iv), $M_L \times M_L$ splits into six irreducible 6-dimensional irreducible $[g_0g_0]_L$-modules, three of weight π_1 and three of weight π_5.

The corresponding highest weights of g_L are ω_1 and ω_5 (more generally, $\omega_1 + k\omega_6$, $\omega_5 + k\omega_6$), with restrictions α to t_0. With t_0 acting by α, we thus see that the irreducible g_0-module $M \times M$ is the unique α-admissible g_0-module.

iii) H: In the homogeneous case of iii), we need only consider the cases $M = A$, $M = A' = A^{op}$, as in §3, with $[M:E] = 9$, so $[M \times M:F] = 36$. Here C (for either M) is a central simple algebra of dimension 36 over F, so C_L has six orthogonal (isomorphic) idempotents, and $M_L \times M_L$ is a direct sum of six isomorphic 6-dimensional $[g_0g_0]_L$-submodules. Because $M = A$ and $M = A'$ are $[g_0g_0]$-dual, an irreducible $[g_0g_0]_L$-submodule of $(A \times A)_L$ is dual to one in $(A' \times A')_L$. As before, each must be of highest weight π_1 or π_5, and these two weights are the highest weights of dual modules.

Now it follows as before that: There are two non-isomorphic α-admissible g_0-modules. They are realized as an irreducible $[g_0g_0]$-submodule of $A \times A$, and as an irreducible $[g_0g_0]$-submodule of $A^{op} \times A^{op}$, in each case with t_0 operating by α.

i) From §2, we have $[M \times M:F] = 12$, $[C:F] = 12$, with the center of C being the cubic extension J_0 of F, so $[C:J_0] = 4$, C a quaternion algebra over J_0. Over the splitting field L, J_{0L} is the product of three copies of L, and C_L is the product of three matrix algebras $M_2(L)$, thus has 6 orthogonal idempotents, isomorphic in pairs. It follows that $M_L \times M_L$ splits into a sum of six non-trivial $[g_0g_0]_L$-modules, each of dimension two (hence irreducible) and consisting of three non-isomorphic pairs of isomorphic modules. From our splitting, these must be the highest t_0-weight spaces for irreducible g_L-modules of highest weights ω_1, ω_3, ω_4. Moreover, irreducible $[g_0g_0]$-submodules of $M \times M$, with t_0 acting by α, are the only α-admissible g_0-modules. From the structure of C, we see that either $M \times M$ is an irreducible $[g_0g_0]$-module, or it is the sum of two isomorphic irreducible $[g_0g_0]$-submodules. In either case, there is a unique α-admissible g_0-module, namely any irreducible g_0-submodule of $M \times M$.

§7. Completeness. Summary.

To complete our constructive program for the modules for the "super-exceptional" central simple Lie algebras of relative rank one, we must show how to generate the λ-admissible g_0-modules for $\lambda = k\alpha$ with values of k as determined in §5. These values result from restriction of highest weights of g_L in two ways: Either as restrictions of fundamental weights, or as restrictions of composite weights, i.e. as $\sum_{i=1} m_i \omega_i$, where $\Sigma m_i > 1$. For composite weights, once we have in hand admissible g_0-modules for which the corresponding irreducible g-modules R have the property that the irreducible g_L-constituents of these R_L exhaust all irreducible g_L-modules with <u>fundamental</u> highest weights $\omega_1,\ldots,\omega_\ell$, we consider irreducible g_0-submodules of appropriate tensor products of the given set of admissible g_0-modules. By the considerations of §2.3; this set will include all irreducible g_0-modules W such that W_L decomposes into highest t_0-weight spaces for our composite highest-weight g_L-modules with highest weight ω of given restriction to t_0.

The process of the last paragraph is just what we have called "Cartan multiplication" in the present context (§2.3; [Se3], Chapter 2). The reduction it effects limits our problem to finding all λ-admissible g_0-modules Q for λ as listed in the individual cases below, subject to the further restriction that, in the corresponding irreducible g-module R, some irreducible constituent of R_L have a fundamental highest weight. (From Galois-theoretical considerations it follows that this will then be the case for <u>all</u> irreducible constituents of R_L.)

We now tabulate according to our previous notation:

i) ω_1, ω_3, ω_4: We have seen that these have restriction α to t_0, and that their highest t_0-weight spaces occur in L-extensions of the module $M \times M$ of §2, with t_0 acting by α. This module is either irreducible, of dimension 12, or splits into two isomorphic 6-dimensional irreducible submodules. In either case, we have a unique α-admissible irreducible submodule, and a unique irreducible g-module of highest t_0-weight α. Its dimension is either 48 or 24.

ω_2: The g_L-module here is the adjoint module, with trivial $[g_0 g_0]_L$-module as highest weight space, and highest t_0-weight 2α. The corresponding 2α-admissible g_0-module is thus the trivial $[g_0 g_0]$-module. Because the unique α-admissible module must be self-dual, the trivial module occurs in its tensor square.

We therefore conclude:

<u>Theorem</u> 8.1. <u>In case</u> i), <u>where</u> g <u>is a form of</u> D_4 <u>constructed from a central</u> 8-<u>dimensional Allison division algebra, there is a unique</u> α-<u>admissible</u> g_0-<u>module</u> W, <u>constructed as in</u> §2. <u>For any</u> k, <u>the</u> kα-<u>admissible</u> g_0-<u>modules are the irreducible</u> g_0-<u>constituents of</u> $\otimes^k W$.

iii) H ω_1, ω_5: These have restrictions α to t_0. Their highest t_0-weight spaces occur in L-extensions of the modules $M \times M$ of §3, with $M = A$ or A^{op}, our associative central E-division algebra of dimension 9, with t_0 acting by α. Thus $M \times M$ is 36-dimensional in either case, and is irreducible by §3. The corresponding irreducible g-module of highest t_0-weight α has in each case dimension $6 \cdot 27$, and these two g-modules are duals of one another.

ω_6: The g_L-module is the adjoint module, with trivial $[g_0g_0]_L$-module as highest weight space and highest weight 2α. Thus the trivial $[g_0g_0]$-module is the corresponding 2α-admissible g_0-module. It is clearly a submodule of the tensor product $W_1 \otimes W_2$, where W_1, W_2 are the dual α-admissible g_0-modules, associated with ω_1 and ω_5, of the preceding paragraph.

ω_2, ω_4: The highest t_0-weight space of the corresponding irreducible g_L-module is an irreducible $[g_0g_0]_L$-module, of highest weight π_2 resp. π_4. This $[g_0g_0]_L$-module is just the exterior square $\Lambda^2(V_1)$ or $\Lambda^2(V_5)$, where V_i is the irreducible $[g_0g_0]_L$-module of highest weight π_i. Accordingly, $\Lambda^2(W_1)$ and $\Lambda^2(W_2)$, $[g_0g_0]$-modules of t_0-weight 2α, contained in $\otimes^2 W_1$ and $\otimes^2 W_2$, contain as irreducible g_0-submodules 2α-admissible submodules Y_1, Y_5 such that $(Y_i)_L \geq \Lambda^2(V_i)$ as g_{0L}-submodule. Together with our earlier remarks, this yields that $\otimes^2 W_1, \otimes^2 W_2, W_1 \otimes W_2$ <u>contain all</u> 2α-<u>admissible</u> g_0-<u>modules</u>.

ω_3: As for ω_2, ω_4, the $[g_0g_0]_L$-module is the exterior cube $\Lambda^3(V_1) \approx \Lambda^3(V_5)$, so occurs in both $(\otimes^3 W_1)_L$ and in $(\otimes^3 W_2)_L$. From this and the previous remarks, we see that $\otimes^3 W_1, \otimes^3 W_2, W_1 \otimes W_1 \otimes W_2, W_1 \otimes W_2 \otimes W_2$ contain all 3α-admissible g_0-modules. We have also identified this particular 3α-admissible module with the space X in §1, where we saw that it is absolutely irreducible (of dimension 20, with corresponding irreducible g-module of highest t_0-weight 3α and dimension 2925). In other words, <u>the irreducible</u> g_L-<u>module of highest weight</u> ω_3 <u>is defined over</u> F. (It is also easy to see that $\Lambda^2(\text{ad } g) \approx \text{ad } g \oplus$ this module.)

In our format, this is summarized as follows:

<u>Theorem</u> 8.2. <u>In case</u> iii)H, <u>where</u> g <u>is a form of</u> E_6 <u>constructed from a central 20-dimensional Allison division algebra with associated Jordan algebra</u> J <u>a central division algebra</u> A <u>of degree</u> 3 <u>over</u> E, <u>and where the map</u> w <u>is canonically associated with an automorphism of special simple</u> J-<u>bimodules, there are two non-isomorphic</u> α-<u>admissible</u> g_0-<u>modules</u>, W_1, W_2, <u>constructed as in</u> §3. <u>For any</u> k, <u>the</u> $k\alpha$-<u>admissible</u> g_0-<u>modules are the irreducible</u> g_0-<u>constituents of</u> k-<u>fold tensor products</u> $W_{i_1} \otimes \ldots \otimes W_{i_k}$, <u>with</u> $i_j = 1$ <u>or</u> 2.

<u>Remark</u>. Based on the classification scheme of Tits (cf. [T2], diagrams ${}^1E_6$ on p. 58), it appears that there should be no Allison division algebras satisfying the hypotheses of Theorem 8.2. This would follow if we could display a quaternion algebra inside C, which could be shown to be split, as in the case $A = M_3(E)$ of Proposition 8.4.

iii) NH ω_1, ω_5: Again these are the only weights of g_L with restrictions α to t_0. We have seen that $M \times M$, of dimension 12 or 36, is an α-admissible g_0-module, which splits over g_L so as to contain both the highest t_0-weight modules of irreducible g_L-modules of highest weights ω_1 and ω_5. The corresponding irreducible g-module has dimension $2 \cdot 27$ or $6 \cdot 27$.

ω_6: Again the situation is as in iii)H above, the g_L-module being the adjoint module. Here the unique α-admissible W is self-dual as $[g_0 g_0]$-module, so the trivial 2α-admissible g_0-module is a submodule of $W \otimes W$.

ω_2, ω_4: As in iii)H, the irreducible $[g_0 g_0]_L$-modules which are the highest t_0-weight spaces for modules of these highest weights are contained in $\Lambda^2(W)_L \subset (\otimes^2 W)_L$. It follows that $W \otimes W$ contains all 2α-admissible g_0-modules.

ω_3: From iii)H, we see that the corresponding irreducible $[g_0 g_0]_L$-module is contained in $\Lambda^3(W)_L \subset (\otimes^3 W)_L$, and therefore that $\otimes^3 W$ contains all 3α-admissible g_0-modules.

Our summarizing theorem here reads:

Theorem 8.3. *In case iii)NH, where g is a form of E_6 constructed from a central 20-dimensional Allison division algebra with associated Jordan algebra J a central division algebra A of degree 3 over E or a matrix algebra $A = M_3(E)$, and where the map w is canonically associated with a map s connecting the two non-isomorphic special simple J-bimodules, there is a unique α-admissible g_0-module $W = M \times M$, constructed as in §§2 and 3. For any k, the $k\alpha$-admissible g_0-modules are the irreducible g_0-constituents of $\otimes^k W$.*

iv) ω_1, ω_5: As in iii)NH above, any irreducible constituent W of the module $M \times M$ is the unique α-admissible g_0-module, its extension to L containing the highest t_0-weight modules of the irreducible g_L-modules of highest weights ω_1, ω_5.

$\omega_6, \omega_2, \omega_4, \omega_3$: These considerations are exactly as in the cases iii).

Our summary theorem now is

Theorem 8.4. *In case iv), where g is a form of E_6 constructed from a central 20-dimensional Allison division algebra with associated Jordan algebra J the fixed elements for an involution of second kind over E in a central simple algebra of degree 3 over a quadratic extension of E, there is a unique α-admissible g_0-module W, realized as (any) irreducible submodule of the g_0-module $M \times M$ of §§2 or 3. For any k, the $k\alpha$-admissible g_0-modules are the irreducible g_0-constituents of $\otimes^k W$.*

v) ω_1: With M the unique special simple bimodule for our Jordan algebra J of dimension 15, we let t_0 act on the $[g_0 g_0]$-module $M \times M$ by α, and we let W be any irreducible g_0-submodule of $M \times M$. It follows as before from what we have done that W is the unique α-admissible g_0-module.

ω_6: The corresponding g_L-module is the adjoint module, and the trivial $[g_0 g_0]$-module, with t_0-weight 2α, is the associated 2α-admissible g_0-module. From the fact that W is self-dual (because the representation of g_L of highest weight ω_1 is self-dual), it follows that this 2α-admissible g_0-module is contained in $\otimes^2 W$.

ω_2: Here we saw at the beginning of §6 that the adjoint $[g_0g_0]$-module with t_0 acting by 2α, is a form of the highest t_0-weight space of this module for g_L. The adjoint $[g_0g_0]$-module is clearly a submodule of $\mathrm{End}_F(W) = W^* \otimes W = W \otimes W$.

ω_5: We have seen at the beginning of §6 that the 32-dimensional absolutely irreducible g_0-module X, with t_0 acting by 3α, is a form of the highest t_0-weight space for the irreducible g_L-module of highest weight ω_5. As a half-spin module for $[g_0g_0]_L$, it cannot occur in any tensor power of W.

ω_7: Let W and X be the irreducible $[g_0g_0]$-modules as above, and consider the $[g_0g_0]$-module $W \otimes X$. Over L, this $[g_0g_0]_L$-module decomposes as does the tensor product of modules of highest weights π_1 and π_5 (normally with higher multiplicities). By our considerations on Clifford algebras of §5.3, extended to the case of 12-dimensional spaces with quadratic form, the irreducible $[g_0g_0]_L$-module of highest weight π_6 occurs with multiplicity one in this tensor product of irreducible $[g_0g_0]_L$-modules, with all other irreducible constituents having dimension greater than 32. It follows that there is an irreducible $[g_0g_0]$-submodule Y of $W \otimes X$ having this module as $[g_0g_0]_L$-submodule of Y_L, and all irreducible submodules of Y_L are isomorphic.

From our theory of modules for generalized even Clifford algebras in Chapter 3 we know that the $[g_0g_0]_L$-modules of highest weights π_5 and π_6 occur in $[g_0g_0]$-modules that are realized as irreducible modules for the Clifford algebra A_d, and that A_d is either simple with center a quadratic extension of F or is the product of two central simple algebras over F. The dimension of A_d is 2^{11}, so A_d can only have the 32-dimensional absolutely irreducible module X if A_d is the product of two algebras of dimension 2^{10}. Now Y is an irreducible A_d-module, not isomorphic to X, so is isomorphic to a minimal right ideal in the "other factor" of A_d. Giving Y the structure of g_0-module by letting t_0 operate by 2α, we see that Y_L contains the highest t_0-weight module of the irreducible g_L-module of highest weight ω_7. Thus Y is the missing 2α-admissible g_0-module.

ω_3, ω_4: By consideration of the split case, the corresponding 3α- and 4α-admissible g_0-modules are contained in $\Lambda^3(W) \subset \otimes^3 W$ resp. $\Lambda^4(W) \subset \otimes^4 W$.

Our summary theorem in this case is a little more complicated:

Theorem 8.5. In case v), where g is a form of E_7 constructed from a central 32-dimensional Allison division algebra with associated Jordan algebra J the fixed elements for an involution of first kind and symplectic type in a central simple algebra of degree 6 over E, there is a unique α-admissible g_0-module W, realized as (any) irreducible submodule of the g_0-module $M \times M$ of §2 or §3. The 3α-admissible module X of §2 is absolutely irreducible, and is one of two non-isomorphic irreducible modules for the Clifford algebra A_d associated with $[g_0g_0]$ (i.e., with a central simple associative F-algebra with degree 12 and involution of first kind, othogonal type, with $[g_0g_0]$ as skew elements). The other irreducible A_d-module $\mathcal{Y}$ becomes a g_0-module with t_0 acting by 2α, and is 2α-admissible.

For any non-negative integer k, the $k\alpha$-admissible g_0-modules are the irreducible g_0-modules of tensor products $(\otimes^r \mathcal{W}) \otimes (\otimes^s \mathcal{Y}) \otimes (\otimes^t \mathcal{X})$, where $k = r + 2s + 3t$.

vi) ω_1: Here the trivial $[g_0g_0]$-module, with t_0 acting by 2α, is the highest t_0-weight space of the adjoint representation. This representation is absolutely irreducible, with ω_1 as its highest weight for g_L. Thus we have the trivial 2α-admissible module F.

ω_7: The irreducible g_L-module of highest weight ω_7 has as highest t_0-weight space of t_0-weight 2α, the adjoint module for $[g_0g_0]_L$, an algebra of type E_7. Thus the adjoint module $[g_0g_0]$, with t_0 acting by 2α, is an absolutely irreducible 2α-admissible g_0-module. The dimension of the corresponding (absolutely) irreducible g-module is 3875. By our calculation on restrictions to t_0 of highest weights for g_L, this module and the previous one exhaust the 2α-admissible g_0-modules.

ω_2: We have seen that the $[g_0g_0]$-module $\mathcal{X}$, of dimension 56, is absolutely irreducible and is 3α-admissible when t_0 is made to act by 3α. The module $\mathcal{X}_L$ is then the highest t_0-weight space of the irreducible g_L-module of highest weight ω_2.

ω_3: Consider the $[g_0g_0]$-module $\Lambda^2\mathcal{X}$, of dimension 1540. Over L, one sees that this $[g_0g_0]_L$-module splits into the sum of a trivial module and an irreducible module of highest weight π_2. With t_0 acting by 4α, this latter module, of dimension 1539, is the highest t_0-weight space of the irreducible g_L-module of highest weight ω_3. Now all of this is rational over F; the alternate $[g_0g_0]$-invariant bilinear form yielding the trivial $[g_0g_0]$-module as homomorphic image of $\Lambda^2 \mathcal{X}$, hence as submodule with absolutely irreducible complement $\mathcal{Y}$, of dimension 1539, may be taken to be

$$<(\xi,x),(\eta,y)> = \frac{\xi\bar{\eta} - \eta\bar{\xi}}{\xi} + \frac{t(yx^W) - t(xy^W)}{\xi} \quad .$$

Thus $\mathcal{Y}$, with t_0 acting by 4α, is a 4α-admissible g_0-module corresponding to ω_3. The corresponding irreducible g-module is absolutely irreducible, of dimension 2,450,240.

ω_6: Here we consider the $[g_0g_0]$-module $\Lambda^2(\text{ad } [g_0g_0])$ of dimension 8778. The Lie bracket in $[g_0g_0]$ shows that this module contains $\text{ad}[g_0g_0]$ with a complementary submodule $\mathcal{W}$, of dimension 8645. Over L, we find that $\Lambda^2(\text{ad } [g_0g_0]_L)$ decomposes into $\text{ad}[g_0g_0]_L$ and the irreducible $[g_0g_0]_L$-module of highest weight π_5. With t_0 acting by 4α, as it does in $\otimes^2(\text{ad } [g_0g_0])$ when $\text{ad}[g_0g_0]$ is as under "ω_7" above, this g_{0L}-module is the highest t_0-weight module of the irreducible g_L-module of highest weight ω_6. Accordingly, we have an absolutely irreducible 4α-admissible g_0-module $\mathcal{W}$, associated with ω_6, and $\mathcal{W} \subset \otimes^2 \mathcal{A}$, where $\mathcal{A}$ is the 2α-admissible module $[g_0g_0]$, as under "ω_7". The corresponding g-module is absolutely irreducible, of dimension 6,696,000. (From our restrictions to t_0 of highest weights for g_L, and from remarks above, we see that all 4α-admissible g_0-modules are constructible from modules in the list so far developed.)

ω_4: The $[g_0g_0]_L$-module $\Lambda^3 X_L$ is seen to decompose into an irreducible module of dimension 56, which must be X_L, and an irreducible module of highest weight π_3, of dimension 27,664. Thus $\Lambda^3 X$ has a decomposition into absolutely irreducible modules which are forms of these. We let B be the 27,664-dimensional absolutely irreducible submodule, and we make B into a g_0-module by letting t_0 operate by 5α. Then B is an F-form of the highest t_0-weight space of the irreducible g_L-module of highest weight ω_4, and accordingly is 5α-admissible. The corresponding g-module is absolutely irreducible, of dimension 146,325,270, according to Tits' tables [T4]; these are our source for dimensions in general.

ω_5: The $[g_0g_0]_L$-module $\Lambda^4 X_L$ again contains an irreducible submodule of dimension 365,750 and highest weight π_4, with a complement of dimension 1540. As in the case of ω_4, there is a unique irreducible $[g_0g_0]$-submodule C of $\Lambda^4 X$ of dimension 365,750, which is 6α-admissible if we let t_0 act by 6α. This module is better realized as a submodule of $\Lambda^3(\mathrm{ad}[g_0g_0])$ for our purpose, because then it is a submodule of the tensor power $\otimes^3$ (ad $[g_0g_0]$), where ad$[g_0g_0]$ is the 2α-admissible g_0-module associated with ω_7, and as such is superfluous for our constructions by Cartan multiplication.

Now it is easy to see that $\Lambda^3(\mathrm{ad}[g_0g_0]_L)$, a $[g_0g_0]_L$-module of dimension 383,306, contains a highest weight vector of weight π_4, thus an irreducible $[g_0g_0]_L$-submodule of dimension 365,750, which must be unique. It follows as above that $\Lambda^3(\mathrm{ad}[g_0g_0])$ contains an F-form of this module, i.e., contains the module C. As g_0-modules, C is a submodule of $\otimes^3(\mathrm{ad}[g_0g_0])$, and is 6α-admissible. The corresponding representation of g is absolutely irreducible, of dimension 6,899,079,264.

ω_8: I am indebted to Alex Feingold and Steve Kass for making me aware of the extensive calculations of McKay, Moody and Patera (not yet published) on the decomposition of tensor products of irreducible modules for the split algebra E_8. In our notation, one of the simplest of these results is that the tensor product of the modules of highest weights ω_2 and ω_7 decomposes in such a way that the module of highest weight ω_8 occurs with multiplicity one, and no other irreducible module has the dimension of that of highest weight ω_8.

The implications for our context are that if one forms the tensor product of the absolutely irreducible g-modules associated above with ω_7 and ω_2, this tensor product contains a g-submodule M such that M_L is the irreducible g_L-module of highest weight ω_8. (Thus in particular M is absolutely irreducible.) By our calculations with restrictions, the highest t_0-weight space of M is a 3α-admissible (absolutely irreducible) g_0-module, the module that induces our absolutely irreducible g-module M of dimension 147,250.

The theorem summarizing these observations is

Theorem 8.6. *When g is a form of E_8 constructed from a central 56-dimensional Allison division algebra associated with an exceptional simple Jordan algebra, there are two* 2α-*admissible* g_0-*modules, described under* "ω_1" *and* "ω_7" *above. As* $[g_0 g_0]$-*modules, they are the trivial and the adjoint modules. There are two* 3α-*admissible* g_0-*modules described under* "ω_2" *and* "ω_8" *above. There is one* 4α-*admissible* g_0-*module, described under* "ω_3" *above, which does not occur in a tensor product of* 2α-*admissible modules and one* 5α-*admissible module, described under* "ω_4", *which does not occur in a tensor product of* 2α- *and* 3α-*admissible modules. These along with all other fundamental modules, are absolutely irreducible, and they generate all irreducible modules in a sense analogous to that of the last statement of Theorem* 8.5.

CHAPTER 9: COMPLEMENTS

§1. Coordinatization Theorems.

The construction of Chapter 7 and those of Chapter 5 for relative rank one are part of a larger family of results about "coordinatization" of (usually simple) Lie algebras containing suitable 3-dimensional simple subalgebras, the latter usually split. These results do not in general require a ground field of characteristic zero, although small prime characteristics are generally excluded, and some results do not require finite-dimensionality of the Lie algebra.

The first of these is commonly known as the Koecher [Ko] - Tits [T1] construction. As Tits formulates the setting, one has a subalgebra $\mathfrak{s} \simeq \mathfrak{sl}(2)$ in the finite-dimensional Lie algebra $\mathfrak{g}$, about which Tits makes assumptions slightly weaker than simplicity, and one assumes that, as adjoint $\mathfrak{s}$-module, $\mathfrak{g}$ is the sum of copies of the trivial module and the 3-dimensional module isomorphic to the adjoint module $\mathfrak{s}$. That is, $\mathfrak{g} \cong \mathcal{D} \oplus (\mathfrak{s} \otimes A)$, as $\mathfrak{s}$-module, where $\mathcal{D}$ and A are trivial $\mathfrak{s}$-modules. This isomorphism enables one to transport the structure of Lie algebra from $\mathfrak{g}$ to $\mathcal{D} \oplus (\mathfrak{s} \otimes A)$. Now Tits appeals to the structure of tensor products of the irreducible $\mathfrak{s}$-modules involved to deduce the existence of a commutative multiplication in A and a skew-symmetric pairing $A \times A \to \mathcal{D}$, together with a pairing $A \times \mathcal{D} \to A$; of course, $\mathcal{D}$, the centralizer of $\mathfrak{s}$, is a subalgebra. The inclusion of $\mathfrak{s}$ in $\mathfrak{g}$ gives A a unit element, annihilating A in the pairing to $\mathcal{D}$ and $\mathcal{D}$ in the pairing $A \times \mathcal{D} \to A$. From the Jacobi identity one then deduces that the pairing $A \times A \to A$ makes A into a Jordan algebra, on which $\mathcal{D}$ acts by derivations, and the pairing $A \times A \to \mathcal{D}$ is just the assignment to $a, a' \in A$ the inner derivation $D_{a,a'}$. If $\mathfrak{g}$ is simple, so is A, and $\mathcal{D}$ is the Lie algebra of inner derivations of A. If a maximal torus in $\mathfrak{s}$ is assumed to be a maximal split toral subalgebra of $\mathfrak{g}$, then A must be a Jordan division algebra.

For example, when A is an exceptional, central simple Jordan algebra, $\mathcal{D}$ is central simple of type F_4, and the construction of Lie algebra structure as above on $\mathfrak{g} = \mathcal{D} \oplus (\mathfrak{s} \otimes A)$ results in a central simple Lie algebra $\mathfrak{g}$ of type E_7. One finds in [J5], §IX.12, constructions, due to Tits, of the exceptional central division algebras A (the ground field must be suitable; there are no such over number fields, for instance), and thus of exotic rank-one forms of E_7. It will be clear from Chapters 5 and 7, as well as from Chapter III of [Sel], how much our approach owes to Tits' example.

Other authors have introduced compositions directly in the weight spaces, a procedure discussed in a more comprehensive setting in what follows. We cite the arguments of [A4] as typical of this approach.

In that work, Allison assumes that one has a Lie algebra g over a field F of characteristic not 2, 3 or 5, of finite dimension, with a subalgebra $s \approx s\ell(2)$ such that g is a completely reducible s-module with (only) irreducible summands of dimension 1, 3, 5, and with no non-zero ideal centralized by s. With a canonical basis {e,f,h} for s, ad h has eigenvalues among 0, ±2, ±4, and an automorphism of s inducing the Weyl reflection on Fh is induced by a linear bijection S of g, conjugation by S stabilizing $\mathrm{ad}_g\, s$. The weight space A belonging to the eigenvalue 2 for ad h then decomposes into $A = A_3 \oplus A_5$, where A_i is the intersection of A with the sum of the i-dimensional irreducible s-submodules of g, and S maps A to the (-3)-eigenspace of ad h, stabilizing each isotypic component. Writing $a = a_3 + a_5$, Allison sets $\bar{a} = a_3 - a_5 \in A$. For $a,b \in A$,

$$ab = \frac{1}{4}([[a^S, \bar{b}]e] - [[a,\bar{b}^S]e] - 2[[a\ \bar{b}]f])$$

is again in A, and this product and involution $(a \to \bar{a})$ make (A,-) into an Allison algebra ([A4], Theorem 4) from which the Lie algebra g is recovered, to within trivial changes in parametrization, as in the construction of Chapter 7. (The centralizer of s must be contained in the derivations of (A,-), and must contain the inner derivations, and (A,-) need not be simple, nor have skew elements of dimension one.)

Other authors have shown that it is possible to relax the requirement of finite-dimensionality for g. Thus Benkart [Be] considers a setting as in the last paragraph, where the characteristic is different from 2, 3 and where g is a direct sum of irreducible s-submodules of dimensions 1, 2, 3, but without finiteness assumptions. When 2-dimensional constituents are absent, one has the Koecher-Tits context, and the 2-eigenspace of ad h carries the structure of Jordan algebra J. This last is still the case in general, but then the 1-eigenspace of ad h carries the structure of special J-module. This analysis is applied by Benkart to Lie algebras <u>with minimum condition on inner ideals</u>, as in the next paragraph.

For characteristics ≠ 2, an <u>inner ideal</u> in a Lie algebra g is a subspace m such that for each $x \in m$, $(\mathrm{ad}\ x)^2$ maps g into m. (A similar notion and a structure theory based on it have been given for Jordan algebras by Jacobson and McCrimmon - cf. [J4], Chapter IV - and these enter into Benkart's study.) Those $x \neq 0$ for which $(\mathrm{ad}\ x)^2 = 0$ each span trivial one-dimensional inner ideals; such x are called absolute zero-divisors, and the (finite dimensional) Lie algebras of prime characteristic containing absolute zero divisors are what Kostrikin calls <u>strongly degenerate</u>. Simple Lie algebras of this kind exist (only in prime characteristic), but are excluded from consideration by an explicit axiom of Benkart.

When g has no absolute zero divisors, but has an element $e \neq 0$ such that $(\mathrm{ad}\ e)^3 = 0$ and satisfying conditions which, in the presence of minimum condition on inner ideals, assure that e is contained in a subalgebra s = {e,f,h} isomorphic to $\mathfrak{sl}(2)$, then, for (a more general requirement than) characteristics $\neq 2,3$, Benkart proves:

The algebra $\mathfrak{g}$ decomposes as $\mathfrak{s}$-module into a direct sum of 1, 2 and 3-dimensional irreducible submodules, and the sum of the highest weight spaces of three-dimensional modules has the structure of Jordan algebra A with minimum condition on inner ideals, the sum of the highest weight spaces for two-dimensional modules being a "special unitary module" for A. Thus the situation is completely analogous to that in the finite-dimensional case previously discussed. The assumption of simplicity for $\mathfrak{g}$ is equivalent to simplicity for the Jordan algebra A, to which the Jacobson-McCrimmon theory may then be applied.

A more comprehensive Jordan theory has been developed by Zelmanov, and has been similarly applied in [Z]. In Benkart's grading, the eigenvalues 0, ±1, ±2 of ad h give $\mathfrak{g}$ a grading $\mathfrak{g} = \sum_{i=-2}^{2} \mathfrak{g}_i$ by the corresponding eigenspaces. Any element x of $\mathfrak{g}_2$ satisfies the condition $(\mathrm{ad}\ x)^3 = 0$, so the presence of the grading already assures the existence of candidates for copies of $\mathfrak{sl}(2)$. Thus Zelmanov considers Lie algebras with findings $\mathfrak{g} = \sum_{i=-n}^{n} \mathfrak{g}_i$, where the $\mathfrak{g}_i$ may have infinite dimension. He further assumes $\mathfrak{g}$ to be simple, and that $\mathfrak{g} \neq \mathfrak{g}_0$. Both in order to apply the techniques and in order to exclude known simple finite-dimensional algebras of characteristic p to which the results do not apply (e.g., the Jacobson-Witt algebras [J2]), Zelmanov adds the hypothesis that the ground field does not have prime characteristic less than $4n + 1$.

In the previous cases, the introduction of a Jordan structure on a suitable subspace ($\mathfrak{g}_1$ or $\mathfrak{g}_2$ in the current setting) really resulted from an identification of this subspace with its "opposite". In place of taking this as his starting point, Zelmanov notes that if $|n|$ is maximal with $\mathfrak{g}_n \neq 0$, then $\mathfrak{g}_{-n} + \mathfrak{g}_n$ has the structure of "Jordan pair", directly from the operations of the Lie algebra, and he bases his analysis of simple $\mathfrak{g}$ as above on a general theory that he has developed for Jordan pairs. The conclusions are not surprising, involving only infinite-dimensional analogues of classical Lie algebras and of the Koecher-Tits construction (for which he adjoins I. L. Kantor to the list of those deserving credit).

§2. Rational Representations.

The Galois-theoretic approach to rational representations has been referred to at points in the discussion. It is presented in [BT] and [T5] for the case of algebraic groups defined over F, where usually no restriction is made on the characteristic of F, but where F is sometimes assumed to be perfect. The results are easily translated to Lie algebras of characteristic zero; one may use [Sel] as a basic dictionary.

In [T5] the main stress is on the absolutely irreducible representations and their description. This is a somewhat more delicate question than that of parametrizing all F-irreducible representations, a class that from the point of view of [BT] and [T5] is parametrized as follows:

Let F_s be the separable closure of F, $\Gamma = \mathrm{Gal}(F_s/F)$. If g is (for us) central simple over F, Γ acts on the split simple Lie algebra $F_s \otimes g$ by semi-automorphisms, and these stabilize $F_s \otimes h$, where h is a given Cartan subalgebra of g. Accordingly, each $\sigma \in \Gamma$ effects a permutation of the roots of $F_s \otimes g$ relative to the splitting Cartan subalgebra $F_s \otimes h$. There is a unique w_σ in the Weyl group W of $F_s \otimes h$ such that σw_σ^{-1} stabilizes the set of simple roots (an ordering assumed fixed). One thus obtains an action of Γ on the set of simple roots (or on the Dynkin diagram), which extends by $\mathbb{Q}$-linearity to the lattice of weights. Then the irreducible F-representations are parametrized by Γ-orbits of dominant integral functions on $F_s \otimes h$. If λ is such a function, so is each λ^σ, $\sigma \in \Gamma$. Denoting by V_μ the irreducible $F_s \otimes g$-module of highest weight μ, one may try to make Γ act on $\sum_{\mu=\lambda^\sigma,\ \text{some}\ \sigma\in\Gamma} \oplus V_\mu$, in such a way that the fixed elements are an irreducible g-module. There are some difficulties to be overcome with this program as stated; for example, if Γ acts trivially on the diagram, no action of Γ on V_λ is evident from the above. Our point of view would be to note that a finite subextension K of F already splits g, so that V_λ is defined over K, say $V_\lambda = F_s \otimes_K M_\lambda$. Now M_λ is a finite-dimensional g-module, all of whose irreducible summands are isomorphic, and the F-irreducible g-module corresponding to the Γ-orbit $\{\lambda\}$ of λ is any one of these. To achieve an analogous result in the case of reductive groups requires more elaborate machinery, in particular for reduction of the base field.

It is a consequence of the analysis in [T5] that the representation of the centralizer of a maximal F-split torus on the highest relative weight space determines the F-irreducible module for the given reductive algebraic group defined over F. This was also observed at about the same time by Parshall (unpublished, except in his 1971 Yale thesis, "Rational Representations and Subgroups of Algebraic Groups"), and other aspects are treated in [St].

To give the Galois approach a "constructive" character by our standards, it would be sufficient to carry out the following program: Given g and h, find a "small" finite Galois extension K of F splitting these data, and determine the associated action of $\mathrm{Gal}(K/F)$ on the Dynkin diagram. Using explicit descriptions of the fundamental K-irreducible modules for $K \otimes g$, one then can proceed as indicated in the next-to-last paragraph above, replacing M_λ by the sum of the M_{λ^σ}, λ^σ in the $\mathrm{Gal}(K/F)$-orbit of the (fundamental) weight λ. This program would be applicable in the anisotropic case as well.

Therein we see one difficulty, namely that over general fields of characteristic zero we do not have assurance that we are "given" all central simple g, when g is anisotropic central simple of dimension 28, 78, 133 or 248. Another is that, even in classical cases, we cannot be sure as to what kind of Galois splitting extension is available. For example, if a central F-division algebra D is a crossed product, and if E is a Galois maximal subfield, then the elements of trace zero in E are a Cartan subalgebra of the derived Lie algebra D', and we may take K = E; on the other hand, if D is not a crossed product we may have a quite uncertain situation.

A rather systematic study on splitting fields, aiming at finding smallest ones, was undertaken by Weisfeiler [We] (at that time transliterated "Veisfeiler"). This work also has some bearing on the question of smallest field extensions over which g becomes isotropic, and thus on the limitations of the constructions of Chapter 7, beginning with an anisotropic 3-dimensional algebra. (Over a quadratic extension, one already acquires nilpotent elements.) Unfortunately, as Weisfeiler told the author, his paper contains errors and uncertainties, and he is no longer among us to point them out. Still it is recommended for careful study and reworking in connection with the Galois approach to problems treated here.

§3. When Are All Modules Rational?

In several cases it has been shown that, if K is a splitting field for the maximal torus h in the central simple Lie algebra g, then all the fundamental irreducible g_K-modules are defined over F, i.e., they have F-forms as g-modules. Of course this must be the case if g is to have the property that every finite-dimensional irreducible g_K-module is defined over F. In fact, this necessary condition is also sufficient. For if

$$\omega = \sum_{i=1}^{\ell} m_i\omega_i$$

is the highest weight of an irreducible g_K-module M, with respect to h_K, where the ω_i are the fundamental weights and the m_i non-negative integers, suppose M_i is an irreducible g-module such that $(M_i)_K$ is the irreducible g_K-module of highest weight ω_i $(1 \le i \le \ell)$. Then M occurs with multiplicity one in the tensor product

$$(\otimes^{m_1}(M_1)_K) \otimes (\otimes^{m_2}(M_2)_K) \otimes \dots \otimes (\otimes^{m_\ell}(M_\ell)_K), \tag{1}$$

all tensor products being taken over K, and every other irreducible submodule has highest weight λ satisfying $\lambda < \omega$, in the usual partial ordering. Now the tensor product (1) is isomorphic to

$$((\otimes^{m_1}M_1) \otimes \dots \otimes (\otimes^{m_\ell}M_\ell))_K, \tag{2}$$

where the tensor products are taken over F, and where we may assume K/F is finite galois.

The action, resulting from (2), of an element σ of the Galois group G of K/F on (1) maps M to another irreducible g_K-submodule M^σ. If w_σ is an inner automorphism of g_K stabilizing h_K such that σw_σ^{-1} stabilizes the set of positive roots of g_K with respect to h_K, then the highest weight λ of M^σ can only satisfy $\lambda \leq \omega$ if $\lambda = \omega$. Thus $M^\sigma = M$ for all $\sigma \in G$, and the G-fixed elements of M are an F-form of M.

Reviewing the results of these notes and those of [Se3] case-by-case, we see that the answer to our title-question is affirmative, that is, that all modules are defined over F, when:

1) g is split of any type.
2) g_K is of type B or D, and g is the Lie algebra of skew transformations of an F-vector space V with respect to a non-degenerate symmetric F-bilinear form, provided that the center of the even Clifford algebra of V is split, along with each of its (at most two) simple ideals.

Moreover 1) and 2) are the only cases of classical central simple Lie algebras (those of Chaps. 1 and 3) where all modules are rational. (See [Se3], Prop. V. 3, Th. VI.1, Chap. VII, as well as noting that if V is a vector space over a division algebra $D \neq F$, where g acts as in Chaps. 1 and 3, then V is a g-module whose irreducible constituents are not absolutely irreducible.)

3) If g_K is of type G_2 or F_4, then classical results of Jacobson and Tomber show that $g \cong \mathrm{Der}(A)$, where A is an octonion algebra resp. exceptional simple Jordan algebra. In each case the space A_0 of elements of trace zero of A is an absolutely irreducible g-module, as is the adjoint module g. When A is an octonion algebra, these are F-forms of the two fundamental g_K-modules, so all g_K-modules are rational. When A is an exceptional Jordan algebra, these are F-forms of two fundamental g_K-modules, and $(\wedge^2 A_0)_K$ resp. $(\wedge^2 g)_K$ contain the remaining fundamental modules, each with multiplicity one, and with no other constituents having the dimension of $\wedge^2 A_0$ resp. $\wedge^2 g$. This is sufficient to answer our question affirmatively.
4) In the remaining isotropic cases, investigated in [Se3] and here, the results are as follows:

When $g = s\ell(3,O)$, $sp(6,O)$, or $F_4(O)$, with O an octonionic division algebra, g_K is of respective type E_6, E_7, E_8, and our answer is <u>affirmative</u> ([Se3], Chap. IX).

When $g = s\ell(2,J)$, J an exceptional Jordan division algebra, g_K is of type E_6, and our answer is <u>affirmative</u>. ([Se3], Chap. X).

When g is of relative type G_2, our answer is affirmative if and only if g is split or is coordinatized by an exceptional Jordan division algebra, in which case g_K is of type E_8 ([Se3], Chap. XI).

For the exceptional algebras of this volume, we see by examining the results of Chapter 6, in particular in the cases where the highest weight space is a Clifford module for g_0, that, among the g_K-types E_6, E_7, E_8, the only ones of Chapter 6 yielding an affirmative answer are those of type E_8 (cf. Prop. E_8 of §6.5 and Prop. $E_{8,2}$ of §6.6).

When g is as in Chaps. 7 and 8, we follow the labeling i)-vi) used there. In case i), where g_K is of type D_4, the g_K-modules of highest weight $\omega_1,\omega_3,\omega_4$ have no F-forms, so our answer is negative. The same applies in case iii) H, to the g_K-modules of highest weights ω_1 and ω_5, and likewise to iii) NH (see Theorems 8.2 and 8.3). Likewise for iv), where the action of g_0 on "$M \times M$" is centralized by the quadratic field extension. The action of the algebra C of §8.6 on the g_0-module "$M \times M$" shows that the answer is negative in case v). Finally, we see from Theorem 8.6 that the answer is affirmative in case vi), where g_K is of type E_8. One is tempted to conjecture that all modules are rational whenever g is a form of E_8, isotropic or not.

Indeed the work of McKay, Moody and Patera cited in connection with Theorem 8.6 shows that this is the case. For starting with the adjoint module g, we find that the module $g \otimes g$ contains forms of the irreducible g_K-modules of highest weights ω_2 and ω_7 (cf. also [Sel], §A.2). The cited work of the three authors shows that the tensor product of these last two contains forms of the irreducible modules of highest weights ω_3,ω_6 and ω_8. (The precise result reads:

$$M_{\omega_2} \otimes M_{\omega_7} = M_{\omega_2+\omega_7} \oplus M_{\omega_1+\omega_8} \oplus M_{\omega_6} \oplus M_{\omega_1+\omega_2} \oplus M_{\omega_3} \oplus 2M_{\omega_1+\omega_7} \oplus M_{\omega_8} \oplus M_{2\omega_1} \oplus$$

$$\oplus M_{\omega_7} \oplus M_{\omega_1}).$$

Finally we see in rather straightforward fashion that $\wedge^4 M_{\omega_1}$ and $\wedge^5 M_{\omega_1}$ contain M_{ω_4} resp. M_{ω_5}, with multiplicity one in each case. Thus all fundamental g_k-modules have F-forms, and we are done.

BIBLIOGRAPHY

[Ab] Albert, A. A., Structure of Algebras. Amer. Math. Soc. Colloquium Publs. No. XXIV, AMS, New York, 1939.

[A1] Allison, B. N., Lie algebras of type BC_1. Trans. Amer. Math. Soc. 224 (1976), 75-86.

[A2] ______, A construction of Lie algebras from J-ternary algebras. Amer. Jour. Math. 98 (1976), 285-294.

[A3] ______, A class of nonassociative algebras with involution containing the class of Jordan algebras. Math. Annalen 237 (1978), 133-156.

[A4] ______, Models of isotropic Lie algebras. Communications in Algebra 7 (1979), 1835-1875.

[A5] ______, Structurable division algebras and relative rank one simple Lie algebras. Canad. Math. Soc. Conf. Proceedings 5 (1986), 139-156.

[AF] Allison, B. N., and Faulkner, J. R., A Cayley-Dickson process for a class of structurable algebras. Trans. Amer. Math. Soc. 283 (1984), 185-210.

[AH] Allison, B. N. and Hein, W., Isotopes of some nonassociative algebras with involution. Jour. Algebra 69 (1981), 120-142.

[Be] Benkart, G. M., On inner ideals and ad-nilpotent elements of Lie algebras. Trans. Amer. Math. Soc. 232 (1977), 61-81.

[BT] Borel, A., and Tits, J., Groupes réductifs. Publ. Math. I.H.E.S. 27 (1965), 55-151.

[Bo] Bourbaki, N., Groupes et algèbres de Lie, Chaps. IV-VI. Éléments de Math., Fasc. XXXIV, Hermann, Paris, 1968.

[BW] Brauer, R. and Weyl, H., Spinors in n dimensions. Amer. Jour. of Math. 57 (1935), 425-449.

[Br] Brown, R. B., Groups of type E_7. Jour. f. reine angew. Math. 236 (1969), 79-102.

[C] Chevalley, C., The Algebraic Theory of Spinors. Columbia Univ. Press, New York, 1954.

[Fa] Faulkner, J. R., A construction of Lie algebras from a class of ternary algebras. Trans. Amer. Math. Soc. 155 (1971), 397-408.

[FF] Faulkner, J. R. and Ferrar, J. C., On the structure of symplectic ternary algebras. Nederl. Akad. Wetensch. Proc. Ser. A 75 (= Indag. Math. 34) (1972), 247-256.

[Fr] Freudenthal, H., Sur le groupe exceptionnel E_7. Nederl. Akad. Wetensch. Proc. Ser. A. 56 (= Indag. Math. 15) (1953), 81-89.

[He] Hein, W., A construction of Lie algebras by triple systems. Trans. Amer. Math. Soc. 205 (1975), 79-95.

[Ho] Howe, R., θ-series and invariant theory. Proc. Symp. Pure Math. XXXIII (1979), Amer. Math. Soc., Providence. Part I, pp. 275-285.

[Hu] Humphreys, J. E., Introduction to Lie Algebras and Representation Theory. Springer-Verlag, New York, 1972.

[J1] Jacobson, N., Rational methods in the theory of Lie algebras. Ann. of Math. 36 (1935), 875-881.

[J2] ______, Completely reducible Lie algebras of linear transformations. Proc. Amer. Math. Soc. 2 (1951), 105-113.

[J3] ______, Lie Algebras. Wiley-Interscience, New York 1962. Reprinted Dover Publs., New York, 1979.

[J4] ______, Structure and Representation of Jordan Algebras. A.M.S. Colloquium Publ. v. XXXIX. Amer. Math. Soc., Providence, 1968.

[J5] ______, Basic Algebra II. Freeman, San Francisco, 1980.

[Ka] Kantor, I. L., Models of exceptional Lie algebras. (Russian) Dokl. Akad. Nauk SSSR 208 (1973), 1276-1279. (English trans.) Soviet Math. Doklady 14 (1973), 254-258.

[Ko] Koecher, M., Imbedding of Jordan algebras in Lie algebras. I. Amer. Jour. Math. 89 (1967), 787-815.

[Mc1] McCrimmon, K., The Freudenthal-Springer-Tits constructions of exceptional Jordan algebras. Trans. Amer. Math. Soc. 139 (1969), 495-510.

[Mc2] ______, The Freudenthal-Springer-Tits constructions revisited. Trans. Amer. Math. Soc. 148 (1970), 293-314.

[Me] Meyberg, K., Eine Theorie der Freudenthalschen Tripelsysteme I, II. Nederl. Akad. Wetensch Proc. Ser. A. 71 (= Indag. Math. 30) (1968), 162-190.

[O'M] O'Meara, O. T., Introduction to Quadratic Forms. Grundlehren der Math. Wiss., v. 117. Springer, 1963.

[Sa] Saltman, D. J., The Brauer group is torsion. Proc. Amer. Math. Soc. 81 (1981), 385-387.

[St] Satake, I., Symplectic representations of algebraic groups satisfying a certain analyticity condition. Acta Math. 127 (1967), 215-279.

[Se1] Seligman, G. B., Rational Methods in Lie Algebras. Lect. Notes in Pure and Applied Math., v. 17. Marcel Dekker, New York, 1976.

[Se2] ______, Mappings into symmetric powers. Jour. of Algebra 62 (1980), 455-472.

[Se3] ______, Rational constructions of modules for simple Lie algebras. Contemp. Math., v. 5. Amer. Math. Society, Providence, 1981.

[Se4] ______, Higher even Clifford algebras. Sem. d'algèbre P. Dubreil et M.-P. Malliavin, Paris, 1982. Lecture Notes in Math., v. 1029, pp. 159-191. Springer, 1983.

[Se5] ______, Generalized even Clifford algebras. Jour. of Algebra 82 (1983), 398-458.

[Se6] ______, On automorphisms of Lie algebras of classical type. III. Trans. A.M.S. 97 (1960), 286-316.

[Sp] Springer, T. A., Characterization of a class of cubic forms. Nederl. Akad. Wetensch. Proc. Ser. A. 65 (= Indag. Math. 24) (1962), 259-265.

[T1] Tits, J., Une classe d'algèbres de Lie en relation avec les algèbres de Jordan. Nederl. Akad. Wetensch. Proc. Ser. A. 65 (= Indag. Math. 24) (1962), 530-535.

[T2] ______, Classification of algebraic semi-simple groups. Proc. Symp. Pure Math. v. 9. Amer. Math. Soc., Providence, 1966, pp. 33-62.

[T3] ______, Algèbres alternatives, algèbres de Jordan et algèbres de Lie exceptionnelles. I. Nederl. Akad. Wetensch. Proc. Ser. A. 69 (= Indag. Math. 28) (1966), 223-237.

[T4] ______, Tabellen zu den einfachen Lie Gruppen und ihren Darstellungen. Lect. Notes in Math. Bd. 40. Springer-Verlag, 1967.

[T5] ______, Représentations linéaires irréductibles d'un groupe réductif sur un corps quelconque. Jour. f. reine angew. Math. 247 (1971), 196-220.

[We] Weisfeiler, B. J. (Veisfeiler, B. Ju.), Some properties of singular semisimple algebraic groups over nonclosed fields. (English translation) Trans. Moscow Math. Soc. for the Year 1969. Vol. 20 (Amer. Math. Soc., Providence, 1971), pp. 109-134. Russian original: Trudy Moscow Math. Soc. 20 (1969), 111-136.

[Z] Zelmanov, E. I., Lie algebras with a finite grading. (Russian) Mat. Sbornik 124 (166) (1984), 353-392. (English translation) Math. of the USSR-Sbornik 52 (1985), 347-385.

Index

LECTURE NOTES IN MATHEMATICS
Edited by A. Dold and B. Eckmann

Some general remarks on the publication of monographs and seminars

In what follows all references to monographs, are applicable also to multiauthorship volumes such as seminar notes.

1. Lecture Notes aim to report new developments - quickly, informally, and at a high level. Monograph manuscripts should be reasonably self-contained and rounded off. Thus they may, and often will, present not only results of the author but also related work by other people. Furthermore, the manuscripts should provide sufficient motivation, examples and applications. This clearly distinguishes Lecture Notes manuscripts from journal articles which normally are very concise. Articles intended for a journal but too long to be accepted by most journals, usually do not have this "lecture notes" character. For similar reasons it is unusual for Ph.D. theses to be accepted for the Lecture Notes series.

   Experience has shown that English language manuscripts achieve a much wider distribution.

2. Manuscripts or plans for Lecture Notes volumes should be submitted either to one of the series editors or to Springer-Verlag, Heidelberg. These proposals are then refereed. A final decision concerning publication can only be made on the basis of the complete manuscripts, but a preliminary decision can usually be based on partial information: a fairly detailed outline describing the planned contents of each chapter, and an indication of the estimated length, a bibliography, and one or two sample chapters - or a first draft of the manuscript. The editors will try to make the preliminary decision as definite as they can on the basis of the available information.

3. Lecture Notes are printed by photo-offset from typed copy delivered in camera-ready form by the authors. Springer-Verlag provides technical instructions for the preparation of manuscripts, and will also, on request, supply special staionery on which the prescribed typing area is outlined. Careful preparation of the manuscripts will help keep production time short and ensure satisfactory appearance of the finished book. Running titles are not required; if however they are considered necessary, they should be uniform in appearance. We generally advise authors not to start having their final manuscripts specially tpyed beforehand. For professionally typed manuscripts, prepared on the special stationery according to our instructions, Springer-Verlag will, if necessary, contribute towards the typing costs at a fixed rate.

   The actual production of a Lecture Notes volume takes 6-8 weeks.

.../...

4. Final manuscripts should contain at least 100 pages of mathematical text and should include

   - a table of contents
   - an informative introduction, perhaps with some historical remarks. It should be accessible to a reader not particularly familiar with the topic treated.
   - subject index; this is almost always genuinely helpful for the reader.

5. Authors receive a total of 50 free copies of their volume, but no royalties. They are entitled to purchase further copies of their book for their personal use at a discount of 33 1/3 %, other Springer mathematics books at a discount of 20 % directly from Springer-Verlag.

   Commitment to publish is made by letter of intent rather than by signing a formal contract. Springer-Verlag secures the copyright for each volume.

Vol. 1145: G. Winkler, Choquet Order and Simplices. VI, 143 pages. 1985.

Vol. 1146: Séminaire d'Algèbre Paul Dubreil et Marie-Paule Malliavin. Proceedings, 1983–1984. Edité par M.-P. Malliavin. IV, 420 pages. 1985.

Vol. 1147: M. Wschebor, Surfaces Aléatoires. VII, 111 pages. 1985.

Vol. 1148: Mark A. Kon, Probability Distributions in Quantum Statistical Mechanics. V, 121 pages. 1985.

Vol. 1149: Universal Algebra and Lattice Theory. Proceedings, 1984. Edited by S. D. Comer. VI, 282 pages. 1985.

Vol. 1150: B. Kawohl, Rearrangements and Convexity of Level Sets in PDE. V, 136 pages. 1985.

Vol 1151: Ordinary and Partial Differential Equations. Proceedings, 1984. Edited by B.D. Sleeman and R.J. Jarvis. XIV, 357 pages. 1985.

Vol. 1152: H. Widom, Asymptotic Expansions for Pseudodifferential Operators on Bounded Domains. V, 150 pages. 1985.

Vol. 1153: Probability in Banach Spaces V. Proceedings, 1984. Edited by A. Beck, R. Dudley, M. Hahn, J. Kuelbs and M. Marcus. VI, 457 pages. 1985.

Vol. 1154: D.S. Naidu, A.K. Rao, Singular Pertubation Analysis of Discrete Control Systems. IX, 195 pages. 1985.

Vol. 1155: Stability Problems for Stochastic Models. Proceedings, 1984. Edited by V.V. Kalashnikov and V.M. Zolotarev. VI, 447 pages. 1985.

Vol. 1156: Global Differential Geometry and Global Analysis 1984. Proceedings, 1984. Edited by D. Ferus, R.B. Gardner, S. Helgason and U. Simon. V, 339 pages. 1985.

Vol. 1157: H. Levine, Classifying Immersions into $\mathbb{R}^4$ over Stable Maps of 3-Manifolds into $\mathbb{R}^2$. V, 163 pages. 1985.

Vol. 1158: Stochastic Processes – Mathematics and Physics. Proceedings, 1984. Edited by S. Albeverio, Ph. Blanchard and L. Streit. VI, 230 pages. 1986.

Vol. 1159: Schrödinger Operators, Como 1984. Seminar. Edited by S. Graffi. VIII, 272 pages. 1986.

Vol. 1160: J.-C. van der Meer, The Hamiltonian Hopf Bifurcation. VI, 115 pages. 1985.

Vol. 1161: Harmonic Mappings and Minimal Immersions, Montecatini 1984. Seminar. Edited by E. Giusti. VII, 285 pages. 1985.

Vol. 1162: S.J.L. van Eijndhoven, J. de Graaf, Trajectory Spaces, Generalized Functions and Unbounded Operators. IV, 272 pages. 1985.

Vol. 1163: Iteration Theory and its Functional Equations. Proceedings, 1984. Edited by R. Liedl, L. Reich and Gy. Targonski. VIII, 231 pages. 1985.

Vol. 1164: M. Meschiari, J.H. Rawnsley, S. Salamon, Geometry Seminar "Luigi Bianchi" II – 1984. Edited by E. Vesentini. VI, 224 pages. 1985.

Vol. 1165: Seminar on Deformations. Proceedings, 1982/84. Edited by J. Ławrynowicz. IX, 331 pages. 1985.

Vol. 1166: Banach Spaces. Proceedings, 1984. Edited by N. Kalton and E. Saab. VI, 199 pages. 1985.

Vol. 1167: Geometry and Topology. Proceedings, 1983–84. Edited by J. Alexander and J. Harer. VI, 292 pages. 1985.

Vol. 1168: S.S. Agaian, Hadamard Matrices and their Applications. III, 227 pages. 1985.

Vol. 1169: W.A. Light, E.W. Cheney, Approximation Theory in Tensor Product Spaces. VII, 157 pages. 1985.

Vol. 1170: B.S. Thomson, Real Functions. VII, 229 pages. 1985.

Vol. 1171: Polynômes Orthogonaux et Applications. Proceedings, 1984. Edité par C. Brezinski, A. Draux, A.P. Magnus, P. Maroni et A. Ronveaux. XXXVII, 584 pages. 1985.

Vol. 1172: Algebraic Topology, Göttingen 1984. Proceedings. Edited by L. Smith. VI, 209 pages. 1985.

Vol. 1173: H. Delfs, M. Knebusch, Locally Semialgebraic Spaces. XVI, 329 pages. 1985.

Vol. 1174: Categories in Continuum Physics, Buffalo 1982. Seminar. Edited by F.W. Lawvere and S.H. Schanuel. V, 126 pages. 1986.

Vol. 1175: K. Mathiak, Valuations of Skew Fields and Projective Hjelmslev Spaces. VII, 116 pages. 1986.

Vol. 1176: R.R. Bruner, J.P. May, J.E. McClure, M. Steinberger, H_∞ Ring Spectra and their Applications. VII, 388 pages. 1986.

Vol. 1177: Representation Theory I. Finite Dimensional Algebras. Proceedings, 1984. Edited by V. Dlab, P. Gabriel and G. Michler. XV, 340 pages. 1986.

Vol. 1178: Representation Theory II. Groups and Orders. Proceedings, 1984. Edited by V. Dlab, P. Gabriel and G. Michler. XV, 370 pages. 1986.

Vol. 1179: Shi J.-Y. The Kazhdan-Lusztig Cells in Certain Affine Weyl Groups. X, 307 pages. 1986.

Vol. 1180: R. Carmona, H. Kesten, J.B. Walsh, École d'Été de Probabilités de Saint-Flour XIV – 1984. Édité par P.L. Hennequin. X, 438 pages. 1986.

Vol. 1181: Buildings and the Geometry of Diagrams, Como 1984. Seminar. Edited by L. Rosati. VII, 277 pages. 1986.

Vol. 1182: S. Shelah, Around Classification Theory of Models. VII, 279 pages. 1986.

Vol. 1183: Algebra, Algebraic Topology and their Interactions. Proceedings, 1983. Edited by J.-E. Roos. XI, 396 pages. 1986.

Vol. 1184: W. Arendt, A. Grabosch, G. Greiner, U. Groh, H.P. Lotz, U. Moustakas, R. Nagel, F. Neubrander, U. Schlotterbeck, One-parameter Semigroups of Positive Operators. Edited by R. Nagel. X, 460 pages. 1986.

Vol. 1185: Group Theory, Beijing 1984. Proceedings. Edited by Tuan H.F. V, 403 pages. 1986.

Vol. 1186: Lyapunov Exponents. Proceedings, 1984. Edited by L. Arnold and V. Wihstutz. VI, 374 pages. 1986.

Vol. 1187: Y. Diers, Categories of Boolean Sheaves of Simple Algebras. VI, 168 pages. 1986.

Vol. 1188: Fonctions de Plusieurs Variables Complexes V. Séminaire, 1979–85. Edité par François Norguet. VI, 306 pages. 1986.

Vol. 1189: J. Lukeš, J. Malý, L. Zajíček, Fine Topology Methods in Real Analysis and Potential Theory. X, 472 pages. 1986.

Vol. 1190: Optimization and Related Fields. Proceedings, 1984. Edited by R. Conti, E. De Giorgi and F. Giannessi. VIII, 419 pages. 1986.

Vol. 1191: A.R. Its, V.Yu. Novokshenov, The Isomonodromic Deformation Method in the Theory of Painlevé Equations. IV, 313 pages. 1986.

Vol. 1192: Equadiff 6. Proceedings, 1985. Edited by J. Vosmansky and M. Zlámal. XXIII, 404 pages. 1986.

Vol. 1193: Geometrical and Statistical Aspects of Probability in Banach Spaces. Proceedings, 1985. Edited by X. Fernique, B. Heinkel, M.B. Marcus and P.A. Meyer. IV, 128 pages. 1986.

Vol. 1194: Complex Analysis and Algebraic Geometry. Proceedings, 1985. Edited by H. Grauert. VI, 235 pages. 1986.

Vol.1195: J.M. Barbosa, A.G. Colares, Minimal Surfaces in $\mathbb{R}^3$. X, 124 pages. 1986.

Vol. 1196: E. Casas-Alvero, S. Xambó-Descamps, The Enumerative Theory of Conics after Halphen. IX, 130 pages. 1986.

Vol. 1197: Ring Theory. Proceedings, 1985. Edited by F.M.J. van Oystaeyen. V, 231 pages. 1986.

Vol. 1198: Séminaire d'Analyse, P. Lelong – P. Dolbeault – H. Skoda. Seminar 1983/84. X, 260 pages. 1986.

Vol. 1199: Analytic Theory of Continued Fractions II. Proceedings, 1985. Edited by W.J. Thron. VI, 299 pages. 1986.

Vol. 1200: V.D. Milman, G. Schechtman, Asymptotic Theory of Finite Dimensional Normed Spaces. With an Appendix by M. Gromov. VIII, 156 pages. 1986.

Vol. 1201: Curvature and Topology of Riemannian Manifolds. Proceedings, 1985. Edited by K. Shiohama, T. Sakai and T. Sunada. VII, 336 pages. 1986.

Vol. 1202: A. Dür, Möbius Functions, Incidence Algebras and Power Series Representations. XI, 134 pages. 1986.

Vol. 1203: Stochastic Processes and Their Applications. Proceedings, 1985. Edited by K. Itô and T. Hida. VI, 222 pages. 1986.

Vol. 1204: Séminaire de Probabilités XX, 1984/85. Proceedings. Edité par J. Azéma et M. Yor. V, 639 pages. 1986.

Vol. 1205: B.Z. Moroz, Analytic Arithmetic in Algebraic Number Fields. VII, 177 pages. 1986.

Vol. 1206: Probability and Analysis, Varenna (Como) 1985. Seminar. Edited by G. Letta and M. Pratelli. VIII, 280 pages. 1986.

Vol. 1207: P.H. Bérard, Spectral Geometry: Direct and Inverse Problems. With an Appendix by G. Besson. XIII, 272 pages. 1986.

Vol. 1208: S. Kaijser, J.W. Pelletier, Interpolation Functors and Duality. IV, 167 pages. 1986.

Vol. 1209: Differential Geometry, Peñíscola 1985. Proceedings. Edited by A.M. Naveira, A. Ferrández and F. Mascaró. VIII, 306 pages. 1986.

Vol. 1210: Probability Measures on Groups VIII. Proceedings, 1985. Edited by H. Heyer. X, 386 pages. 1986.

Vol. 1211: M.B. Sevryuk, Reversible Systems. V, 319 pages. 1986.

Vol. 1212: Stochastic Spatial Processes. Proceedings, 1984. Edited by P. Tautu. VIII, 311 pages. 1986.

Vol. 1213: L.G. Lewis, Jr., J.P. May, M. Steinberger, Equivariant Stable Homotopy Theory. IX, 538 pages. 1986.

Vol. 1214: Global Analysis – Studies and Applications II. Edited by Yu.G. Borisovich and Yu.E. Gliklikh. V, 275 pages. 1986.

Vol. 1215: Lectures in Probability and Statistics. Edited by G. del Pino and R. Rebolledo. V, 491 pages. 1986.

Vol. 1216: J. Kogan, Bifurcation of Extremals in Optimal Control. VIII, 106 pages. 1986.

Vol. 1217: Transformation Groups. Proceedings, 1985. Edited by S. Jackowski and K. Pawalowski. X, 396 pages. 1986.

Vol. 1218: Schrödinger Operators, Aarhus 1985. Seminar. Edited by E. Balslev. V, 222 pages. 1986.

Vol. 1219: R. Weissauer, Stabile Modulformen und Eisensteinreihen. III, 147 Seiten. 1986.

Vol. 1220: Séminaire d'Algèbre Paul Dubreil et Marie-Paule Malliavin. Proceedings, 1985. Edité par M.-P. Malliavin. IV, 200 pages. 1986.

Vol. 1221: Probability and Banach Spaces. Proceedings, 1985. Edited by J. Bastero and M. San Miguel. XI, 222 pages. 1986.

Vol. 1222: A. Katok, J.-M. Strelcyn, with the collaboration of F. Ledrappier and F. Przytycki, Invariant Manifolds, Entropy and Billiards; Smooth Maps with Singularities. VIII, 283 pages. 1986.

Vol. 1223: Differential Equations in Banach Spaces. Proceedings, 1985. Edited by A. Favini and E. Obrecht. VIII, 299 pages. 1986.

Vol. 1224: Nonlinear Diffusion Problems, Montecatini Terme 1985. Seminar. Edited by A. Fasano and M. Primicerio. VIII, 188 pages. 1986.

Vol. 1225: Inverse Problems, Montecatini Terme 1986. Seminar. Edited by G. Talenti. VIII, 204 pages. 1986.

Vol. 1226: A. Buium, Differential Function Fields and Moduli of Algebraic Varieties. IX, 146 pages. 1986.

Vol. 1227: H. Helson, The Spectral Theorem. VI, 104 pages. 1986.

Vol. 1228: Multigrid Methods II. Proceedings, 1985. Edited by W. Hackbusch and U. Trottenberg. VI, 336 pages. 1986.

Vol. 1229: O. Bratteli, Derivations, Dissipations and Group Actions on C*-algebras. IV, 277 pages. 1986.

Vol. 1230: Numerical Analysis. Proceedings, 1984. Edited by J.-P. Hennart. X, 234 pages. 1986.

Vol. 1231: E.-U. Gekeler, Drinfeld Modular Curves. XIV, 107 pages. 1986.

Vol. 1232: P.C. Schuur, Asymptotic Analysis of Soliton Problems. VIII, 180 pages. 1986.

Vol. 1233: Stability Problems for Stochastic Models. Proceedings, 1985. Edited by V.V. Kalashnikov, B. Penkov and V.M. Zolotarev. VI, 223 pages. 1986.

Vol. 1234: Combinatoire énumérative. Proceedings, 1985. Edité par G. Labelle et P. Leroux. XIV, 387 pages. 1986.

Vol. 1235: Séminaire de Théorie du Potentiel, Paris, No. 8. Directeurs: M. Brelot, G. Choquet et J. Deny. Rédacteurs: F. Hirsch et G. Mokobodzki. III, 209 pages. 1987.

Vol. 1236: Stochastic Partial Differential Equations and Applications. Proceedings, 1985. Edited by G. Da Prato and L. Tubaro. V, 257 pages. 1987.

Vol. 1237: Rational Approximation and its Applications in Mathematics and Physics. Proceedings, 1985. Edited by J. Gilewicz, M. Pindor and W. Siemaszko. XII, 350 pages. 1987.

Vol. 1238: M. Holz, K.-P. Podewski and K. Steffens, Injective Choice Functions. VI, 183 pages. 1987.

Vol. 1239: P. Vojta, Diophantine Approximations and Value Distribution Theory. X, 132 pages. 1987.

Vol. 1240: Number Theory, New York 1984–85. Seminar. Edited by D.V. Chudnovsky, G.V. Chudnovsky, H. Cohn and M.B. Nathanson. V, 324 pages. 1987.

Vol. 1241: L. Gårding, Singularities in Linear Wave Propagation. III, 125 pages. 1987.

Vol. 1242: Functional Analysis II, with Contributions by J. Hoffmann-Jørgensen et al. Edited by S. Kurepa, H. Kraljević and D. Butković. VII, 432 pages. 1987.

Vol. 1243: Non Commutative Harmonic Analysis and Lie Groups. Proceedings, 1985. Edited by J. Carmona, P. Delorme and M. Vergne. V, 309 pages. 1987.

Vol. 1244: W. Müller, Manifolds with Cusps of Rank One. XI, 158 pages. 1987.

Vol. 1245: S. Rallis, L-Functions and the Oscillator Representation. XVI, 239 pages. 1987.

Vol. 1246: Hodge Theory. Proceedings, 1985. Edited by E. Cattani, F. Guillén, A. Kaplan and F. Puerta. VII, 175 pages. 1987.

Vol. 1247: Séminaire de Probabilités XXI. Proceedings. Edité par J. Azéma, P.A. Meyer et M. Yor. IV, 579 pages. 1987.

Vol. 1248: Nonlinear Semigroups, Partial Differential Equations and Attractors. Proceedings, 1985. Edited by T.L. Gill and W.W. Zachary. IX, 185 pages. 1987.

Vol. 1249: I. van den Berg, Nonstandard Asymptotic Analysis. IX, 187 pages. 1987.

Vol. 1250: Stochastic Processes – Mathematics and Physics II. Proceedings 1985. Edited by S. Albeverio, Ph. Blanchard and L. Streit. VI, 359 pages. 1987.

Vol. 1251: Differential Geometric Methods in Mathematical Physics. Proceedings, 1985. Edited by P.L. García and A. Pérez-Rendón. VII, 300 pages. 1987.

Vol. 1252: T. Kaise, Représentations de Weil et GL_2 Algèbres de division et GL_n. VII, 203 pages. 1987.

Vol. 1253: J. Fischer, An Approach to the Selberg Trace Formula via the Selberg Zeta-Function. III, 184 pages. 1987.

Vol. 1254: S. Gelbart, I. Piatetski-Shapiro, S. Rallis. Explicit Constructions of Automorphic L-Functions. VI, 152 pages. 1987.

Vol. 1255: Differential Geometry and Differential Equations. Proceedings, 1985. Edited by C. Gu, M. Berger and R.L. Bryant. XII, 243 pages. 1987.

Vol. 1256: Pseudo-Differential Operators. Proceedings, 1986. Edited by H.O. Cordes, B. Gramsch and H. Widom. X, 479 pages. 1987.

Vol. 1257: X. Wang, On the C*-Algebras of Foliations in the Plane. V, 165 pages. 1987.

Vol. 1258: J. Weidmann, Spectral Theory of Ordinary Differential Operators. VI, 303 pages. 1987.

Vol. 1259: F. Cano Torres, Desingularization Strategies for Three-Dimensional Vector Fields. IX, 189 pages. 1987.

Vol. 1260: N.H. Pavel, Nonlinear Evolution Operators and Semigroups. VI, 285 pages. 1987.

Vol. 1261: H. Abels, Finite Presentability of S-Arithmetic Groups. Compact Presentability of Solvable Groups. VI, 178 pages. 1987.

Vol. 1262: E. Hlawka (Hrsg.), Zahlentheoretische Analysis II. Seminar, 1984–86. V, 158 Seiten. 1987.

Vol. 1263: V.L. Hansen (Ed.), Differential Geometry. Proceedings, 1985. XI, 288 pages. 1987.

Vol. 1264: Wu Wen-tsün, Rational Homotopy Type. VIII, 219 pages. 1987.

Vol. 1265: W. Van Assche, Asymptotics for Orthogonal Polynomials. VI, 201 pages. 1987.

Vol. 1266: F. Ghione, C. Peskine, E. Sernesi (Eds.), Space Curves. Proceedings, 1985. VI, 272 pages. 1987.

Vol. 1267: J. Lindenstrauss, V.D. Milman (Eds.), Geometrical Aspects of Functional Analysis. Seminar. VII, 212 pages. 1987.

Vol. 1268: S.G. Krantz (Ed.), Complex Analysis. Seminar, 1986. VII, 195 pages. 1987.

Vol. 1269: M. Shiota, Nash Manifolds. VI, 223 pages. 1987.

Vol. 1270: C. Carasso, P.-A. Raviart, D. Serre (Eds.), Nonlinear Hyperbolic Problems. Proceedings, 1986. XV, 341 pages. 1987.

Vol. 1271: A.M. Cohen, W.H. Hesselink, W.L.J. van der Kallen, J.R. Strooker (Eds.), Algebraic Groups Utrecht 1986. Proceedings. XII, 284 pages. 1987.

Vol. 1272: M.S. Livšic, L.L. Waksman, Commuting Nonselfadjoint Operators in Hilbert Space. III, 115 pages. 1987.

Vol. 1273: G.-M. Greuel, G. Trautmann (Eds.), Singularities, Representation of Algebras, and Vector Bundles. Proceedings, 1985. XIV, 383 pages. 1987.

Vol. 1274: N. C. Phillips, Equivariant K-Theory and Freeness of Group Actions on C*-Algebras. VIII, 371 pages. 1987.

Vol. 1275: C.A. Berenstein (Ed.), Complex Analysis I. Proceedings, 1985–86. XV, 331 pages. 1987.

Vol. 1276: C.A. Berenstein (Ed.), Complex Analysis II. Proceedings, 1985–86. IX, 320 pages. 1987.

Vol. 1277: C.A. Berenstein (Ed.), Complex Analysis III. Proceedings, 1985–86. X, 350 pages. 1987.

Vol. 1278: S.S. Koh (Ed.), Invariant Theory. Proceedings, 1985. V, 102 pages. 1987.

Vol. 1279: D. Ieşan, Saint-Venant's Problem. VIII, 162 Seiten. 1987.

Vol. 1280: E. Neher, Jordan Triple Systems by the Grid Approach. XII, 193 pages. 1987.

Vol. 1281: O.H. Kegel, F. Menegazzo, G. Zacher (Eds.), Group Theory. Proceedings, 1986. VII, 179 pages. 1987.

Vol. 1282: D.E. Handelman, Positive Polynomials, Convex Integral Polytopes, and a Random Walk Problem. XI, 136 pages. 1987.

Vol. 1283: S. Mardešić, J. Segal (Eds.), Geometric Topology and Shape Theory. Proceedings, 1986. V, 261 pages. 1987.

Vol. 1284: B.H. Matzat, Konstruktive Galoistheorie. X, 286 pages. 1987.

Vol. 1285: I.W. Knowles, Y. Saitō (Eds.), Differential Equations and Mathematical Physics. Proceedings, 1986. XVI, 499 pages. 1987.

Vol. 1286: H.R. Miller, D.C. Ravenel (Eds.), Algebraic Topology. Proceedings, 1986. VII, 341 pages. 1987.

Vol. 1287: E.B. Saff (Ed.), Approximation Theory, Tampa. Proceedings, 1985–1986. V, 228 pages. 1987.

Vol. 1288: Yu. L. Rodin, Generalized Analytic Functions on Riemann Surfaces. V, 128 pages, 1987.

Vol. 1289: Yu. I. Manin (Ed.), K-Theory, Arithmetic and Geometry. Seminar, 1984–1986. V, 399 pages. 1987.

Vol. 1290: G. Wüstholz (Ed.), Diophantine Approximation and Transcendence Theory. Seminar, 1985. V, 243 pages. 1987.

Vol. 1291: C. Mœglin, M.-F. Vignéras, J.-L. Waldspurger, Correspondances de Howe sur un Corps p-adique. VII, 163 pages. 1987

Vol. 1292: J.T. Baldwin (Ed.), Classification Theory. Proceedings, 1985. VI, 500 pages. 1987.

Vol. 1293: W. Ebeling, The Monodromy Groups of Isolated Singularities of Complete Intersections. XIV, 153 pages. 1987.

Vol. 1294: M. Queffélec, Substitution Dynamical Systems – Spectral Analysis. XIII, 240 pages. 1987.

Vol. 1295: P. Lelong, P. Dolbeault, H. Skoda (Réd.), Séminaire d'Analyse P. Lelong – P. Dolbeault – H. Skoda. Seminar, 1985/1986. VII, 283 pages. 1987.

Vol. 1296: M.-P. Malliavin (Ed.), Séminaire d'Algèbre Paul Dubreil et Marie-Paule Malliavin. Proceedings, 1986. IV, 324 pages. 1987.

Vol. 1297: Zhu Y.-l., Guo B.-y. (Eds.), Numerical Methods for Partial Differential Equations. Proceedings. XI, 244 pages. 1987.

Vol. 1298: J. Aguadé, R. Kane (Eds.), Algebraic Topology, Barcelona 1986. Proceedings. X, 255 pages. 1987.

Vol. 1299: S. Watanabe, Yu.V. Prokhorov (Eds.), Probability Theory and Mathematical Statistics. Proceedings, 1986. VIII, 589 pages. 1988.

Vol. 1300: G.B. Seligman, Constructions of Lie Algebras and their Modules. VI, 190 pages. 1988.